SYSTEM MODELING AND OPTIMIZATION

IFIP – The International Federation for Information Processing

IFIP was founded in 1960 under the auspices of UNESCO, following the First World Computer Congress held in Paris the previous year. An umbrella organization for societies working in information processing, IFIP's aim is two-fold: to support information processing within its member countries and to encourage technology transfer to developing nations. As its mission statement clearly states,

> *IFIP's mission is to be the leading, truly international, apolitical organization which encourages and assists in the development, exploitation and application of information technology for the benefit of all people.*

IFIP is a non-profitmaking organization, run almost solely by 2500 volunteers. It operates through a number of technical committees, which organize events and publications. IFIP's events range from an international congress to local seminars, but the most important are:

- The IFIP World Computer Congress, held every second year;
- Open conferences;
- Working conferences.

The flagship event is the IFIP World Computer Congress, at which both invited and contributed papers are presented. Contributed papers are rigorously refereed and the rejection rate is high.

As with the Congress, participation in the open conferences is open to all and papers may be invited or submitted. Again, submitted papers are stringently refereed.

The working conferences are structured differently. They are usually run by a working group and attendance is small and by invitation only. Their purpose is to create an atmosphere conducive to innovation and development. Refereeing is less rigorous and papers are subjected to extensive group discussion.

Publications arising from IFIP events vary. The papers presented at the IFIP World Computer Congress and at open conferences are published as conference proceedings, while the results of the working conferences are often published as collections of selected and edited papers.

Any national society whose primary activity is in information may apply to become a full member of IFIP, although full membership is restricted to one society per country. Full members are entitled to vote at the annual General Assembly, National societies preferring a less committed involvement may apply for associate or corresponding membership. Associate members enjoy the same benefits as full members, but without voting rights. Corresponding members are not represented in IFIP bodies. Affiliated membership is open to non-national societies, and individual and honorary membership schemes are also offered.

SYSTEM MODELING AND OPTIMIZATION

Proceedings of the 21st IFIP TC7 Conference held in July 21st-25th, 2003, Sophia Antipolis, France

Edited by

John Cagnol
Pôle Universitaire Léonard de Vinci,
Paris, France

Jean-Paul Zolésio
CNRS/INRIA
Sophia Antipolis, France

KLUWER ACADEMIC PUBLISHERS

John Cagnol
Pôle Universitaire Léonard de Vinci
Courbevoie, FRANCE

Jean-Paul Zolésio
CNRS/INRIA
Sophia Antipolis, FRANCE

Library of Congress Cataloging-in-Publication Data

A C.I.P. Catalogue record for this book is available from the Library of Congress.

SYSTEM MODELING AND OPTIMIZATION / Edited by John Cagnol, John-Paul Zolésio.
p.cm. —(The International Federation for Information Processing)
Includes bibliographical references..

ISBN: (HC) 1-4020-7760-2/ (eB00K) ISBN: 0-387-23467-5
Printed on acid-free paper.

Printed in the United States of America.

9 8 7 6 5 4 3 2 1 SPIN 11054290 (HC) / 11333548 (eBook)
springeronline.com

Contents

Foreword

This volume comprises selected papers from the 21st Conference on System Modeling and Optimization that took place from July 21st to July 25th, 2003, in Sophia Antipolis, France. This event is part of a series of conferences that meet every other year and bring together the seventh Technical Committee of the International Federation for Information Processing (IFIP). It has been co-organized by three institutions: Institut National de Recherche en Informatique et Automatique (INRIA), Pôle Universitaire Léonard de Vinci and Ecole des Mines de Paris. It was chaired by Jean-Paul Zolésio and co-chaired by John Cagnol.

IFIP is a multinational federation of professional and technical organizations concerned with information processing. The Federation is organized into the IFIP Council, the Executive Board, and the Technical Assembly. The Technical Assembly is divided into eleven Technical Committees of which TC 7 is one. The TC 7 on system modeling and optimization aims to provide an international clearing house for computational, as well as related theoretical, aspects of optimization problems in diverse areas and to share computing experience gained on specific applications. It also aims to promote the development of importants high-level theory to meet the needs of complex optimization problems and establish appropriate cooperation with the International Mathematics Union and similar organizations. In addition, IFIP fosters interdisciplinary activity on optimization problems spanning the various areas such as Economics, including Business Administration and Management, Biomedicine, Meteorology, etc. in cooperation with associated international bodies. The technical committee is composed of seven working groups and is chaired by Irena Lasiecka. It was founded by A.V. Balakrishnan, J.L. Lions and M. Marchuk.

System modeling and optimization are two disciplines arising from many spheres of scientific activities. Their fields include, but are not limited to: bioscience, environmental science, optimal design, transport and telecommunications, control in electromagnetics, image analysis, multiphisics systems that are coupled by moving interfaces, free boundary problems, non cylindrical evolution control, etc... The emergence of smart materials allows the existence of new actuators and new configurations, and thus we are required to revisit many classical settings. For example, the dynamical systems involved are often non autonomous. The uncertainty in the modeling and the robustness (or the lack thereof) results in stochastic modeling. In addition, intrinsic geometry is increasing in control theory since the boundaries are moving and minimal regularity is sought. The coupling of fluid and structural mechanics leads to the superposition of Eulerian and Lagrangian representations. The coupling of several physical models such as fluid (wind, blood, solar flux,...), structures (elastic shell, elasto-plastic crash, airfoils, arteries,...), electromagnetism (antennas, dynamical frequency assignations, nerves and heart control), thermal effects (rheology, boundary conditions, damping,...), acoustics (supersonic plane, sound control, helicopter cabin noise reduction...), and chemical effects (climate, pollution, ionisation,....) lead to hierarchical modeling associated with multiscale control theory and computation. Optimization and optimal control of such systems include inverse problems and topological identification analysis for applications to non destructive control such as cracks and surface identifications. Many of these problems lead to non linear, non quadratic control problems.

The editors would like to acknowledge the contributions of the many members of the IFIP Program Committee who have given valuable advice. They would like to thank George Avalos, Arunabha Bagchi, Francesca Bucci, Dan Dolk, Hitohsi Furuta, Irena Lasiecka, Catherine Lebiedzik, Guenter Leugering, Zdzislaw Naniewicz, Vyacheslav Maksimov, Luciano Pandolfi, Mike Polis, Hans-Jürgen Sebastian, Irina Sivergina, Jan Sokolowski, Marc Thiriet and Fredi Troeltzsch for accepting to organize minisymposia.

The editors would also like to thank Michel Cosnard, head of INRIA-Sophia Antipolis who made possible the organization of the conference and Yves Laboureur head of the Sophia-Antipolis branch of the Ecole des Mines for hosting the conference.

Finally, the chair and co-chair would like to thank their wives Monique and Bethany for their support and help during the organization of the conference.

John Cagnol and Jean-Paul Zolésio

Organizing Institutions

INRIA, the National Institute for Research in Computer Science and Control, was created in 1967 at Rocquencourt near Paris. INRIA is a public scientific and technological establishment under the joint supervision of the Research Ministry and the Ministry of Economy, Treasury and Industry. INRIA's mission is to be a world player, a research institute at the heart of the information society. INRIA aims to network skills and talents from the fields of information and computer science and technology from the entire French research system. This network allows scientific excellence to be used for technological progress, for creating employment, and prosperity and for finding renewed applications in response to socio-economic needs. Its decentralized organization (six research units), small autonomous teams, and regular evaluation enable INRIA to develop partnerships with 95 research projects shared with universities, Grandes Ecoles and research organizations.

Pôle Universitaire Léonard de Vinci is a private university founded in 1995, and located in Paris La Défense, France. It includes an accredited engineering school with several departments: Scientific Computation, Computer Science, Financial Engineering and Computational Mechanics. These programs have received very positive feedback from the industry.

The Ecole des Mines de Paris was founded in 1783 by Louis XVI. It was originally a mining school. The exploitation and processing of raw materials formed the basis of the development of Europe's economy. The art of mining in particular was one area in which scientific thinking had to be applied. Naturally, the focus of the School closely followed industrial development and the Ecole des Mines now studies, develops and teaches a wide range of sciences and techniques of value to engineers, including economic and social sciences. Today the Ecole des Mines de Paris is split into four locations: Paris, Fontainebleau, Evry and Sophia Antipolis.

Contributing Authors

Arunabha Bagchi
University of Twente, The Netherlands

A.V. Balakrishnan
Flight Systems Research Center, UCLA, U.S.A.

Bradley M. Bell
University of Washington, U.S.A.

Marcelo M. Cavalcanti
State University of Maringá, Brazil

Valéria N. Domingos Cavalcanti
State University of Maringá, Brazil

Andrea De Gaetano
Università Cattolica del Sacro Cuore, Italy

Michel C. Delfour
CRM, Université de Montréal, Canada

Zdzisław Denkowski
Jagiellonian University, Poland

Domenico Di Martino
IASI-CNR, Italy

Nicolas Doyon
Université Laval, Canada

Hitoshi Furuta
Kansai University, Japan

Alfredo Germani
Università degli Studi dell'Aquila, Italy

Mrinal K. Ghosh
Indian Institute of Science, India

Thibault Henri
INSA de Rennes, France

Jacques Henry
INRIA, France

Naoki Katoh
Kyoto University, Japan

Mitsuo Kawatani
Kobe University, Japan

Chul-Woo Kim
Kobe University, Japan

Kazuhiro Koyama
Kansai University, Japan

A.B. Kurzhanski
Moscow State University, Russia

B. Louro
Universidade Nova de Lisboa, Portugal

Vyacheslav Maksimov
Russian Academy of Sciences, Russia

Costanzo Manes
Università degli Studi dell'Aquila, Italy

Stanisław Migórski
Jagiellonian University, Poland

Serguei A. Nazarov
Russian Academy of Sciences, Russia

Yury Orlov
CICESE Research Center, Mexico

Luciano Pandolfi
Politecnico di Torino, Italy

Gianluigi Pillonetto
University of Padova, Italy

Enrico Priola
Università di Torino, Italy

Mauro L. Santos
Universidade Federal do Pará, Brazil

Klaus Schittkowski
University of Bayreuth, Germany

Victor Shutyaev
Russian Academy of Sciences, Russia

M.C. Soares
Universidade Nova de Lisboa, Portugal

Jan Sokolowski
Université Henri Poincaré Nancy I, France

M.M. Sorokina
Moscow State University, Russia

P. Thoft-Christensen
Aalborg University, Denmark

Roberto Triggiani
University of Virginia, U.S.A.

Jean-Pierre Yvon
INSA de Rennes, France

Christian Zillober
University of Bayreuth, Germany

Jean-Paul Zolésio
CNRS and INRIA, France

TOWARD A MATHEMATICAL THEORY OF AEROELASTICITY

A.V. Balakrishnan*
Flight Systems Research Center
UCLA
bal@ee.ucla.edu

Abstract This paper initiates a mathematical theory of aeroelasticity centered on the canonical problem of the flutter boundary — an instability endemic to aircraft that limits attainable speed in the subsonic regime. We develop a continuum mathematical model that exhibits the known flutter phenomena and yet is amenable to analysis — non-numeric theory. Thus we model the wing as a cantilever beam and limit the aerodynamics to irrotational, isentropic so that we work with the quasi-linear Transonic Small Disturbance Equations with the attached flow and Kutta-Joukowsky boundary conditions. We can obtain a Volterra expansion for the solution showing in particular that the stability is determined by the linearized model consistent with the Hopf Bifurcation Theory. Specializing to linear aerodynamics, the time domain version of the aeroelastic problem is shown to be a convolution-evolution equation in a Hilbert space. The aeroelastic modes are shown to be the eigenvalues of the infinitesimal generator of a semigroup, which models the combined aerostructure state space dynamics. We are also able to define flutter boundary in terms of the "root locus" — the modes as a function of the air speed U. We are able to track the dependence of the flutter boundary on the Mach number — a crucial problem in aeroelasticity — but many problems remain for Mach numbers close to one. The model and theory developed should open the way to Control Design for flutter boundary expansion.

Introduction

To a mathematician specializing in the problems of stability and control for partial differential equations, Aeroelasticity offers a fertile, if challenging, field of application. Currently, however, to a mathemati-

*Research supported in part by NASA Grant NCC4-157

cian — even an applied mathematician — Aeroelasticity (to paraphrase Richard E. Meyer, in *Introduction to Mathematical Fluid Dynamics* [1]) "appears to be built on a quicksand of assorted intuitions" — plus numerical approximations. This paper is a first halting step toward a "mathematical theory of Aeroelasticity."

The canonical problem of Aeroelasticity is flutter. It is an instability endemic to aircraft wings that occurs at high enough airspeed in subsonic flight and thus limits the attainable speed. The purpose of Control Design is to "expand" this "flutter boundary."

Control Design, however, requires a mathematical model that is simple enough for non-numeric analysis and yet displays the phenomena of interest — in this case flutter. In contrast almost all the extant work on this problem has been computational (see the review paper by Friedmann [2]). Computational techniques despite their success and universal use, require that numerical parameters be specified and thus cannot contribute to Control Design. The lack of a faithful enough mathematical model is undoubtedly one reason why all attempts at flutter control have failed so far. As we shall show, the kind of models needed require crucially recent advances in boundary control of partial differential equations. Even then many purely mathematical questions relating to the model are unanswered as yet.

We begin in Section 2 with the wing model, incorporating in addition a model for self-straining actuators. Section 3 is devoted to the aerodynamic model where we derive the TSD Equation from the Full Potential Equation clarifying the many assumptions made, and allowing for nonzero angle of attack. We linearize the TSD Equation and show it can be solved by the Possio Integral Equation, generalized to include nonzero angle of attack. We also develop a solution to the Linear Nonhomogeneous TSD Equation for zero initial and boundary conditions. Using these results we show how to construct a power series expansion — actually a Volterra kernel series expansion for the solution of the nonlinear TSD Equation. We are then able to obtain what is perhaps the most significant result — that the stability of the system is determined by the stability of the linear system — consistent with the Hopf Bifurcation Theory.

In Section 4 we go on to the abstract or time domain formulation of the flutter control problem. It turns out to be convolution-evolution equation in a Hilbert Space for the structure state — which is not quite the full state for which we used to go to a Banach Space formulation, enabling us to identify the aeroelastic modes as eigenvalues of the infinitesimal generator of the Banach Space semigroup. Of primary interest on the practical side is the calculation of these modes. This in turn leads to the "root locus" — the modes as a function of U — and the definition of flutter speed. The dependence of the flutter speed on M is an important unresolved issue here.

1. The Wing Model

The wing is modelled as a flexible structure — the flexibility is of course the key feature — as a "straight" uniform rectangular plate. Identifying the modes of the wing structure is one of the standard activities (vibration testing) in flight centers. The structure model must have the ability to conform to the first few measured modes at least. Following the model initiated by Goland [3] in 1954 we allow two degrees of freedom — plunge (displacement) and pitch (angle) about the elastic axis. Let

$$x(s,t) = \begin{vmatrix} h(s,t) \\ \theta(s,t) \end{vmatrix}, \quad 0 \le t, \ 0 \le s \le \ell, \tag{1}$$

where ℓ is the wing span (one sided). Then the Goland model is:

$$M_s\,\ddot{x}(s,t) + D_s\,\dot{x}(s,t) + K_s\,x(s,t) = \begin{vmatrix} L(s,t) \\ M(s,t) \end{vmatrix}, \quad 0 < s < \ell, \tag{2}$$

where K_s is the differential operator

$$K_s = \begin{vmatrix} EI\,\frac{d^4}{ds^4} & 0 \\ 0 & -GJ\,\frac{d^2}{ds^2} \end{vmatrix}$$

$$M_s = \begin{vmatrix} m & S \\ S & I_\theta \end{vmatrix}$$

$$\det M_s > 0$$

$$D_s = 0$$

and $L(s,t)$, $M(s,t)$ denote the aerodynamic lift and moment. We are thus modelling the structure as a beam which would imply that the spread $2b$ ("chord length") is "small" compared to the span ℓ. Following Goland the beam is a cantilever clamped at the root $s = 0$ and free at the tip $s = \ell$, so that we have the end conditions: at the root:

$$\theta(0,t) = 0; \quad h(0,t) = h'(0,t) = 0 \tag{3}$$

and at the tip:

$$\theta'(\ell,t) = 0; \quad h''(\ell,t) = h'''(\ell,t) = 0 \tag{4}$$

where the super primes denote derivative with respect to s and the superdots denote time derivatives, in the usual notation.

We will need to change the tip conditions to:

$$\left.\begin{aligned} EI\,h''(\ell,t) \;+\; g_h\,\dot{h}'(\ell,t) &= 0 \\ GJ\,\theta'(\ell,t) \;+\; g_\theta\,\dot{\theta}(\ell,t) &= 0 \end{aligned}\right\} \tag{4a}$$

if we wish to include a generally accepted model for self-straining actuators, with $g_h\,, g_\theta \geq 0$ being the gains.

2. The Aerodynamic Model

The aerodynamics is far the more complicated part. To comply with space limitation, the presentation will need to be quite compressed with minimal details of proofs.

To begin with, we shall assume the flow to be non-viscous. Next we will assume that it is isentropic and that the Perfect Gas Law applies. In this case, as shown in [4], the flow can be described by a velocity potential $\phi(x, y, z, t)$ which satisfies the so-called Full Potential Equation given by:

$$\begin{aligned} \frac{\partial^2\phi}{\partial t^2} + \frac{\partial}{\partial t}\,|\nabla\phi|^2 \\ = a_\infty^2\,\nabla^2\phi\left(1 + \frac{\gamma-1}{a_\infty^2}\left(\frac{|q_\infty|^2}{2} - \frac{\partial\phi}{\partial t} - \frac{|\nabla\phi|^2}{2}\right)\right) \\ - \nabla\phi\cdot\nabla\left(\tfrac{1}{2}\,|\nabla\phi|^2\right), \quad -\infty < x, y, z < \infty,\ 0 \leq t \end{aligned} \tag{5}$$

where q_∞ is the free stream (far-field) velocity and a_∞ is the free stream (far-field) speed of sound, ∇ denotes gradients in the usual notation, and

$$M_\infty = \frac{|q_\infty|}{a_\infty},$$

the far stream Mach number assumed ≤ 1, and γ is the ratio of specific heats.

This equation would appear to be complex but fortunately can be simplified since our primary concern is stability. Hence we go one level down to the Transonic Small Disturbance Equation — there are various versions [6], [10] but we shall follow Nixon [5] — see also [4]. Thus we assume that

$$\varphi = \frac{\phi - \phi_\infty}{U} \tag{6}$$

is "small" (see below for how it is used) where ϕ_∞ is the undisturbed or far stream potential:

$$\phi_\infty = (xq_1 + yq_2 + zq_3)\,U,$$

$$U^2 = |q_\infty|^2, \quad q_1^2 + q_2^2 + q_3^2 = 1.$$

We have then the TSD Equation for φ (see [7, equation 2.22]):

$$\begin{aligned}
\frac{\partial^2\varphi}{\partial t^2} + 2U\left(q_1\,\frac{\partial^2\varphi}{\partial x\partial t} + q_2\,\frac{\partial^2\varphi}{\partial y\partial t} + q_3\,\frac{\partial^2\varphi}{\partial z\partial t}\right) \\
= a_\infty^2\left(1 - M_\infty^2 q_1^2 - (1+\gamma)M_\infty^2 q_1\,\frac{\partial\varphi}{\partial x}\right)\frac{\partial^2\varphi}{\partial x^2} \\
+ a_\infty^2\left(1 - M_\infty^2 q_2^2 - (1+\gamma)M_\infty^2 q_2\,\frac{\partial\varphi}{\partial y}\right)\frac{\partial^2\varphi}{\partial y^2} \\
+ a_\infty^2\left(1 - M_\infty^2 q_3^2 - (1+\gamma)M_\infty^2 q_3\,\frac{\partial\varphi}{\partial z}\right)\frac{\partial^2\varphi}{\partial z^2}
\end{aligned} \qquad (7)$$

Note that this is a quasi-linear equation with the right hand side neither elliptic nor hyperbolic, studied by Tricomi [6], Bers [7], Guderley [8], extensively, specialized to the stationary case.

2.1 The Aeroelastic Problem

Our interests are different in that we need to go beyond Transonic Aerodynamics to Transonic Aeroelasticity, as reflected in our preoccupation with the boundary conditions:

i) Flow Tangency Condition:

$$\left.\frac{\partial\varphi}{\partial z}\right|_{z=0} = w_a(x,y,t) \; : -b < s < b; \; 0 < y < \ell, \qquad (8)$$

where $w_a(x,y,t)$ the "downwash" is the normal velocity of the structure. For our structure model of zero thickness, with $z(x,y,t)$ denoting the instantaneous displacement of the wing along the z-axis, we can calculate that:

$$\begin{aligned}
w(x,y,t) &= \frac{Dz(x,y,t)}{Dt} \\
&= (-1)\left[\dot{h}(y,t) + (x-ab)\dot{\theta}(y,t) + Uq_1\theta(y,t)\right], \\
&\qquad -b < x < b,\; 0 < y < \ell,\; |a| < 1,
\end{aligned} \qquad (9)$$

where $x = ab$ locates the elastic axis of the wing in the xy plane.

ii) Kutta-Joukowsky Conditions:

"Zero pressure jump off the wing and at the trailing edge" (10)

Now from [4] we have that pressure $p(x,y,z,t)$ can be expressed as

$$p(x,y,z,t) = \frac{\rho_\infty a_\infty^2}{\gamma}\left(\frac{\rho}{\rho_\infty}\right)^\gamma$$

where $\rho(x,y,z,t)$ is the density, and

$$\rho = \rho_\infty \left[1 + \frac{\gamma-1}{a_\infty^2}\left(\tfrac{1}{2}U^2 - \psi(x,y,z,t)\right)\right]^{\frac{1}{\gamma-1}}$$

where $\psi(x,y,z,t)$ is the acceleration potential

$$\psi = \frac{\partial\phi}{\partial t} + \tfrac{1}{2}|\nabla\phi|^2$$

$$\psi_\infty = \tfrac{1}{2}U^2.$$

Now consistent with our small disturbance assumption,

$$\tfrac{1}{2}U^2 - \psi(x,y,z,t)$$

can be approximated as

$$\tilde{\psi} = \psi_\infty - \psi = U\left[-q_\infty \cdot \nabla\varphi - \frac{\partial\varphi}{\partial t}\right].$$

Further

$$\left(1 + \frac{\gamma-1}{a_\infty^2}(\psi_\infty - \psi)\right)^{\frac{1}{\gamma-1}} = \left(1 + \frac{\gamma-1}{a_\infty^2}\,U\left(-q_\infty\cdot\nabla\varphi - \frac{\partial\varphi}{\partial t}\right)\right)^{\frac{1}{\gamma-1}}$$

as in [4] can be approximated:

$$= 1 + \frac{U}{a_\infty^2}\left[-q_\infty \cdot \nabla\varphi - \frac{\partial\varphi}{\partial t}\right].$$

Hence finally

$$\left(\frac{\rho}{\rho_\infty}\right)^\gamma = \left(1 + \frac{\gamma}{a_\infty^2}\,\tilde{\psi}\right).$$

Hence the pressure jump

$$\delta p = \left.(p(x,y,z,t) - p(x,y,-z,t))\right|_{z=0}$$

$$= \rho_\infty\,\delta\tilde{\psi}.$$

Hence we may express the Kutta-Joukowsky condition as:

$$\left.\begin{aligned} \delta\tilde{\psi} &= 0, \quad |x| > b,\ 0 < y < \ell \\ &= 0, \quad x = b-,\ 0 < y < \ell \end{aligned}\right\} \tag{11}$$

where

$$\tilde{\psi} = \left(-q_\infty \cdot \nabla\varphi - \frac{\partial\varphi}{\partial t}\right) U. \tag{12}$$

The lift $L(s,t)$ in (1) is then given by

$$L(y,t) = \rho_\infty \int_{-b}^{b} \delta\tilde{\psi}\, dx, \quad 0 < y < \ell, \tag{13}$$

and so the moment $M(y,t)$ in (1) is given by

$$M(y,t) = \rho_\infty \int_{-b}^{b} (x - ab)\, \delta\tilde{\psi}\, dx, \quad 0 < y < \ell. \tag{14}$$

We have thus completed our aeroelastic model, simplified to "small disturbance" theory. As Nixon notes in his review paper: The TSD (7) is the "minimum complexity equation that should be used for transonic flow prediction..." The first mathematical question is of course that of existence and uniqueness of solution for (7) subject to the stipulated boundary conditions. We may and do take the initial conditions to be zero, since we are interested only in the question of stability. Note that we have a "boundary-input" problem — the input being the normal velocity of the wing $w_a(x,y,t)$ and the "output" may be taken as $\tilde{\psi}(x,y,z,t)$. For arbitrary $w_a(x,y,t)$, we have thus a pure aerodynamic problem which we need to solve first. For the aeroelastic problem the function $w_a(x,y,t)$ is linear in the structure state:

$$\begin{vmatrix} X(y,t) \\ \dot{X}(y,t) \end{vmatrix} = Z(y,t)$$

$$w_a(x,y,t) = [\ell_1\,,\, Z(y,t)] + x[\ell_2\,,\, Z(y,t)], \quad |x| < b.$$

Heuristically then, by invoking physical realizability or Duhamel's principle we would have

$$L(\cdot,t) = \int_0^t \mathcal{L}(Z(\cdot,\sigma),\, t-\sigma)\, d\sigma$$

$$M(\cdot,t) = \int_0^t \mathcal{M}(Z(\cdot,\sigma),\, t-\sigma)\, d\sigma$$

where $\mathcal{L}(\cdot,t)$, $\mathcal{M}(\cdot,t)$ are nonlinear operators. This would make (1) a nonlinear "integro-differential" equation. We are primarily interested in the stability properties as a function of U for fixed M, particularly in the transonic case for $M > 0.8$.

2.2 The Incompressible Case

Before we proceed to the general problem there is one special limiting case — the "incompressible flow" case, corresponding to $M = 0$ which is much the backbone of Aeroelasticity Theory. Thus we divide through by a_∞^2 first in (7) and allow $M_\infty \to 0$, $a_\infty \to \infty$ but keeping

$$U = a_\infty \cdot M_\infty$$

finite. Then (7) (as well as in the Full Potential Equation (5)!) simplifies to

$$\frac{\partial^2\varphi}{\partial x^2} + \frac{\partial^2\varphi}{\partial y^2} + \frac{\partial^2\varphi}{\partial z^2} = 0, \qquad -\infty < x, y, z < \infty$$

so that the flow is incompressible and the field equation is stationary, but of course U enters via the boundary conditions (9) and (10). Unfortunately this "3D problem for a finite wing" is still unsolved at this level of generality and we refer to [9] for a recent treatment.

2.3 High-Aspect Ratio Wings: Typical Section Theory

We now make a further simplifying assumption:

$$\frac{\ell}{b} \approx \infty$$

which is the mathematical equivalent of "high" aspect ratio wing. Or, formulated another way, we drop the dependence on y so that we have a "typical section" theory. Thus (7) becomes

$$\begin{aligned}\frac{\partial^2\varphi}{\partial t^2} &+ 2U\left(\cos\alpha\,\frac{\partial^2\varphi}{\partial x\partial t} + \sin\alpha\,\frac{\partial^2\varphi}{\partial z\partial t}\right)\\ &= a_\infty^2\left[1 - M_\infty^2\cos^2\alpha - (1+\gamma)M_\infty^2\cos\alpha\,\frac{\partial\varphi}{\partial x}\right]\frac{\partial^2\varphi}{\partial x^2}\\ &\quad + a_\infty^2\left[1 - M_\infty^2\sin^2\alpha - (1+\gamma)M_\infty^2\sin\alpha\,\frac{\partial\varphi}{\partial z}\right]\frac{\partial^2\varphi}{\partial z^2}\end{aligned} \tag{7a}$$

with the boundary conditions

$$\left.\frac{\partial\varphi}{\partial z}\right|_{z=0} = w_a(x,s,t), \qquad |x|<b,\ 0<x<\ell$$

$$= (-1)\left[\dot{h}(s,t) + (x-ab)\,\dot{\theta}(s,t) + U\cos\alpha\,\theta(s,t)\right].$$

Note that as far as the aerodynamics is concerned, the span parameter s is fixed. And the Kutta-Joukowsky conditions:

$$\begin{aligned}\delta\tilde{\psi} &= 0, \quad |x|>b\\ &= 0, \quad x=b-\end{aligned}$$

where

$$\tilde{\psi} = U\left(\frac{\partial\varphi}{\partial t} + U\cos\alpha\,\frac{\partial\varphi}{\partial x} + U\sin\alpha\,\frac{\partial\varphi}{\partial z}\right).$$

There are as yet no general existence uniqueness results for this class of problems. This is typical for this area. Here we shall present a general solution technique.

2.4 The Linear TSD Equation: Possio Integral Equation

First however we need to consider the linear (or "linearized," as we shall show below) TSD which is obtained by eliminating the nonlinear or quasi-linear part — that is setting

$$(1+\gamma)M_\infty^2 = 0$$

in (7a), but with the same boundary conditions, which would make the spatial part elliptic. Thus the linear TSD is

$$\frac{\partial^2\varphi}{\partial t^2} + 2U\left(\cos\alpha\,\frac{\partial^2\varphi}{\partial x\partial t} + \sin\alpha\,\frac{\partial^2\varphi}{\partial z\partial t}\right)$$

$$= a_\infty^2\left((1-M_\infty^2\cos^2\alpha)\,\frac{\partial^2\varphi}{\partial x^2} + (1-M_\infty^2\sin^2\alpha)\,\frac{\partial^2\varphi}{\partial z^2}\right). \qquad \text{(7L)}$$

Details of the function space or "abstract" formulation of this problem for $\alpha=0$ are given in [11], and extended to the case $\alpha\neq 0$ in [4]. The technique is to go to the equivalent formulation as an integral equation called the Possio Integral Equation after the initiator Possio [12], after taking Laplace Transforms — (actually Fourier Transforms in the early work as in [12]). It is customary to do this in terms of the function

$$\left.\begin{aligned} A(x,t) &= \frac{\delta P}{\rho_\infty U} = \frac{-\delta\tilde{\psi}}{U}, \quad |x| < b \\ &= 0, \quad |x| > b \end{aligned}\right\}. \tag{15}$$

Note that the Kutta condition requires

$$A(b-,t) = 0.$$

Because of the primary interest in stability, we work with Laplace Transforms. Thus let

$$\hat{A}(x,\lambda) = \int_0^\infty e^{-\lambda t} A(x,t)\, dt, \quad \mathrm{Re}\,\lambda > \sigma_a$$

and

$$\hat{w}_a(x,\lambda) = \int_0^\infty e^{-\lambda t} w_a(x,t)\, dt.$$

We have (the Possio Integral Equation valid for nonzero-angle-of-attack) normalizing b to 1 and λ to $k = \frac{\lambda b}{U}$ (see [4]).

$$\hat{w}_a(x,\lambda) = \int_{-1}^{1} \hat{P}(x-\xi,\lambda)\,\hat{A}(\xi,\lambda)\, d\lambda, \quad |x| < 1, \tag{16}$$

where

$$\int_{-\infty}^{\infty} \hat{P}(x,\lambda)\, e^{-i\omega x}\, dx$$

$$= \frac{1}{k + i\omega\cos\alpha}\left(\frac{M^2k^2 + 2M^2ki\omega\cos\alpha + \omega^2(1-M^2\cos^2\alpha)}{\sqrt{M^2k^2 + 2M^2ki\omega\cos\alpha + \omega^2(1-M^2)}}\right)$$

$$-\infty < \omega < \infty, \tag{17}$$

and M_∞ is relaced by M, and

$$\hat{w}_a(x,\lambda) = (-U)\left[k\,\hat{h}(s,\lambda) + (1-ak)\hat{\theta}(s,\lambda) + xk\,\hat{\theta}(s,\lambda)\right], \quad |x| < 1. \tag{18}$$

For existence and uniqueness and abstract forulation of this problem see [11]. Here we shall simply assume this, so that in turn the linear TSD has a unique solution calculated via $\hat{A}(\cdot,\lambda)$ — as in [11]. For the aeroelastic problem the solution $\hat{A}(\cdot,\lambda)$ suffices.

Next we need to consider:

2.5 The Linear Nonhomogeneous TSD Equation

In developing the solution to the nonlinear TSD Equation (7), we need to continue with the linear equation (7L) but now the nonhomogeneous case — nonzero right hand side. Thus we need to consider:

$$\frac{\partial^2\varphi}{\partial t^2} + 2U\left(\cos\alpha\,\frac{\partial^2\varphi}{\partial x\partial t} + \sin\alpha\,\frac{\partial^2\varphi}{\partial z\partial t}\right)$$
$$- a_\infty^2\left(1 - M^2\cos^2\alpha\right)\frac{\partial^2\varphi}{\partial x^2} - a_\infty^2\left(1 - M^2\sin^2\alpha\right)\frac{\partial^2\varphi}{\partial z^2}$$
$$= f(x,z,t), \quad -\infty < x, z < \infty;\ 0 < t, \qquad \text{(7LNH)}$$

with zero initial conditions:

$$\begin{aligned}\varphi(x,z,0) &= 0\\ \dot\varphi(x,z,0) &= 0\end{aligned}$$

and zero boundary conditions

$$\left.\frac{\partial\varphi}{\partial z}\right|_{z=0} = 0, \quad |x| < 1,$$

$$\delta\,\tilde\psi = 0, \quad -\infty < x < \infty$$

and zero far-field conditions. In that case we can show that (7LNH) has a unique solution given in fact in terms of a Green's function:

$$\varphi(x,z,t) = \int_0^t\int_{-\infty}^{\infty}\int_{-\infty}^{\infty} g(x-\xi, y-\eta, t-\sigma)\, f(\xi,\eta,\sigma)\; d\xi\; d\eta\; d\sigma,$$
$$-\infty < x, z < \infty, \qquad (19)$$

for f in the same space as $\varphi(\cdot)$. We skip the details.

Let us use the notation:

$$\varphi = \mathcal{L} f \qquad (20)$$

where $\mathcal{L}$ is a linear bounded transformation.

Let us now return to the TSD (7). We shall outline a general technique of solution. The function space choices are described in [11]. We assume first that we have a solution which is analytic in the input $w_a(\cdot,\cdot)$, the initial conditions φ and $\dot\varphi$ being zero at $t = 0$. In other words we assume a solution $\varphi(\lambda)$ corresponding to the input $\lambda\, w_a(\cdot,\cdot)$ which is analytic in the complex variable λ. From the physical point of view, where we assume that the model also is, this goes without saying. Then $\varphi(\lambda)$ can be expanded in a power series in λ about $\lambda = 0$, valid for any λ in the finite part of the plane:

$$\varphi(\lambda) = \sum_0^\infty \lambda^k \varphi_k \tag{21}$$

where

$$\varphi(0) = 0$$

and

$$\varphi_k = \left. \frac{\frac{d^k}{d\lambda^k}\varphi(\lambda)}{k!} \right|_{\lambda=0} . \tag{22}$$

Now by (7), where $\varphi(\lambda)$ is more explicitly

$$\varphi(x, z, t;\ \lambda),$$

we have

$$\begin{aligned} &\frac{\partial^2 \varphi(\lambda)}{\partial t^2} + 2U\left(\cos\alpha\,\frac{\partial^2\varphi(\lambda)}{\partial x \partial t} + \sin\alpha\,\frac{\partial^2\varphi(\lambda)}{\partial z\partial t}\right) \\ &- a_\infty^2(1 - M^2\cos^2\alpha)\,\frac{\partial^2\varphi(\lambda)}{\partial x^2} - a_\infty^2(1 - M^2\sin^2\alpha)\,\frac{\partial^2\varphi(\lambda)}{\partial z^2} \\ &= U^2(1+\gamma)\cos\alpha\,\frac{\partial\varphi(\lambda)}{\partial x}\,\frac{\partial^2\varphi(\lambda)}{\partial x^2} U^2(1+\gamma)\sin\alpha\,\frac{\partial\varphi(\lambda)}{\partial z}\,\frac{\partial^2\varphi(\lambda)}{\partial z^2} \end{aligned} \tag{23}$$

and

$$U\left.\frac{\partial\varphi(\lambda)}{\partial z}\right|_{z=0} = \lambda\, w_a(t, x), \quad |x| < 1. \tag{24}$$

Hence differentiating with respect to λ in (24):

$$U\,\frac{\partial}{\partial z}\left.\frac{\partial\varphi(\lambda)}{\partial\lambda}\right|_{z=0} = w_a(t, x), \quad |x| < 1 \tag{25}$$

$$U\,\frac{\partial}{\partial z}\left.\frac{\partial^k\varphi(\lambda)}{\partial\lambda^k}\right|_{z=0} = 0, \qquad |x| < 1,\ k \geq 2 \tag{26}$$

and in (23)

$$\begin{aligned} &\frac{\partial^2\varphi_k}{\partial t^2} + 2U\left(\cos\alpha\,\frac{\partial^2\varphi_k}{\partial x\partial t} + \sin\alpha\,\frac{\partial^2\varphi_k}{\partial z\partial t}\right) \\ &\quad - a_\infty^2(1 - M^2\cos^2\alpha)\,\frac{\partial^2\varphi_k}{\partial x^2} - a_\infty^2(1 - M^2\sin^2\alpha)\,\frac{\partial^2\varphi_k}{\partial z^2} \end{aligned}$$

$$= k!\,U^2(1+\gamma)\cos\alpha\;\frac{\partial^k}{\partial\lambda^k}\left[\frac{\partial\varphi(\lambda)}{\partial x}\,\frac{\partial^2\varphi(\lambda)}{\partial x^2}\right]\Bigg|_{\lambda=0}$$

$$+\,k!\,U^2(1+\gamma)\sin\alpha\;\frac{\partial^k}{\partial\lambda^k}\left[\frac{\partial\varphi(\lambda)}{\partial z}\,\frac{\partial^2\varphi(\lambda)}{\partial z^2}\right]\Bigg|_{\lambda=0}. \tag{27}$$

Hence, for $k = 1$ we see that (27) reduces to the linear TSD Equation (7L) with the associated boundary conditions. The solution φ_1 is then uniquely determined via the corresponding nonzero-angle-of-attack Possio Integral Equation (17).

Next we see that φ_2 satisfies

$$\frac{\partial^2\varphi_2}{\partial t^2} + 2U\left(\cos\alpha\,\frac{\partial^2\varphi_2}{\partial x\partial t} + \sin\alpha\,\frac{\partial^2\varphi_2}{\partial z\partial t}\right)$$

$$-\,a_\infty^2(1 - M^2\cos^2\alpha)\,\frac{\partial^2\varphi_2}{\partial x^2} - a_\infty^2(1 - M^2\sin^2\alpha)\,\frac{\partial^2\varphi_2}{\partial z^2}$$

$$= 2U^2(1+\gamma)\cos\alpha\left(2\,\frac{\partial\varphi_1}{\partial x}\,\frac{\partial^2\varphi_1}{\partial x^2}\right)$$

$$+\,2U^2(1+\gamma)\sin\alpha\left(2\,\frac{\partial\varphi_1}{\partial z}\,\frac{\partial^2\varphi_1}{\partial z^2}\right) \tag{28}$$

with

$$\frac{\partial\varphi_1}{\partial z}\bigg|_{z=0} = 0, \quad |x| < 1. \tag{29}$$

But this is the linear nonhomogeneous equation (7LNH) we have already treated. With f_2 denoting the right side of (28), we have that

$$\varphi_2 = \mathcal{L}\,f_2\,.$$

More generally, with

$$f_k = k!\,U^2(1+\gamma)\cos\alpha\;\frac{\partial^k}{\partial\lambda^k}\left(\frac{\partial\varphi(\lambda)}{\partial x}\frac{\partial^2\varphi(\lambda)}{\partial x^2}\right)\bigg|_{\lambda=0}$$

$$+\,k!\,U^2(1+\gamma)\sin\alpha\;\frac{\partial^k}{\partial\lambda^k}\left(\frac{\partial\varphi(\lambda)}{\partial z}\frac{\partial^2\varphi(\lambda)}{\partial z^2}\right)\bigg|_{\lambda=0},$$

we have that

$$\varphi_k = \mathcal{L}\,f_k\,, \quad k \geq 2.$$

Hence we obtain:

$$\varphi(\lambda) = \lambda\varphi_1 + \sum_{2}^{\infty} \lambda^k \mathcal{L} f_k \tag{30}$$

or, taking $\lambda = 1$, the solution to (5) is given by

$$\varphi = \varphi_1 + \sum_{2}^{\infty} \mathcal{L} f_k . \tag{31}$$

Our main interest in (31) is that of stability.

THEOREM 1 *Suppose the linear solution φ_1 is stable. That is, denoting the dependence on t by $\varphi_1(\cdot, t)$ suppose*

$$\varphi_1(\cdot, t) \to 0 \quad \textit{as } t \to \infty$$

then so does $\varphi(\cdot, t)$. Also suppose $\varphi_1(\cdot, t)$ is periodic in t, then so is $\varphi(\cdot, t)$ with the same period.

REMARK We note that these statements are consistent with the central result of the more general Hopf Bifurcation Theory — as treated for example in [14]. In particular stability or instability is determined by the linearized equation.

PROOF *These results are easily deduced from the expansion* (31). *Thus if φ_1 is stable so is $\mathcal{L}(f_k)$ for each k. Similarly if φ_1 is periodic so is each f_k and then also $\mathcal{L}(f_k)$ with the same period.* QED.

REMARK We can show that the series (30) converges at a power series in λ and that the function so defined satisfies (7).

Since our primary interest is in stability as a function of U and by the Hopf Bifurcation Theory it is determined by the linear or linearized equation (7L) we shall now consider the linear problem in more detail, combining (1) and (7L). (It would appear that R. Triggiani in the paper presented at this conference (the 21st IFIP TC-7 Conference) pursues a similar idea.)

3. Time-Domain Formulation of Control Problem

We now turn to the time-domain formulation of (1) with the aerodynamic lift and moment terms determined by the linearized equation (7L) (specializing to the typical section aerodynamics). We use the term "time-domain formulation" because in the aeroelastic literature, going back to the classic treatises [15], [16] only the Laplace Transform (or actually the Fourier Transform) theory is considered consistent with the primary interest in stability. Moreover abstract formulation as here is totally new.

We need first to calculate the lift $L(\cdot, t)$ and moment $M(\cdot, t)$, for which we use the solution to the Possion Equation (16). Here we take advantage of the speical form of $\hat{w}(\cdot, \lambda)$ in (18) and introduce the functions

$$\begin{aligned} f_1(x) &= 1, \quad |x| < 1 \\ f_2(x) &= x, \quad |x| < 1. \end{aligned}$$

Let $\hat{A}_i(\cdot, \lambda)$ denote the solution to the Possio Equation:

$$\int_{-1}^{1} \hat{P}(x - \xi)\, \hat{A}(\xi, \lambda)\, d\xi = f_i(x), \quad |x| < 1$$

and let

$$\hat{w}_{ij}(M, k) = \int_{-1}^{1} f_i(x)\, \hat{A}_j(x, \lambda)\, dx$$

as in [13]. Then

$$\begin{aligned} \hat{L}(s, \lambda) &= \int_0^\infty e^{-\lambda t}\, L(s, t)\, dt \\ &= -\rho b U^2 \Big[k\, \hat{w}_{11}\, \hat{h}(s, \lambda) + (k\, \hat{w}_{12} + (1-ak)\hat{w}_{11})\, \hat{\theta}(s, \lambda)\Big] \end{aligned} \tag{32}$$

$$\begin{aligned} \hat{M}(s, \lambda) &= \int_0^\infty e^{-\lambda t}\, M(s, t)\, dt \\ &= -\rho b^2 U^2 \Big[k(\hat{w}_{21} - \hat{w}_{11})\, \hat{h}(s, \lambda) \\ &\quad + (k\, \hat{w}_{22} + (1-ak)\hat{w}_{21} - ak\, \hat{w}_{12} - a(1-ak)\hat{w}_{11})\, \hat{\theta}(s, \lambda)\Big] \end{aligned} \tag{33}$$

Correspondingly, the inverse Laplace Transforms may be expressed

$$L(s, t) = \int_0^t \ell(t - \sigma)\, y(s, \sigma)\, d\sigma$$

$$M(s, t) = \int_0^t m(t - \sigma)\, y(s, \sigma)\, d\sigma$$

where

$$y(s,t) = \begin{vmatrix} X(s,t) \\ \dot{X}(s,t) \end{vmatrix}.$$

A minor complication in inverse Laplace Transforming is that we may have to allow for δ function and its derivative in $\ell(\cdot)$ and $m(\cdot)$, especially at $t = 0$. This is illustrated by the case $M = 0$ treated in [17]. Hence we shall write

$$\begin{vmatrix} L(s,t) \\ M(s,t) \end{vmatrix} = M_a\,\ddot{x}(t) + D_a\,\dot{x}(t) + K_a\,x(t)$$

$$+ \begin{vmatrix} \int_0^t \ell_{00}(t-\sigma)\,y(s,\sigma)\,d\sigma + \int_0^t \ell_{01}(t-\sigma)\,\dot{y}(s,\sigma)\,d\sigma \\ \int_0^t \ell_{10}(t-\sigma)\,y(s,\sigma)\,d\sigma + \int_0^t \ell_{11}(t-\sigma)\,\dot{y}(s,\sigma)\,d\sigma \end{vmatrix}$$

(which only proves the efficiency of Laplace Transforms!) where we may assume (in the absence of proof here!) that

$$M = M_s - M_a > 0 \quad \text{(nonsingular and nonnegative definite).}$$

To proceed to the abstract formulation we need next to take care of the end conditions due to the possibility of self-training actuators. This means "including the boundary value as part of the state," initiated in [18], [19]. Thus let

$$\begin{aligned} \mathcal{H}_b &= L_2p[0,\ell] \times E^1 \\ \mathcal{H}_p &= L_2p[0,\ell] \times E^1 \\ \mathcal{H} &= \mathcal{H}_b \times \mathcal{H}_p\,. \end{aligned}$$

Define the linear operator

$$A_s = \begin{vmatrix} A_h & 0 \\ 0 & A_\theta \end{vmatrix}$$

with domain and range in $\mathcal{H}$.

$$\mathcal{D}(A_h) = \left[\begin{vmatrix} h(\cdot) \\ c \end{vmatrix}, \quad \begin{matrix} h'''' \in L_2[0,\ell] \\ h(0) = h'(0) = 0; \quad h'''(\ell) = 0 \\ c = h'(\ell) \end{matrix} \right] \tag{34}$$

$$A_h \begin{vmatrix} h(\cdot) \\ h'(\ell) \end{vmatrix} = \begin{vmatrix} EIh''''(\cdot) \\ EIh''(\ell) \end{vmatrix} \tag{35}$$

$$\mathcal{D}(A_\theta) = \left[\begin{vmatrix} \theta(\cdot) \\ c \end{vmatrix}, \quad \begin{matrix} \theta'' \in L_2[0,\ell] \\ \theta(0) = 0 \\ c = \theta(\ell) \end{matrix} \right]$$

$$A_\theta \begin{vmatrix} \theta(\cdot) \\ \theta(\ell) \end{vmatrix} = \begin{vmatrix} -GJ\theta''(\cdot) \\ GJ\theta'(\ell) \end{vmatrix}. \tag{36}$$

Thus defined, A_s is self-adjoint, nonnegative definite with dense domain. Let $x \in \mathcal{D}(A_s)$. Then

$$[A_s x, x] = EI \int_0^\ell |h''(s)|^2 \, ds + GJ \int_0^\ell |\theta'(s)|^2 \, ds$$

where

$$x = \begin{vmatrix} h(\cdot) \\ h'(\ell) \\ \theta(\cdot) \\ \theta(\ell) \end{vmatrix}$$

and, in particular, we see that

$$A_s x = 0 \quad \text{implies} \quad x = 0.$$

(There are no rigid-body modes.)

With $\sqrt{A_s}$ denoting the positive square root, we can verify that if $x \in \mathcal{D}\left(\sqrt{A_s}\right)$ we must have that

$$x = \begin{vmatrix} h(\cdot) \\ h'(\ell) \\ \theta(\cdot) \\ \theta(\ell) \end{vmatrix} \quad \text{is such that} \quad \begin{matrix} h''(\cdot) \in L_2[0,\ell] \\ \theta'(\cdot) \in L_2[0,\ell] \\ h(0) = h'(0) = 0 \\ \theta(0) = 0 \end{matrix}.$$

Also, $\sqrt{A_s}$ has a bounded inverse.

Next we define the Hilbert space (energy space):

$$\mathcal{H}_E = \mathcal{D}\left(\sqrt{A_s}\right) \times L_2[0,\ell]^2$$

with inner product

$$[Y, Z]_E = \left[\sqrt{A_s}\, x_1, \sqrt{A_s}\, x_2\right] + [M z_1, z_2]$$

where

$$Y = \begin{vmatrix} x_1 \\ z_1 \end{vmatrix}, \quad Z = \begin{vmatrix} x_2 \\ z_2 \end{vmatrix}$$

$$x_1 = \begin{vmatrix} h_1 \\ h_1'(\ell) \\ \theta_1 \\ \theta_1(\ell) \end{vmatrix}, \quad z_1 = \begin{vmatrix} h_2(\cdot) \\ \theta_2(\cdot) \end{vmatrix}$$

$$x_2 = \begin{vmatrix} h_3 \\ h_3'(\ell) \\ \theta_3 \\ \theta_3(\ell) \end{vmatrix}, \quad z_2 = \begin{vmatrix} h_4(\cdot) \\ \theta_4(\cdot) \end{vmatrix}$$

where

$$M = (M_s - M_a) > 0.$$

Define $\mathcal{A}_s$ with domain and range in $\mathcal{H}_E$ by:

$$\mathcal{D}(\mathcal{A}_s) = \left[Y = \begin{vmatrix} x_1 \\ z_1 \end{vmatrix}, \; x_1 = \begin{vmatrix} h_1 \\ h_1'(\ell) \\ \theta_1 \\ \theta_1(\ell) \end{vmatrix} \in \mathcal{D}(\mathcal{A}_s), \; z_1 = \begin{vmatrix} h_2(\cdot) \\ \theta_2(\cdot) \end{vmatrix} \right.$$

$$\left. \text{and} \quad \begin{vmatrix} h_2 \\ -EI\frac{h_1''(\ell)}{g_h} \\ \theta_2 \\ -GJ\frac{\theta_1'(\ell)}{g_\theta} \end{vmatrix} \in \mathcal{D}\left(\sqrt{\mathcal{A}_s}\right) \right].$$

(The last condition implies in particular that

$$h_2'(\ell) = \frac{-EIh_1''(\ell)}{g_h}$$

$$\theta_2(\ell) = \frac{-GJ\theta_1'(\ell)}{g_\theta} \Bigg)$$

$$\mathcal{A}_s Y = \begin{vmatrix} h_2 \\ \frac{-EIh_1''(\ell)}{g_h} \\ \theta_2 \\ \frac{-GJ\theta_1'(\ell)}{g_\theta} \\ -M^{-1} \begin{vmatrix} EIh_1''''(\cdot) \\ -GJ\theta_1''(\cdot) \end{vmatrix} \end{vmatrix}.$$

The defined $\mathcal{A}_s$ is closed with dense domain, and compact resolvent. Moreover

$$\begin{aligned}\operatorname{Re}[\mathcal{A}_s Y, Y] &= \frac{-1}{g_h}(EI)^2|h_1''(\ell)|^2 - \frac{1}{g_\theta}(GJ)^2|\theta_1'(\ell)|^2 \\ &= -g_h|h_1'(\ell)|^2 - g_\theta|\theta_1(\ell)|^2. \end{aligned} \tag{37}$$

REMARK The last relation allows extension to the "limiting" cases:

$$\begin{aligned} g_h &= 0 \quad \text{by adding the condition } h_1''(\ell) = 0 \\ g_\theta &= 0 \quad \text{by adding the condition } \theta_1'(\ell) = 0. \end{aligned}$$

Next we define the linear operators D and K on $\mathcal{H}_E$ into $\mathcal{H}_E$:

$$Y = \begin{vmatrix} h_1 \\ h_1'(\ell) \\ \theta_1 \\ \theta_1(\ell) \\ h_2 \\ \theta_2 \end{vmatrix}; \quad DY = \begin{vmatrix} 0 \\ 0 \\ 0 \\ 0 \\ M^{-1}D_a \begin{vmatrix} h_2 \\ \theta_2 \end{vmatrix} \end{vmatrix}; \quad KY = \begin{vmatrix} 0 \\ 0 \\ 0 \\ 0 \\ M^{-1}K_a \begin{vmatrix} h_1 \\ \theta_1 \end{vmatrix} \end{vmatrix}$$

Thus defined, D and K are bounded linear operators, their precise bounds being not of interest. Define

$$\mathcal{A} = \mathcal{A}_s + UD + U^2 K. \tag{38}$$

Then $\mathcal{A}$ generates a C_0-semigroup; denote it $S(t)$, $t \geq 0$. Also

$$\operatorname{Re}[\mathcal{A}Y, Y] = \operatorname{Re}[\mathcal{A}_s Y, Y] + \pi\rho U^2 \operatorname{Re}[\theta_1, \theta_2]. \tag{39}$$

The semigroup $S(\cdot)$ is thus not necessarily a contraction for nonzero U. But the resolvent of $\mathcal{A}$, $R(\lambda, \mathcal{A})$, is compact and there are no eigenvalues in the half plane

$$\operatorname{Re}\lambda > \sigma_a$$

where σ_a is the growth bound of the semigroup generated by $\mathcal{A}$, which of course depends on U.

Next for each $t \geq 0$, define the linear operators $L_0(t)$, $L_1(t)$ on $\mathcal{H}_E$ into $\mathcal{H}_E$ by

$$L_0(t)\,Y = \begin{vmatrix} 0 \\ 0 \\ 0 \\ 0 \\ M^{-1} \begin{vmatrix} \ell_{00}(t)\,V \\ \ell_{10}(t)\,V \end{vmatrix} \end{vmatrix} \tag{40}$$

$$L_1(t)\,Y \;=\; \begin{vmatrix} 0 \\ 0 \\ 0 \\ 0 \\ M^{-1}\begin{vmatrix} \ell_{01}(t)\,V \\ \ell_{11}(t)\,V \end{vmatrix} \end{vmatrix} \tag{41}$$

where

$$V \;=\; \begin{vmatrix} h_1 \\ \theta_1 \\ h_2 \\ \theta_2 \end{vmatrix}.$$

Then our abstraction version (or time domain formulation) is the evolution-convolution equation in a Hilbert space:

$$\dot{Y}(t) \;=\; \mathcal{A}Y(t) \;+\; \int_0^t L_0(t-\sigma)\,Y(\sigma)\;d\sigma \;+\; \int_0^t L_1(t-\sigma)\,\dot{Y}(\sigma)\;d\sigma \tag{42}$$

where we shall need to impose some additional properties of the operators $L_0(\cdot)$, $L_1(\cdot)$, such as

$$\int_0^\infty \|L_0(t)\|e^{-\sigma t} \;+\; \int_0^\infty \|L_1(t)\|e^{-\sigma t} \;<\; 0, \quad \sigma > 0$$

and

$L_1(t)$ coverges strongly to $L_0(\infty)$, linear bounded as $t \to \infty$

$L_0(t)$ coverges strongly to $L_1(\infty)$, linear bounded as $t \to \infty$.

Equations like (42) have been discussed in the pure mathematics literature (e.g., [20]) but unfortunately are much too abstract to provide answers to the questions of interest to us here. Following [17] we start by taking Laplace Transforms in (42). Defining

$$\hat{Y}(\lambda) \;=\; \int_0^\infty e^{-\lambda t}\,Y(t)\;dt, \quad \operatorname{Re}\lambda > \sigma_a$$

we have

$$(\lambda I - \mathcal{A})\hat{Y}(\lambda) \;-\; \hat{L}_0(\lambda)\hat{Y}(\lambda) \;-\; \hat{L}_1(\lambda)\left(\lambda\hat{Y}(\lambda) - Y(0)\right) \;=\; Y(0)$$

or

$$\left(\lambda I - \mathcal{A} - \hat{L}_0(\lambda) - \lambda\hat{L}_1(\lambda)\right)\hat{Y}(\lambda) \;=\; (I \;-\; \hat{L}_1(\lambda))Y(0).$$

We shall refer to

$$\left(\lambda I - \mathcal{A} - \hat{L}_0(\lambda) - \lambda \hat{L}_1(\lambda)\right) Y = Z \tag{43}$$

as the generalized resolvent equation and

$$\left(\lambda I - \mathcal{A} - \hat{L}_0(\lambda) - \lambda \hat{L}_1(\lambda)\right)^{-1} \tag{44}$$

the generalized resolvent. We may then state: (cf [19]) without proof:

THEOREM 2 *For any λ, either*

$$\left(\lambda I - \mathcal{A} - \hat{L}_0(\lambda) - \lambda \hat{L}_1(\lambda)\right) Y = 0$$

for $\|Y\| \neq 0$, or

$$\left(\lambda I - \mathcal{A} - \hat{L}_0(\lambda) - \lambda \hat{L}_1(\lambda)\right)$$

has a bounded inverse.

THEOREM 3 *Call λ such that*

$$\left(\lambda I - \mathcal{A} - \hat{L}_0(\lambda) - \lambda \hat{L}_1(\lambda)\right) Y = 0, \quad \|Y\| \neq 0 \tag{45}$$

an "aeroelastic mode." The aeroelastic modes are countable in number for each fixed M, U and only a finite number can have positive real part.

We can show that the aeroelastic modes are precisely the eigenvalues of the infinitesimal generator of a semigroup. We can also view this another way. We make a state space formuation of the linear system represented by the evolution-convolution equation (42). By state space representation we mean the representation

$$\left.\begin{aligned} \dot{Z}(t) &= \mathcal{A}_c Z(t) + \mathcal{B} u(t) \\ Y(t) &= \mathcal{C} z(t) \end{aligned}\right\} \tag{46}$$

where $\mathcal{A}_c$ is the infinitesimal generator of a C_0-semigroup, $\mathcal{B}$ is linear bounded and $\mathcal{C}$ is closed linear. In the present case $\mathcal{B} = 0$, and the controls are included already in "feedback" form. The state space needs to be a Banach Space. Such a representation for the case $L_0(\cdot) = 0$ is given in [21], and is readily generalized to the present case. We should note that the representation (46) allows us to check controllability and stabilizability for any given control scheme. $Y(\cdot)$ represents only the structure state and $Z(\cdot)$ includes a "stand in" for the aerodynamic state.

3.1 Calculation of Aeroelastic Modes: Flutter Speed

From the practical point of view perhaps the most important problem is to track the aeroelastic modes as a function of U, for fixed M, in order to determine stability. For $\alpha = 0$ this is carried out in [13] and the extension to nonzero α is straightforward.

Thus we need to start with "unwinding" (45), returning to the Laplace Transform version of (1). We show that the modes are the zeros of a function

$$d(M, \lambda, U)$$

which is analytic in λ except for a logarithmic singularity for $\lambda \leq 0$. We need to define the roots as a single-valued function of U — define the "root locus." For $U = 0$ we obtain the structure modes — two sequences — the "bending" modes and the "pitching" modes. We begin with:

$$\lambda_k(\lambda, M, 0) = i\omega_k$$

the structure modes, and we show that

$$\frac{\partial}{\partial \lambda} d(M, \lambda, U)\bigg|_{\lambda = i\omega_k} \neq 0$$

so that we can via the usual implicit function theory define the roots $\lambda_k(M, U)$ as a function of U with

$$-\frac{\frac{\partial d}{\partial U}}{\frac{\partial d}{\partial \lambda}} = \frac{\partial \lambda(M, U)}{\partial U} .$$

See [13] for the details. If, for example,

$$\lambda_k(M, 0) = i\omega_k$$

is the the kth bending mode we keep calling $\lambda_k(M, U)$ the root locus of the kth bending mode. Let

$$\sigma_k(M, U) = \operatorname{Re} \lambda_k(M, U).$$

Then

$$\sigma_k(M, 0) = 0.$$

The curve of $\sigma_k(M, U)$ is called the "stability curve." We show that

$$\frac{\partial \sigma_K(M, U)}{\partial U}\bigg|_{U=0} = \text{constant} \left(\frac{-1}{M}\right)$$

for all k with the constant depending on whether it is a pitching mode or a bending mode. This enables us to define flutter speed $U_F(M)$ as the first time

$$\sigma_k(M,U) = 0, \quad \frac{\partial \sigma_k(M,U)}{\partial U} > 0.$$

We are also able [13] to deduce a good many properties of $U_F(M)$ as a function of M, but not nearly enough! Much work still remains to be done especially for M close to 1, being in particular not continuous at $M = 1$.

As shown in [4] the theory can help explain the occurrence of the Transonic Dip due to nonzero angle of attack — and should also explain that due to camber observed in computations [22], [23]. The point is that even though the system is nonlinear the stability as we have seen is still determined by the linearized model.

The solution Y_k of the modal equation corresponding to the kth aeroelastic mode $\lambda_k(M,U)$ is called the "mode shape" and we can express the time-domain ("unsteady") solution of the aeroelastic convolution-evolution equation in terms of the elements in Y_k even though it is not an eigenfunction expansion — see [17], [19]. This is not of much use in application to the flutter problem except to indicate the nature of the instability, since the aerodynamic initial conditions can never be determined. It is nevertheless of mathematical interest.

References

[1] Meyer, R.E., *Introduction to Mathematical Fluid Mechanics.* Dover Publications, 1982.

[2] Friedmann, P.P. "The Renaissance of Aeroelasticity and Its Future." *Journal of Aircraft*, Vol. 36, No. 1 (1999), pp. 105–121.

[3] Goland, M. "The Flutter of a Uniform Cantilever Wing." *Journal of Applied Mechanics, ASME*, Vol. 12, No. 4 (1954), pp. A197–A208.

[4] Balakrishnan, A.V. "On the Transonic Small Disturbance Potential Equation." Submitted to *AIAA Journal.*

[5] Nixon, D. "Basic Equations for Unsteady Transonic Flow." Chapter 2 in: *Unsteady Transonic Aerodynamics.* Progress in Astronautics and Aeronautics Series, Vol. 120. Edited by David Nixon. American Institute of Astronautics and Aeronautics, 1989. Pp. 57–73.

[6] Ferrari, C. and Tricomi, F.G. *Transonic Aerodynamics.* Translated by Raymond H. Cramer. Translation of *Aerodinamica transonica.* Academic Press, New York, 1968.

[7] Bers, L. *Mathematical Aspects of Subsonic and Transonic Gas.* Surveys in Applied Mathematics, Vol. 3. John Wiley and Sons, New York, 1958.

[8] Guderley, K.G. *The Theory of Transonic Flow.* Pergamon Press, 1962.

[9] Balakrishnan, A.V. "On the (Non-numeric) Mathematical Foundations of Linear Aeroelasticity." In: *Fourth International Conference on Nonlinear Problems in Aviation and Aerospace.* Edited by Seenith Sivasundaram. European Conference Publications, 2003. Pp. 11–41.

[10] Cole, J.D. and Cook, L.P. *Transonic Aerodynamics.* North-Holland Series in Applied Mathematics and Mechanics, Vol. 30. Elsevier Science Pub. Co., 1986.

[11] Balakrishnan, A.V. "Possio Integral Equation of Aeroelasticity Theory." *Journal of Aerospace Engineering,* Vol. 16, No. 4 (2003).

[12] Possio, C. "L'azione aerodinamica sul profile oscillante in un fluido compressibile a velocità iposonora." *L'Aerotecnica.* Vol. 18, No. 4 (1938).

[13] Balakrishnan, A.V. and Iliff, K.W. "A Continuum Aeroelastic Model for Inviscid Subsonic Bending-Torsion Wing Flutter." In: *Proceedings of International Forum on Aeroelasticity and Structural Dynamics, June 4–6, 2003, Amsterdam.*

[14] Chorin, A.J. and Marsden, J.E. *A Mathematical Introduction to Fluid Mechanics.* 3rd Edition. Texts in Applied Mathematics, Vol. 4. Springer-Verlag, New York, 1993.

[15] Bisplinghoff, R.L., Ashley, H. and Halfman, R.L. *Aeroelasticity.* Addison-Wesley Publishing Co., 1955.

[16] Fung, Y.C. *An Introduction to the Theory of Aeroelasticity.* Dover, New York, 1983.

[17] Balakrishnan, A.V. "Aeroelastic Control with Self-straining Actuators: Continuum Models." In: *Smart Structures and Materials 1998: Mathematics and Control in Smart Structures.* Edited by Vasundara V. Varadan. Proceedings of SPIE, Vol. 3323. Pp. 44–54.

[18] Balakrishnan, A.V. "Dynamics and Control of Articulated Anisotropic Timoshenko Beams." In: *Dynamics and Controls of Distributed Systems.* Edited by H.S. Tzou and L.A. Bergman. Cambridge University Press, United Kingdom, 1998. Pp. 121–201.

[19] Balakrishnan, A.V. "Subsonic Flutter Suppression Using Self-straining Actuators." *Journal of the Franklin Institute,* Vol. 338 (2001), pp. 149–170.

[20] Prüss, J. *Evolutionary Integral Equations and Applications,* Birkhauser, 1993.

[21] Balakrishnan, A.V. "Representing the Convolution-Semigroup Equation of Aeroelasticity as a Pure Semigroup Equation." Presented at ICNPAA 2000, Third International Conference on Nonlinear Problems in Aviation and Aerospace, Daytona Beach, Florida, May 2000. Unpublished report.

[22] Bendiksen, O.O. "Transonic Flutter." AIAA Paper No. 2002-1488. Presented at: 43rd AIAA/ASME/ASCE/AHS/ASC Structures, Structural Dynamics, and Materials Conference, Denver, Colorado, April 22-25, 2002.

[23] Schultz, S. "Transonic Aeroelastic Simulation of a Flexible Wing Section." Agard Report 822. March 1998.

UNIFORM CUSP PROPERTY, BOUNDARY INTEGRAL, AND COMPACTNESS FOR SHAPE OPTIMIZATION

Michel C. Delfour*
Centre de recherches mathématiques et Département de mathématiques et de statistique Université de Montréal, C.P. 6128, succ. Centre-ville, Montréal (Qc), Canada H3C 3J7
delfour@CRM.UMontreal.CA

Nicolas Doyon
Département de mathématiques et de statistique, Université de Montréal, C. P. 6128, succ. Centre-ville, Montréal (Qc), Canada H3C 3J7
nicodoyon@hotmail.com

Jean-Paul Zolésio
CNRS and INRIA, 2004 route des lucioles, BP 93, FR-06902 Sophia Antipolis, France
Jean-Paul.Zolesio@sophia.inria.fr

Abstract In this paper we consider the family of sets verifying the uniform cusp property introduced in [2] and extended in [4] to cusp functions only continuous at the origin. In the latter case we show that to any extended cusp function, we can associate a continuous, non-negative, and monotone strictly increasing cusp function of the type introduced in [2]. We construct an example of a bounded set in $\mathbf{R}^N$ with a cusp function of the form $c|\theta|^\alpha$, $0 < \alpha < 1$, for which its boundary integral is infinite and the Hausdorff dimension of its boundary is exactly $N - \alpha$. We then give compactness theorems for the family of subsets of a bounded open holdall verifying a uniform cusp property with a uniform bound on either the De Georgi [6] or the γ-density perimeter of Bucur and Zolésio [1]. We also give their *uniform local C^0-graph* versions following [4].

*This research has been supported by National Sciences and Engineering Research Council of Canada discovery grant A-8730 and by a FQRNT grant from the Ministère de l'Éducation du Québec.

This class forms a much larger family than the one of subsets verifying a uniform cone property.

Keywords: Oriented distance function, signed distance function, compactness, cusp property, boundary integral, density perimeter, Hausdorff dimension

Introduction

In this paper we consider the family of sets verifying the uniform cusp property introduced in [2] and extended in [4] to cusp functions only continuous at the origin. In the latter case we show that to any extended cusp function, we can associate a continuous, non-negative, and monotone strictly increasing cusp function of the type originally introduced in [2]. Unlike sets verifying a uniform cone property, such sets do not necessarily have a locally finite boundary integral. This fact is illustrated by constructing an example of a bounded subset of $\mathbf{R}^N$ with cusp function $c|\theta|^\alpha$, $0 < \alpha < 1$, for which the boundary integral is infinite and the Hausdorff dimension of its boundary is exactly $N - \alpha$.

Even without a uniform bound on the perimeter a general compactness theorem was given in [4] for a family of subsets of a bounded hold-all verifying a uniform cusp property with a cusp function only continuous at the origin. In this paper we give compactness theorems for the family of subsets of a bounded open holdall verifying a uniform cusp property with a uniform bound on either the De Georgi [6] or the γ-density perimeter of Bucur and Zolésio [1]. We also give in § 4.3 their *uniform local C^0-graph* versions following [4]. This class of subsets forms a much larger family than the one of subsets verifying a uniform cone property.

1. Preliminaries: Topologies on Families of Sets

We first introduce some notation. Given an integer $N \geq 1$, m_N and H_{N-1} will denote the N-dimensional Lebesgue and $(N-1)$-dimensional Hausdorff measures. The inner product and the norm in $\mathbf{R}^N$ will be written $x \cdot y$ and $|x|$. The *complement* $\{x \in \mathbf{R}^N : x \notin \Omega\}$ and the boundary $\overline{\Omega} \cap \overline{\complement\Omega}$ of a subset Ω of $\mathbf{R}^N$ will be respectively denoted by $\complement\Omega$ or $\mathbf{R}^N \backslash \Omega$ and by $\partial\Omega$ or Γ. The *distance function* $d_A(x)$ from a point x to a subset $A \neq \varnothing$ of $\mathbf{R}^N$ is defined as $\inf\{|y - x| : y \in A\}$.

Recall a few results on metric topologies defined on spaces of equivalence classes of sets constructed from the characteristic function, the distance or the oriented distance functions to a set. Given $\Omega \subset \mathbf{R}^N$, $\Gamma \neq \varnothing$, the *oriented distance function* is defined as

$$b_\Omega(x) \stackrel{\text{def}}{=} d_\Omega(x) - d_{\complement\Omega}(x). \tag{1}$$

It is Lipschitz continuous of constant 1, and ∇b_Ω exists and $|\nabla b_\Omega| \le 1$ almost everywhere in $\mathbf{R}^N$. Thus $b_\Omega \in W^{1,p}_{\text{loc}}(\mathbf{R}^N)$ for all p, $1 \le p \le \infty$. Recall that $b^+_\Omega = d_\Omega$, $b^-_\Omega = d_{\complement\Omega}$, and $|b_\Omega| = d_\Gamma$, and that $\chi_{\text{int}\,\Omega} = |\nabla d_{\complement\Omega}|$, $\chi_{\text{int}\,\complement\Omega} = |\nabla d_\Omega|$, and $\chi_\Gamma = 1 - |\nabla d_\Gamma|$ a.e. in $\mathbf{R}^N$, where χ_A denotes the characteristic function of a subset A of $\mathbf{R}^N$. Given a nonempty subset D of $\mathbf{R}^N$, the family $C_b(D) = \{b_\Omega : \Omega \subset \overline{D} \text{ and } \Gamma \neq \varnothing\}$ is closed in $W^{1,p}(D)$. The following theorem is central. It shows that convergence and compactness in the metric on $C_b(D)$ associated with $W^{1,p}(D)$ will imply the same properties in the other topologies introduced in [2].

THEOREM 1 *Let* $D \subset \mathbf{R}^N$ *be bounded open and* $1 \le p < \infty$. *The maps*

$$b_\Omega \mapsto (b^+_\Omega, b^-_\Omega, |b_\Omega|) = (d_\Omega, d_{\complement\Omega}, d_{\partial\Omega}) : C_b(D) \subset W^{1,p}(D) \to W^{1,p}(D)^3 \quad (2)$$

$$b_\Omega \mapsto (\chi_{\partial\Omega}, \chi_{\text{int}\,\Omega}, \chi_{\text{int}\,\complement\Omega}) : W^{1,p}(D) \to L^p(D)^3 \quad (3)$$

are continuous.

Proof. – They are well-defined from [2] (Chapter 5, Theorem 2.1 (iii), p. 207) for the map (2) and [2] (Chapter 5, Thm 2.2 (iv)-(v), p. 210) for the map (3). They are continuous from [2] (Chapter 5, Thm 5.1). □

2. Extension of the Uniform Cusp Property

The *uniform cusp property* introduced in [2] (Chapter 5, § 11) was specified by a continuous function $h : [0, \rho[\to \mathbf{R}$ such that

$$h(0) = 0, \quad h(\rho) = \lambda, \quad \forall \theta, 0 < \theta < \rho, \quad 0 < h(\theta) < \lambda. \quad (4)$$

Recall that with h of the form $h(\theta) = \lambda\,(\theta/\rho)^\alpha$, $0 < \alpha \le 1$, we recover the *uniform cusp property* for $0 < \alpha < 1$ and the *uniform cone property* for $\alpha = 1$, $\rho = \lambda \tan\omega$ and $h(\theta) = \theta/\tan\omega$ which corresponds to an open cone in 0 of aperture ω, height λ, and axis e_N.

The uniform cusp property was extended in [4] to the family of cusp functions h in the larger space

$$\mathcal{H} \stackrel{\text{def}}{=} \{h : [0, \infty[\to \mathbf{R} \,:\, h(0) = 0 \text{ and } h \text{ is continuous in } 0\} \quad (5)$$

by associating with $h \in \mathcal{H}$, $\rho > 0$, and λ the axi-symmetrical region

$$C(\lambda, h, \rho) \stackrel{\text{def}}{=} \left\{(\zeta', \zeta_N) \in \mathbf{R}^N : |\zeta'| < \rho \text{ and } \limsup_{\xi' \to \zeta'} h(|\xi'|) < \zeta_N < \lambda\right\} \quad (6)$$

around the axis $e_N = (0, \ldots, 0, 1)$ in $\mathbf{R}^N$. Given $\lambda > 0$, $\rho > 0$, $h \in \mathcal{H}$, and a direction $d \in \mathbf{R}^N$, $|d| = 1$, the rotated region from direction e_N to

d is defined as

$$C(\lambda, h, \rho, d) \stackrel{\text{def}}{=} \left\{ y \in \mathbf{R}^N : \begin{array}{l} |P_{H_d}(y)| < \rho \text{ and} \\ \limsup_{z \to y} h(|P_{H_d}(z)|) < y \cdot d < \lambda \end{array} \right\}, \quad (7)$$

where $H_d = \{d\}^\perp$ is the hyperplane through 0 orthogonal to the direction d. Finally, the translation of $C(\lambda, h, \rho, d)$ to the point x will be denoted

$$C_x(\lambda, h, \rho, d) \stackrel{\text{def}}{=} x + C(\lambda, h, \rho, d).$$

LEMMA 2 ([4],[5]) *For all $\lambda > 0$, $\rho > 0$, $h \in \mathcal{H}$, and $x \in \mathbf{R}^N$, the regions $C(\lambda, h, \rho)$ and $C_x(\lambda, h, \rho, d)$ are nonempty and open. Moreover the segment $(x, x + \lambda d)$ is contained in $C_x(\lambda, h, \rho, d)$.*

The function h is referred to as a *cusp function* and the space $\mathcal{H}$ as the *space of cusp functions.* The definition of the uniform cusp property in [2] (Chapter 5, § 11) can now be extended to the larger class $\mathcal{H}$.

DEFINITION 3 *Let Ω be a subset of $\mathbf{R}^N$ such that $\partial\Omega \neq \varnothing$.*

(i) Ω satisfies the local uniform cusp property *if*

$$\begin{array}{ll} \forall x \in \partial\Omega, & \exists h \in \mathcal{H}, \exists \lambda > 0, \exists \rho > 0, \exists r > 0, \exists d \in \mathbf{R}^N, |d| = 1, \\ \textit{such that} & \forall y \in B(x, r) \cap \overline{\Omega}, \quad C_y(\lambda, h, \rho, d) \subset \operatorname{int} \Omega. \end{array}$$

*(ii) Given $h \in \mathcal{H}$, Ω satisfies the h-*local uniform cusp property *if*

$$\begin{array}{ll} \forall x \in \partial\Omega, & \exists \lambda > 0, \exists \rho > 0, \exists r > 0, \exists d \in \mathbf{R}^N, |d| = 1, \\ \textit{such that} & \forall y \in B(x, r) \cap \overline{\Omega}, \quad C_y(\lambda, h, \rho, d) \subset \operatorname{int} \Omega. \end{array}$$

(iii) Ω satisfies the uniform cusp property *for (r, λ, h, ρ) if*

$$\begin{array}{ll} \exists h \in \mathcal{H}, \exists \lambda > 0, \exists \rho > 0, \exists r > 0, & \forall x \in \partial\Omega, \exists d \in \mathbf{R}^N, |d| = 1, \\ \textit{such that} & \forall y \in B(x, r) \cap \overline{\Omega}, \quad C_y(\lambda, h, \rho, d) \subset \operatorname{int} \Omega. \end{array}$$

The three cases of Definition 3 only differ when $\partial\Omega$ is not compact.

THEOREM 4 ([4]) *If $\partial\Omega$ is compact, then the three uniform cusp properties of Definition 3 coincide.*

In fact, when a local uniform cusp property is verified for some cusp function $h \in \mathcal{H}$, it is verified for another cusp function which is continuous, non-negative, and monotone strictly increasing as in (4).

THEOREM 5 *Assume that Ω satisfies the local uniform cusp property in $x \in \partial\Omega$ for some (r, λ, h, ρ), $h \in \mathcal{H}$. Then there exist $(r', \lambda', h', \rho')$,* *with $h' \in \mathcal{H}$ continuous, non-negative, monotone strictly increasing, and $\lambda' = h'(\rho')$, such that Ω satisfies the local uniform cusp property in $x \in \partial\Omega$ for (r, λ', h', ρ').*

Proof. – By continuity of $h \in \mathcal{H}$ in 0,

$$\exists 0 < \theta_0 \leq \rho, \quad \forall 0 \leq \theta \leq \theta_0, \quad |h(\theta)| \leq \lambda/2,$$
$$\forall n \geq 1, \quad \exists 0 < \theta_n < \theta_{n-1}/2, \quad \forall 0 \leq \theta \leq \theta_n, \quad |h(\theta)| \leq \lambda/2^{n+1}.$$

At each step $n \geq 0$ construct the continuous monotone strictly increasing and non-negative function $k_n : [0, \theta_0] \to \mathbf{R}$ defined as follows

$$k_n(\theta) \stackrel{\text{def}}{=} \begin{cases} \frac{\lambda}{2^{j+1}} \frac{\theta_j - \theta}{\theta_j - \theta_{j+1}} + \frac{\lambda}{2^j} \frac{\theta - \theta_{j+1}}{\theta_j - \theta_{j+1}}, & \text{if } \theta_{j+1} < \theta \leq \theta_j, \quad 0 \leq j \leq n-1 \\ \frac{\lambda}{2^{n+1}} \frac{\theta_n - \theta}{\theta_n} + \frac{\lambda}{2^n} \frac{\theta}{\theta_n}, & \text{if } 0 \leq \theta \leq \theta_n. \end{cases}$$

By continuity of h at the origin and the fact that $h(0) = 0$, $\theta_n \to 0$ and $k_n(0) \to 0$. By construction, $0 \leq |h(\theta)| \leq k_{n+1}(\theta) \leq k_n(\theta)$ in $[0, \theta_0]$, $k_{n+1}(\theta) = k_n(\theta)$ in $[\theta_{n+1}, \theta_0]$, and $\|k_{n+1} - k_n\|_{C[0,\theta_{n+1}]} \leq \lambda/2^{n+1}$. Therefore there exists a continuous non-negative and monotone strictly increasing function $k \in C[0, \theta_0]$ such that $k_n \to k$ in $C[0, \theta_0]$, $k(0) = 0$, and $|h(\theta)| \leq k(\theta) \leq \lambda$ in $[0, \theta_0]$. Finally, if $k(\theta_0) = \lambda$, choose ρ' such that $k(\rho') = \lambda$, $\lambda' = \lambda$, and $h' = k$. If $k(\theta_0) < \lambda$, choose $\rho' = \theta_0$, $\lambda' = k(\theta_0)$, and $h' = k$. From the construction, $\rho' \leq \rho$, $\lambda' \leq \lambda$, $h' \geq h$, and hence $C(\lambda', h', \rho') \subset C(\lambda, h, \rho)$. Therefore the local uniform cusp property of Definition 3 is verified with a non-negative, continuous, and monotone strictly increasing cusp function of the form (4). □

We now turn to the compactness theorem. Given a bounded open subset D of $\mathbf{R}^N$, $\rho > 0$, $\lambda > 0$, $r > 0$, and $h \in \mathcal{H}$, consider the family

$$L(D, \lambda, h, \rho, r) \stackrel{\text{def}}{=} \left\{ \Omega \subset \overline{D} : \begin{array}{c} \Omega \text{ satisfies the uniform cusp} \\ \text{property for } (\lambda, h, \rho, r) \end{array} \right\}. \tag{8}$$

The compactness Theorem 11.1 ([2], Chapter 5) readily extends to $\mathcal{H}$.

THEOREM 6 ([4]) *Let D be a nonempty bounded open subset of $\mathbf{R}^N$ and $1 \leq p < \infty$. For $\rho > 0$, $\lambda > 0$, and $h \in \mathcal{H}$ the family*

$$B(D, \lambda, h, \rho, r) \stackrel{\text{def}}{=} \{b_\Omega : \forall \Omega \in L(D, \lambda, h, \rho, r)\}$$

is compact in $C(\overline{D})$ and $W^{1,p}(D)$. As a consequence the families

$$B_d(D, \lambda, h, \rho, r) \stackrel{\text{def}}{=} \{d_\Omega : \forall \Omega \in L(D, \lambda, h, \rho, r)\},$$
$$B_d^c(D, \lambda, h, \rho, r) \stackrel{\text{def}}{=} \{d_{\complement\Omega} : \forall \Omega \in L(D, \lambda, h, \rho, r)\},$$
$$B_d^\partial(D, \lambda, h, \rho, r) \stackrel{\text{def}}{=} \{d_{\partial\Omega} : \forall \Omega \in L(D, \lambda, h, \rho, r)\}$$

are compact in $C(\overline{D})$ *and* $W^{1,p}(D)$*, and the following families are compact in* $L^p(D)$

$$X(D,\lambda,h,\rho,r) \stackrel{\text{def}}{=} \{\chi_\Omega : \forall \Omega \in L(D,\lambda,h,\rho,r)\},$$
$$X^c(D,\lambda,h,\rho,r) \stackrel{\text{def}}{=} \{\chi_{\complement\Omega} : \forall \Omega \in L(D,\lambda,h,\rho,r)\}.$$

3. Extended Uniform Cusp Property and Boundary Integral (Perimeter)

Domains Ω which are locally Lipschitzian epigraphs or, equivalently, satisfy the local uniform cone property enjoy the additional property that the $(N-1)$-Hausdorff measure of their boundary $\partial\Omega$ is locally finite. In general, this is no longer true for domains which are locally Hölderian epigraphs of exponent α, $0<\alpha<1$, but we have an upper bound on the Hausdorff dimension of $\partial\Omega$. We first recall a definition.

DEFINITION 7 *Let* $\Omega \subset \mathbf{R}^N$ *be such that* $\partial\Omega \neq \varnothing$*. The set* Ω *is said to be* locally a $C^{0,\ell}$-epigraph, $0 \le \ell \le 1$*, if for each* $x \in \partial\Omega$ *there exist*

(a) *an open neighborhood* $\mathcal{U}(x)$ *of* x*;*

(b) *a direction* $e_N(x) \in \mathbf{R}^N$, $|e_N(x)| = 1$*;*

(c) *a bounded open neighborhood* $V_{H(x)}$ *of* 0 *in the hyperplane* $H(x) = \{e_N(x)\}^\perp$ *through* 0 *such that*

$$\mathcal{U}(x) \subset \{y \in \mathbf{R}^N : P_{H(x)}(y-x) \in V_{H(x)}\}, \tag{9}$$

where $P_{H(x)}$ *is the orthogonal projection onto* $H(x)$*; and*

(d) *a* $C^{0,\ell}$*-mapping* $a_x \colon V_{H(x)} \to \mathbf{R}$ *such that*

$$\mathcal{U}(x) \cap \partial\Omega = \left\{ x + \zeta' + \zeta_N e_N(x) : \begin{array}{c} \zeta' \in V_{H(x)} \\ \zeta_N = a_x(\zeta') \end{array} \right\} \tag{10}$$

$$\mathcal{U}(x) \cap \text{int}\Omega = \mathcal{U}(x) \cap \left\{ x + \zeta' + \zeta_N e_N(x) : \begin{array}{c} \zeta' \in V_{H(x)} \\ \zeta_N > a_x(\zeta') \end{array} \right\}. \tag{11}$$

THEOREM 8 *If* Ω *in* $\mathbf{R}^N$ *satisfies the uniform cusp property associated with the function* $h(\theta) = \theta^\alpha$, $0<\alpha<1$*, then the Hausdorff dimension of* $\partial\Omega$ *is less or equal to* $N-\alpha$*.*

Proof. – From Theorem 3.3 (i) in [4], Ω is locally a $C^{0,\alpha}$-epigraph and, a fortiori, a C^0-epigraph. Let $r>0$, $\rho>0$, and $\lambda>0$ be the parameters,

$e_N(x) = d_x$ the direction and $H(x)$ the hyperplane through 0 orthogonal to d_x associated with the point $x \in \partial\Omega$. Then there exists $\overline{\rho}$,

$$0 < \overline{\rho} \le r_\lambda \stackrel{\text{def}}{=} \min\{r, \lambda/2\} \tag{12}$$

which is the largest radius such that

$$B_{H(x)}(0, \overline{\rho}) \subset \left\{P_{H(x)}(y - x) \,:\, \forall y \in B(x, r_\lambda) \cap \partial\Omega\right\}.$$

The neighborhoods of Definition 3.2 in [4] or Definition 5.2 in Chapter 2 of [2] that specify the local graph $a_x : V_{H(x)} \to \mathbf{R}$ can be chosen as

$$\begin{aligned} V_{H(x)} &\stackrel{\text{def}}{=} B_{H(x)}(0, \overline{\rho}) \text{ and} \\ \mathcal{U}(x) &\stackrel{\text{def}}{=} B(x, r_\lambda) \cap \left\{y : P_{H(x)}(y - x) \in V_{H(x)}\right\}, \end{aligned} \tag{13}$$

where $B_{H(x)}(0, \overline{\rho})$ is the open ball of radius $\overline{\rho}$ in the hyperplane $H(x)$. For each $\zeta' \in V_{H(x)}$, there exists a unique $y_{\zeta'} \in \partial\Omega \cap \mathcal{U}(x)$ such that $P_{H(x)}(y_{\zeta'} - x) = \zeta'$ and the function

$$\zeta' \mapsto a_x(\zeta') \stackrel{\text{def}}{=} (y_{\zeta'} - x) \cdot d_x : V_{H(x)} \to \mathbf{R}$$

is well-defined, bounded,

$$\forall \zeta' \in V_{H(x)}, \quad |a_x(\zeta')| < r_\lambda, \tag{14}$$

uniformly continuous in $V_{H(x)}$, and

$$\forall \zeta'_1, \zeta'_2 \in V_{H(x)}, \quad \left|a_x(\zeta'_2) - a_x(\zeta'_1)\right| \le c\, \left|\zeta'_2 - \zeta'_1\right|^\alpha. \tag{15}$$

Since $\partial\Omega$ is compact there exists a finite number of points $\{x_i \in \partial\Omega : 1 \le i \le m\}$ such that $\partial\Omega \subset \cup_{i=1}^m \mathcal{U}(x_i)$. Given $\varepsilon < \bar{\rho}$, $\bar{\rho}$ as chosen in (12), let $N_\Omega(\varepsilon)$ be the number of hypercubes of dimension N and side ε required to cover $\partial\Omega$ and let $N_{\Omega,i}(\varepsilon)$ be the number of hypercubes of dimension N and side ε required to cover $\partial\Omega \cap \mathcal{U}(x_i)$.

We have the following estimate

$$N_{\Omega,i}(\varepsilon) \le \left(\frac{r_\lambda}{\varepsilon}\right)^{N-1} \frac{c(\sqrt{N-1}\,\varepsilon)^\alpha}{\varepsilon}.$$

Indeed the neighborhood

$$V_{H(x)} = B_{H(x)}(0, \overline{\rho}) \subset B_{H(x)}(0, r_\lambda)$$

can be covered by $[r_\lambda/\varepsilon]^{N-1}$ $(N-1)$-dimensional hypercubes of side ε. On each $(N-1)$-dimensional hypercube of side ε the variation between the minimum and the maximum of the function a_x is bounded by

$$c\left(\sqrt{(N-1)\varepsilon^2}\right)^\alpha = c\left(\sqrt{N-1}\,\varepsilon\right)^\alpha.$$

So the number of N-dimensional hypercubes of side ε necessary to cover the hypersurface above each $(N-1)$-dimensional hypercube of side ε is

$$\left[\frac{c}{\varepsilon}\left(\sqrt{N-1}\,\varepsilon\right)^{\alpha}\right].$$

Finally

$$\begin{aligned} N_{\Omega,i}(\varepsilon) &\leq \left(\frac{r_\lambda}{\varepsilon}+1\right)^{N-1}\left(\frac{c}{\varepsilon}\left(\sqrt{N-1}\,\varepsilon\right)^{\alpha}+1\right) \\ &\leq \frac{1}{\varepsilon^{N-1}}\frac{1}{\varepsilon^{1-\alpha}}\left(r_\lambda+\varepsilon\right)^{N-1}\left(c\left(\sqrt{N-1}\right)^{\alpha}+\varepsilon^{1-\alpha}\right) \\ &\leq \frac{1}{\varepsilon^{N-\alpha}}\left(r_\lambda+\varepsilon\right)^{N-1}\left(c\left(\sqrt{N-1}\right)^{\alpha}+\varepsilon^{1-\alpha}\right). \end{aligned}$$

As a result for all $\beta > N-\alpha$

$$\begin{aligned} N_{\Omega,i}(\varepsilon) &\leq \sum_{i=1}^{m} N_{\Omega,i}(\varepsilon) \\ &\leq m\,\frac{1}{\varepsilon^{N-\alpha}}\left(r_\lambda+\varepsilon\right)^{N-1}\left(c\left(\sqrt{N-1}\right)^{\alpha}+\varepsilon^{1-\alpha}\right) \\ \Rightarrow\ N_{\Omega}(\varepsilon)\varepsilon^{\beta} &\leq \varepsilon^{\beta-N+\alpha}\, m\left(r_\lambda+\varepsilon\right)^{N-1}\left(c\left(\sqrt{N-1}\right)^{\alpha}+\varepsilon^{1-\alpha}\right) \\ &\Rightarrow \forall\beta > N-\alpha,\quad H_\beta(\partial\Omega)=0. \end{aligned}$$

This means that, by definition, the Hausdorff dimension of $\partial\Omega$ is less or equal to $N-\alpha$. □

It is possible to construct examples of sets verifying the uniform cusp property for which the Hausdorff dimension of the boundary is strictly greater than $N-1$ and hence $H_{N-1}(\partial\Omega)=+\infty$.

EXAMPLE 9 *This following two-dimensional example of an open domain satisfying the uniform cusp condition for the function $h(\theta)=\theta^\alpha$, $0<\alpha<1$, can easily be generalized to an N-dimensional example. Consider the open domain Ω in $\mathbf{R}^2$*

$$\begin{aligned} \Omega \stackrel{\text{def}}{=} &\{(x,y)\ :\ -1 < x \leq 0 \text{ and } 0 < y < 2\} \\ &\cap \{(x,y)\ :\ 0 < x < 1 \text{ and } f(x) < y < 2\} \\ &\cap \{(x,y)\ :\ 1 \leq x < 2 \text{ and } 0 < y < 2\} \end{aligned}$$

where $f:[0,1]\to\mathbf{R}$ is defined as follows

$$f(x) \stackrel{\text{def}}{=} d_C(x)^\alpha, \quad 0 \leq x \leq 1,$$

and C is the Cantor set on the interval $[0,1]$. This function is equal to 0

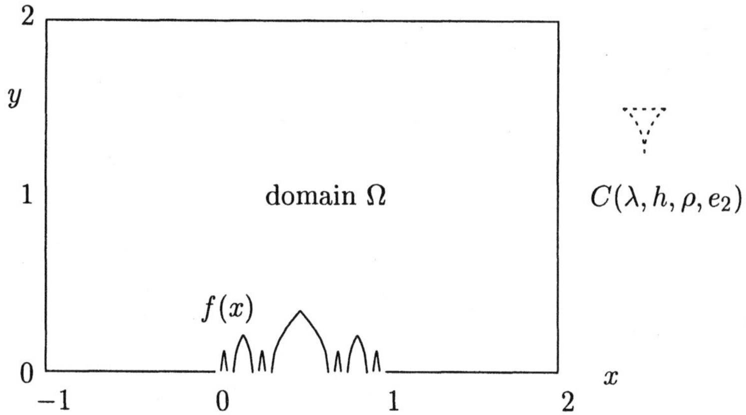

Figure 1. Domain Ω for $N = 2$, $0 < \alpha < 1$, $e_2 = (0, 1)$, $\rho = 1/6$, $\lambda = (1/6)^\alpha$, $h(\theta) = \theta^\alpha$.

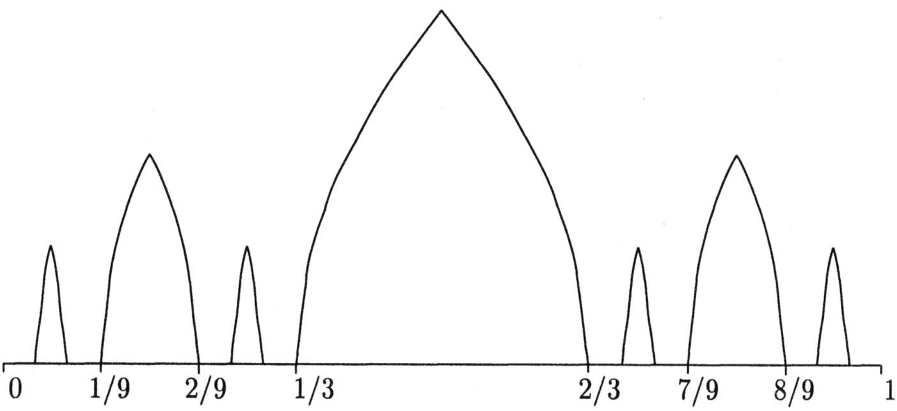

Figure 2. $f(x) = d_C(x)^{1/2}$ constructed on the Cantor set C for $2k + 1 = 3$.

on C. Any point in $[0, 1]\backslash C$ belongs to one of the intervals of length 3^{-k}, $k \geq 1$, which has been deleted from $[0, 1]$ in the sequential construction of the Cantor set. Therefore the distance function $d_C(x)$ is equal to the distance function to the two end points of that interval. In view of this special structure it can be shown that

$$\forall x, y \in [0, 1], \quad |d_C(y)^\alpha - d_C(x)^\alpha| \leq |y - x|^\alpha$$

Denote by Γ the piece of the boundary $\partial\Omega$ specified by the function $f = d_C$. On Γ the uniform cusp condition is verified with $\rho = 1/6$, $\lambda = (1/6)^\alpha$, and $h(\theta) = \theta^\alpha$. Clearly the number $N_\Omega(\varepsilon)$ of hypercubes of *dimension N and side ε required to cover $\partial\Omega$ is greater than the number $N_\Gamma(\varepsilon)$ of hypercubes of dimension N and side ε required to cover Γ. The construction of the Cantor set is done by sequentially deleting intervals. At step $k = 0$ the interval $(1/3, 2/3)$ of width 3^{-1} is removed. At step k a total of 2^k intervals of width $3^{-(k+1)}$ are removed. Thus if we pick $\varepsilon = 3^{-(k+1)}$ the interval $[0, 1]$ can be covered with exactly $3^{(k+1)}$ intervals. Here we are interested in finding a lower bound to the total number of of squares of side ε necessary to cover Γ. For this purpose we only keep the 2^k intervals removed at step k. Vertically it takes*

$$\left\lceil \frac{(2^{-1}3^{-(k+1)})^\alpha}{3^{-(k+1)}} \right\rceil \geq \frac{(2^{-1}3^{-(k+1)})^\alpha}{3^{-(k+1)}} - 1$$

Then we have for $\beta \geq 0$

$$\begin{aligned} N_\Omega(\varepsilon) \geq N_\Gamma(\varepsilon) &\geq 2^k \left(\frac{(2^{-1}3^{-(k+1)})^\alpha}{3^{-(k+1)}} - 1 \right) \\ &\geq 2^{k-\alpha}3^{(k+1)(1-\alpha)} - 2^k = \left(3^{(1-\alpha)}2\right)^k 2^{-\alpha}3^{(1-\alpha)} - 2^k \\ \Rightarrow N_\Omega(\varepsilon)(3^{-k})^{1+\beta} &\geq 3^{-k(1+\beta)} \left(\left(3^{(1-\alpha)}2\right)^k 2^{-\alpha}3^{(1-\alpha)} - 2^k \right) \\ &\geq \left(3^{-(\alpha+\beta)}2\right)^k 2^{-\alpha}3^{(1-\alpha)} - \left(\frac{2}{3^{(1+\beta)}}\right)^k \end{aligned}$$

The second term goes to zero as k goes to infinity. The first term goes to infinity as k goes to infinity if $3^{-(\alpha+\beta)}2 > 1$, that is, $0 < \alpha + \beta < \ln 2/\ln 3$. Under this condition, $H_{1+\beta}(\partial\Omega) = H_{1+\beta}(\Gamma) = +\infty$ for all $0 < \alpha < \ln 2/\ln 3$ and all $0 \leq \beta < \ln 2/\ln 3 - \alpha$. Therefore given $0 < \alpha < \ln 2/\ln 3$

$$\forall \beta,\ 0 \leq \beta + \alpha < \ln 2/\ln 3, \quad H_{1+\beta}(\partial\Omega) = +\infty$$

and the Hausdorff dimension of $\partial\Omega$ is strictly greater than 1.

Given $0 < \alpha < 1$, it is possible to construct an *optimal example* of a set verifying the uniform cusp property for which the Hausdorff dimension of the boundary is exactly $N - \alpha$ and hence $H_{N-1}(\partial\Omega) = +\infty$.

EXAMPLE 10 Optimal example of a set that verifies the uniform cusp property with $h(\theta) = |\theta|^\alpha$, $0 < \alpha < 1$, and whose boundary has Hausdorff dimension exactly equal to $N - \alpha$.

For that purpose, we need a generalization of the Cantor set. Denote by C_1 the Cantor set. Recall that each x, $0 \leq x \leq 1$, can be written *uniquely (if we make a certain convention) as*

$$x = \sum_{j=1}^{\infty} \frac{a_j(3,x)}{3^j}$$

where $a_j(3,x)$ *can be regarded as the jth digit of* x *written in basis 3. From this define the Cantor set is characterized as follows*

$$x \in C_1 \iff \forall j, a_j(3,x) \neq 1.$$

Similarly for an arbitrary integer $k \geq 1$, *each* $x \in [0,1]$ *can be uniquely written in the form*

$$x = \sum_{j=1}^{\infty} \frac{a_j(2k+1,x)}{(2k+1)^j}$$

and we can define the set C_k *as*

$$x \in C_k \iff \forall j, a_j(2k+1,x) \neq k.$$

In a certain sense, if $k_1 > k_2$, C_{k_1} *contains more points than* C_{k_2}. *We now use these sets to construct the family of set* D_k *as follows*

$$x \in D_1 \iff 2x \in C_1$$

and for $k > 1$

$$x \in D_k \iff 2^{k+1}(x - 2^k) \in C_k.$$

Note that, if $k_1 \neq k_2$, $D_{k_1} \bigcap D_{k_2} = \varnothing$ *since the* D_k*'s only contain points from the interval* $[1 - 2^{k-1}, 1 - 2^k]$. *Consider now the following set*

$$D \stackrel{\text{def}}{=} \bigcup_{k=1}^{\infty} D_k$$

and go back to Example 9 with the function f *is replaced by the function*

$$f(x) \stackrel{\text{def}}{=} d_D(x)^\alpha.$$

Again it can be shown that

$$\forall x, y \in [0,1], \quad |d_D(y)^\alpha - d_D(x)^\alpha| \leq |y - x|^\alpha.$$

Note that on the interval $[1 - 2^{k-1}, 1 - 2^k]$ *we have* $d_D(x)^\alpha = d_{D_k}(x)^\alpha$.

Denote by Γ *the piece of boundary* $\partial\Omega$ *specified by the function* $f = d_D$ *and* Γ_k *the part of boundary* $\partial\Omega$ *specified by the function* $f = d_D = d_{D_k}$ *on the interval* $[1 - 2^{k-1}, 1 - 2^k]$. *Once again on* Γ *the uniform cusp property is verified with* $\rho = 1/6$, $\lambda = (1/6)^\alpha$, *and* $h(\theta) = \theta^\alpha$.

Clearly the number $N_\Omega(\varepsilon)$ of hypercubes of dimension N and side ε required to cover $\partial\Omega$ is greater than the number $N_{\Gamma_k}(\varepsilon)$ of hypercubes of dimension N and side ε required to cover Γ_k. The construction of the set C_k is also done sequentially by deleting intervals. At step $j = 0$ the interval $]k/(2k+1), (k+1)/(2k+1)[$ of width $(2k+1)^{-1}$ is removed. At step j a total of 2^j intervals of width $(2j+1)^{-(j+1)}$ are removed. If we consider the intervals that remain at step j, a total of 2^{j+1} nonempty disjoint intervals of width $(\frac{k}{2k+1})^{j+1}$ remain in the set C_k. Each of these intervals contains a gap of length $(\frac{k}{2k+1})^{j+1}\frac{1}{2k+1}$ created at step $j+1$.

If we construct the set D_k in the same way, at step j a total of 2^j nonempty disjoint intervals of width $(\frac{k}{2k+1})^{j+1}\frac{1}{2^k}$ remain in the set D_k. Each of these intervals contains a gap of length $(\frac{k}{2k+1})^{j+1}\frac{1}{2^k(2k+1)}$. Pick

$$\varepsilon = \frac{1}{2^k}\left(\frac{k}{2k+1}\right)^{j+1}$$

and look for a lower bound on the number of squares of side ε necessary to cover Γ_k. For this purpose, only consider the 2^{j+1} nonempty disjoint intervals remaining at step j. As they each contain a gap of length

$$\left(\frac{k}{2k+1}\right)^{j+1}\frac{1}{2^k(2k+1)}$$

vertically it takes

$$\left[\left(\left(\frac{k}{2k+1}\right)^{j+1}\frac{1}{2^{k+1}(2k+1)}\right)^{\alpha} 2^k\left(\frac{2k+1}{k}\right)^{j+1}\right]$$
$$\geq \left(\left(\frac{k}{2k+1}\right)^{j+1}\frac{1}{2^{k+1}(2k+1)}\right)^{\alpha} 2^k\left(\frac{2k+1}{k}\right)^{j+1} - 1$$

ε-cubes. Then we have for $\beta \geq 0$

$$\begin{aligned}
N_\Omega(\varepsilon) &\geq N_\Gamma(\varepsilon)\\
&\geq 2^{j+1}\left(\left(\left(\frac{k}{2k+1}\right)^{j+1}\frac{1}{2^{k+1}(2k+1)}\right)^{\alpha} 2^k\left(\frac{2k+1}{k}\right)^{j+1} - 1\right)\\
&\geq \left(\frac{2(2k+1)k^\alpha}{k(2k+1)^\alpha}\right)^{j+1}\left(\frac{2^k}{2^{\alpha(k+1)}(2k+1)^\alpha}\right) - 2^{j+1}\\
&= (2(2k+1)^{1-\alpha}k^{\alpha-1})^{j+1}\left(\frac{2^{k(1-\alpha)-\alpha}}{(2k+1)^\alpha}\right) - 2^{j+1}
\end{aligned}$$

and hence

$$\varepsilon^{1+\beta} N_\Omega(\varepsilon)$$

$$\geq \left(\frac{1}{2^k}\left(\frac{k}{2k+1}\right)^{j+1}\right)^{1+\beta} \left(\left(2(2k+1)^{1-\alpha}k^{\alpha-1}\right)^{j+1}\left(\frac{2^{k(1-\alpha)-\alpha}}{(2k+1)^\alpha}\right) - 2^{j+1}\right)$$

$$\geq \left(\left(\frac{k}{2k+1}\right)^{\alpha+\beta} 2\right)^{j+1} \frac{2^{k(1-\alpha)-\alpha}}{2^{k(1+\beta)}(2k+1)^\alpha} - 2^{j+1}\left(\frac{1}{2^k}\left(\frac{k}{2k+1}\right)^{j+1}\right)^{1+\beta}$$

The second term goes to zero as j goes to infinity. The first term goes to infinity as j goes to infinity if $(\frac{k}{2k+1})^{\alpha+\beta}2 > 1$ for any integer k, that is, if

$$0 < \alpha + \beta < \frac{\log 2}{\log\left((2k+1)/k\right)}.$$

As k can be chosen arbitrarily large, the former inequality reduces to $0 < \alpha + \beta < 1$. Under this condition there exists an integer k for which $H_{1+\beta}(\partial\Omega) = H_{1+\beta}(\Gamma_k) = +\infty$ for all $0 < \alpha < 1$ and all $0 \leq \beta < 1 - \alpha$. Therefore, given $0 < \alpha < 1$ $\forall\beta, 0 \leq \beta < 1 - \alpha$, $H_{1+\beta}(\partial\Omega) = +\infty$.. This implies that the Hausdorff dimension of $\partial\Omega$ is greater than or equal to $2 - \alpha$ which is the upper bound we obtained in Theorem 8.

4. Compactness under the Uniform Cusp Property and a Bound on the Perimeter

4.1 De Giorgi Perimeter of Caccioppoli Sets

One of the classical notions of perimeter is the one introduced in the context of the problem of minimal surfaces for Caccioppoli sets.

DEFINITION 11 *Let Ω be a measurable subset of $\mathbf{R}^N$. Given an open set D in $\mathbf{R}^N$, Ω is said to have* finite perimeter with respect to *D if $\chi_\Omega \in BV(D)$. This* perimeter *denoted by $P_D(\Omega)$ is given by the expression*

$$P_D(\Omega) \stackrel{\text{def}}{=} \|\nabla\chi_\Omega\|_{M^1(D)^N}, \tag{16}$$

where $BV(D)$ is the space of functions of total bounded variation and $M^1(D)$ is the space of bounded measures on D.

Given a bounded open subset D of $\mathbf{R}^N$, $\rho > 0$, $\lambda > 0$, $r > 0$, $c > 0$, and $h \in \mathcal{H}$, consider the family

$$L(D, \lambda, h, \rho, r, c) \stackrel{\text{def}}{=} \left\{ \Omega \subset \overline{D} : \begin{array}{c} \Omega \text{ satisfies the uniform cusp} \\ \text{property for } (\lambda, h, \rho, r, c) \\ \text{and } P_D(\Omega) \leq c \end{array} \right\}. \tag{17}$$

The compactness Theorem 6 readily extends to this new family.

THEOREM 12 *Let D be a nonempty bounded open subset of $\mathbf{R}^N$ and $1 \le p < \infty$. For $\rho > 0$, $\lambda > 0$, $c > 0$, and $h \in \mathcal{H}$ and assume that $L(D, \lambda, h, \rho, r, c)$ is not empty. Then the family*

$$B(D, \lambda, h, \rho, r, c) \stackrel{\text{def}}{=} \{b_\Omega : \forall \Omega \in L(D, \lambda, h, \rho, r, c)\}$$

is compact in $C(\overline{D})$ and $W^{1,p}(D)$. As a consequence the families

$$\begin{aligned} B_d(D, \lambda, h, \rho, r, c) &\stackrel{\text{def}}{=} \{d_\Omega : \forall \Omega \in L(D, \lambda, h, \rho, r, c)\}, \\ B_d^c(D, \lambda, h, \rho, r, c) &\stackrel{\text{def}}{=} \{d_{\complement\Omega} : \forall \Omega \in L(D, \lambda, h, \rho, r, c)\}, \\ B_d^\partial(D, \lambda, h, \rho, r, c) &\stackrel{\text{def}}{=} \{d_{\partial\Omega} : \forall \Omega \in L(D, \lambda, h, \rho, r, c)\} \end{aligned}$$

are compact in $C(\overline{D})$ and $W^{1,p}(D)$, and the following families are compact in $L^p(D)$

$$\begin{aligned} X(D, \lambda, h, \rho, r, c) &\stackrel{\text{def}}{=} \{\chi_\Omega : \forall \Omega \in L(D, \lambda, h, \rho, r, c)\}, \\ X^c(D, \lambda, h, \rho, r, c) &\stackrel{\text{def}}{=} \{\chi_{\complement\Omega} : \forall \Omega \in L(D, \lambda, h, \rho, r, c)\}. \end{aligned}$$

Proof. – From Theorem 6 there exist Ω in $L(D, \lambda, h, \rho, r)$ and a sequence $\{\Omega_n\}$ in $L(D, \lambda, h, \rho, r, c)$ such that $b_{\Omega_n} \to b_\Omega$ in $W^{1,p}(D)$ and $P_D(\Omega_n) \le c$. In particular, from Theorem 1, $\chi_{\Omega_n} \to \chi_\Omega$ in $L^1(D)$. But, in view of the uniform bound $P_D(\Omega_n) \le c$ on the Ω_n's (cf. [6]), there exist a subsequence $\{\chi_{\Omega_{n_k}}\}$ such that $\chi_{\Omega_{n_k}} \to \chi_{\Omega'}$ in $L^1(D)$ for some Ω' for which $P_D(\Omega') \le c$. But, as a subsequence of $\{\Omega_n\}$,

$$b_{\Omega_{n_k}} \to b_\Omega \text{ in } W^{1,p}(D) \text{ and } \chi_{\Omega_{n_k}} \to \chi_\Omega \text{ in } L^1(D).$$

Hence $\chi_{\Omega'} = \chi_\Omega$, $P_D(\Omega) = P_D(\Omega') \le c$, and $\Omega \in L(D, \lambda, h, \rho, r, c)$. This concludes the proof. □

4.2 Finite γ-density Perimeter

The γ-density perimeter introduced by Bucur and Zolésio [1] is a relaxation of the $(N-1)$-dimensional *upper Minkowski content* which leads to the compactness Theorem 14. We recall the definition and quote the compactness for the $W^{1,p}$-topology under a uniform bound on the γ-density perimeter as revisited in [3].

DEFINITION 13 *Let $\gamma > 0$ be a fixed real and Ω a subset of $\mathbf{R}^N$ with nonempty boundary Γ. Consider the quotient*

$$P_\gamma(\Gamma) \stackrel{\text{def}}{=} \sup_{0<k<\gamma} \frac{m_N(U_k(\Gamma))}{2k}. \tag{18}$$

Whenever $P_\gamma(\Gamma)$ is finite, we say that Ω has a finite γ-density perimeter.

It was shown in [1] that, when $P_\gamma(\Gamma)$ is finite, $m_N(\Gamma) = 0$. The compactness result of [1] can be revisited and established in the $W^{1,p}$-topology from which convergence in all other topologies of Theorem 1 follows.

THEOREM 14 ([3]) *Let $D \neq \varnothing$ be a bounded open subset of $\mathbf{R}^N$ and $\{\Omega_n\}$, $\Gamma_n \neq \varnothing$, be a sequence of subsets of $\overline{D}$. Assume that*

$$\exists \gamma > 0 \text{ and } c > 0 \text{ such that } \forall n, \quad P_\gamma(\Gamma_n) \le c. \tag{19}$$

Then there exist a subsequence $\{\Omega_{n_k}\}$ and Ω, $\Gamma \neq \varnothing$, of $\overline{D}$ such that

$$P_\gamma(\Gamma) \le \liminf_{n\to\infty} P_\gamma(\Gamma_n) \le c \tag{20}$$

$$\forall p,\ 1 \le p < \infty, \quad b_{\Omega_{n_k}} \to b_\Omega \text{ in } W^{1,p}(U_\gamma(D)) \text{ -strong.} \tag{21}$$

The proof of the next result combines Theorem 6 which says that the family $L(D, \lambda, h, \rho, r)$ is compact with Theorem 14 which says that the family of sets verifying (19) is compact in $W^{1,p}(D)$. The intersection of the two families of oriented distance functions is compact in $W^{1,p}(D)$.

THEOREM 15 *For fixed $\gamma > 0$, Theorem 12 remains true when $P_D(\Omega)$ is replaced by the γ-density perimeter $P_\gamma(\Gamma)$.*

4.3 Compactness via Local C^0-graphs

It was shown in [4] (Thm 3.3 and 3.4) that the uniform cusp property is equivalent to conditions on the local C^0-graphs. Thus by adding a condition either on the De Giorgi or the perimeter γ-density perimeter in Theorem 4.1 of [4] we get the analogues of the above Theorems 12 and 15. Recall the definition of the *orthogonal subgroup* of $N \times N$ matrices

$$\mathrm{O(N)} \stackrel{\text{def}}{=} \{A : {}^*\!A\, A = A\, {}^*\!A = I\}, \tag{22}$$

where ${}^*\!A$ is the transposed matrix of A. A direction can be specified either by a matrix (of rotation) $A \in \mathrm{O(N)}$ or the corresponding unit vector $d = Ae_N \in \mathbf{R}^N$.

THEOREM 16 *Let $\rho > 0$ be given and assume that U is a bounded neighborhood of 0 such that*

$$U \subset \{y \in \mathbf{R}^N : P_H(y) \in B_H(0,\rho)\}, \quad V \stackrel{\text{def}}{=} B_H(0,\rho). \tag{23}$$

Let $R > 0$ be such that $B(0, 2R) \subset U$. Given a bounded nonempty subset D of $\mathbf{R}^N$, consider a family $L(D, \rho, U)$ of subsets Ω of $\bar{D}$ with *the following properties: for each* $\Omega \in L(D,\rho,U)$ *and each* $x \in \partial\Omega$, *there exist* $A^\Omega(x) \in \mathrm{O(N)}$ *and a* C^0*-mapping* $a_x^\Omega : V^\Omega(x) \to \mathbf{R}$, *where* $V^\Omega(x) \stackrel{\text{def}}{=} A^\Omega(x)V$ *and* $\mathcal{U}^\Omega(x) \stackrel{\text{def}}{=} x + A^\Omega(x)U$, *such that*

$$\mathcal{U}^\Omega(x) \cap \partial\Omega = \left\{ x + \zeta' + \zeta_N e_N^\Omega(x) : \begin{array}{c} \zeta' \in V^\Omega(x) \\ \zeta_N = a_x^\Omega(\zeta') \end{array} \right\} \tag{24}$$

$$\mathcal{U}^\Omega(x) \cap \mathrm{int}\Omega = \mathcal{U}^\Omega(x) \cap \left\{ x + \zeta' + \zeta_N e_N^\Omega(x) : \begin{array}{c} \zeta' \in V^\Omega(x) \\ \zeta_N > a_x^\Omega(\zeta') \end{array} \right\} \tag{25}$$

where $e_N^\Omega(x) = A^\Omega(x)e_N$.

(i) Assume that there exists $h \in \mathcal{H}$ *and* $c > 0$ *such that*

$$\forall \Omega \in L(D,\rho,U),\ \forall y \in V, \quad \bar{a}_x^\Omega(y) \le h(|y|), \quad P_D(\Omega) \le c \tag{26}$$

where $\bar{a}_x^\Omega = a_x^\Omega \circ A^\Omega(x) : V \to \mathbf{R}$. *Each* Ω *of* $L(D,\rho,U)$ *satisfies the uniform cusp property for the parameters* $(r^\Omega, \lambda^\Omega, \rho^\Omega, h^\Omega) = (R,R,\rho,h)$. *Hence (from Theorem 12) the family*

$$B(D,\rho,U,c) \stackrel{\text{def}}{=} \{b_\Omega : \forall \Omega \in L(D,\rho,U) \text{ and } P_D(\Omega) \le c\}$$

is compact in $C(\overline{D})$ *and* $W^{1,p}(D)$, $1 \le p < \infty$.

(ii) Given $\gamma > 0$, *the results of part (i) remain true with* $P_D(\Omega) \le c$ *in place of* $P_\gamma(\Gamma) \le c$.

References

[1] D. Bucur and J.-P. Zolésio, *Free boundary problems and density perimeter*, J. Differential Equations 126 (1996), 224–243.

[2] M.C. Delfour and J.-P. Zolésio, *Shapes and Geometries: Analysis, Differential Calculus and Optimization*, SIAM series on Advances in Design and Control, Society for Industrial and Applied Mathematics, Philadelphia, USA 2001.

[3] M.C. Delfour and J.-P. Zolésio, *The new family of cracked sets and the image segmentation problem revisited*, CRM Report, May 2003, Université de Montréal, accepted in Communications in Information and Systems.

[4] M.C. Delfour, N. Doyon, and J.-P. Zolésio, *Extension of the uniform cusp property in shape optimization*, in "Control of Partial Differential Equations", G. Leugering, O. Imanuvilov, R. Triggiani, and B. Zhang, eds. Lectures Notes in Pure and Applied Mathematics, Marcel Dekker, May 2003, accepted.

[5] M.C. Delfour, N. Doyon, and J.-P. Zolésio, *The uniform fat segment and cusp properties in shape optimization*, in "Control and Boundary analysis", J. Cagnol and J.-P. Zolésio (Eds.), pp. 85–96, Marcel Dekker 2004.

[6] E. Giusti, *Minimal surfaces and functions of bounded variation*, Birkhäuser, Boston, Basel, Stuttgart, 1984.

INTERIOR AND BOUNDARY STABILIZATION OF NAVIER-STOKES EQUATIONS

Roberto Triggiani*
University of Virginia
Charlottesville, VA 22904 USA
rt7u@virginia.edu

Abstract We report on very recent work on the stabilization of the steady-state solutions to Navier-Stokes equations on an open bounded domain $\Omega \subset R^d$, $d = 2, 3$, by either interior, or else boundary control.

More precisely, as to the interior case, we obtain that the steady-state solutions to Navier-Stokes equations on $\Omega \subset R^d$, $d = 2, 3$, with no-slip boundary conditions, are locally exponentially stabilizable by a finite-dimensional feedback controller with support in an arbitrary open subset $\omega \subset \Omega$ of positive measure. The (finite) dimension of the feedback controller is minimal and is related to the largest algebraic multiplicity of the unstable eigenvalues of the linearized equation.

Second, as to the boundary case, we obtain that the steady-state solutions to Navier-Stokes equations on a bounded domain $\Omega \subset R^d$, $d = 2, 3$, are locally exponentially stabilizable by a boundary closed-loop feedback controller, acting on the boundary $\partial\Omega$, in the Dirichlet boundary conditions. If $d = 3$, the non-linearity imposes and dictates the requirement that stabilization must occur in the space $(H^{\frac{3}{2}+\epsilon}(\Omega))^3$, $\epsilon > 0$, a high topological level. A first implication thereof is that, for $d = 3$, the boundary feedback stabilizing controller *must* be infinite dimensional. Moreover, it generally acts on the entire boundary $\partial\Omega$. Instead, for $d = 2$, where the topological level for stabilization is $(H^{\frac{3}{2}-\epsilon}(\Omega))^2$, the boundary feedback stabilizing controller can be chosen to act on *an arbitrarily small* portion of the boundary. Moreover, still for $d = 2$, it may even be *finite* dimensional, and this occurs if the linearized operator is diagonalizable over its finite-dimensional unstable subspace.

Keywords: Internal stabilization, boundary stabilization, Navier-Stokes Equations.

*Funding provided by NSF grant DMS-0104305 and ARO DAAD19-02-1-0179.

We hereby report on recent joint work on the stabilization of steady-state solutions to Navier-Stokes equations on an open bounded domain $\Omega \subset R^d$, $d = 2, 3$, by either *interior* feedback control or else *boundary* feedback control. The case of interior control is taken from the joint work with V. Barbu in [4]. The case of boundary control is taken from the joint work with V. Barbu and I. Lasiecka in [3]. To enhance readability, we provide independent accounts of each case.

Part I: Interior Control [4]

1. Introduction

The controlled N-S equations. Consider the controlled Navier-Stokes equations (see [6, p. 45], [13, p. 253] for the uncontrolled case $u \equiv 0$) with the non-slip Dirichlet B.C.:

$$\begin{aligned} & y_t(x,t) - \nu\Delta y(x,t) + (y\cdot\nabla)y(x,t) \\ &= m(x)u(x,t) + f_e(x) + \nabla p(x,t) \text{ in } Q = \Omega\times(0,\infty), \qquad (1) \end{aligned}$$

$$\begin{aligned} \nabla\cdot y &= 0 && \text{in } Q; \\ y &= 0 && \text{on } \Sigma = \partial\Omega\times(0,\infty); \\ y(x,0) &= y_0(x) && \text{in } \Omega. \end{aligned}$$

Here, Ω is an open smooth bounded domain of R^d, $d = 2, 3$; m is the characteristic function of an open smooth subset $\omega \subset \Omega$ of positive measure; u is the control input; and $y = (y_1, y_2, \ldots, y_d)$ is the state (velocity) of the system. The function $v = mu$ can be viewed itself as an internal controller with support in $Q_\omega = \omega \times (0,\infty)$. The functions $y_0, f_e \in (L^2(\Omega))^d$ are given, the latter being a body force, while p is the unknown pressure.

Let $(y_e, p_e) \in ((H^2(\Omega))^d \cap V) \times H^1(\Omega)$ be a steady-state solution to equation (1), i.e.,

$$-\nu\Delta y_e + (y_e\cdot\nabla)y_e = f_e + \nabla p_e \text{ in } \Omega; \qquad (2)$$

$$\begin{aligned} \nabla\cdot y_e &= 0 \text{ in } \Omega; \\ y_e &= 0 \text{ on } \partial\Omega. \end{aligned}$$

The steady-state solution is known to exist for $d = 2, 3$, [6, Theorem 7.3, p. 59]. Here [6, p. 9], [13, p. 18]

$$\begin{aligned} V &= \{y\in(H_0^1(\Omega))^d;\ \nabla\cdot y = 0\}, \text{ with norm } \|y\|_V \equiv \|y\| \\ &= \left\{\int_\Omega |\nabla y(x)|^2 d\Omega\right\}^{\frac{1}{2}}. \qquad (3) \end{aligned}$$

Literature. According to some recent results of O. Imanuvilov [9] (see also [1]) any such solution y_e is locally exactly controllable on every interval $[0, T]$ with controller u with support in Q_ω. More precisely, if the distance $\|y_e - y_0\|_{H^2(\Omega)}$ is sufficiently small, then there is a solution (y, p, u) to (1) of appropriate regularity such that $y(T) \equiv y_e$. The steering control is *open-loop* and depends on the initial condition. Subsequently, paper [2] proved that any steady-state solution y_e is locally exponentially stabilizable by means of an *infinite-dimensional* feedback controller, by using the controllability of the linear Stokes equation. In contrast, here we shall prove, via the state decomposition technique of [14], [15], and the first-order stabilization Riccati equation method developed in our previous work [2] (see also [5] still in the parabolic case, as well as [11] in the hyperbolic case), that any steady-state solution y_e is locally exponentially stabilizable by a *finite-dimensional closed-loop* feedback controller of the form

$$u = -\sum_{i=1}^{2K} (R_N(y - y_e), \psi_i)_\omega \psi_i, \tag{4}$$

where $R_N \in \mathcal{L}(\mathcal{D}(A^{\frac{1}{4}})) \cap \mathcal{L}(\mathcal{D}(A^{\frac{1}{2}}); H)$ is the solution of the algebraic Riccati equation (18) below associated with the linearized system (14) below and $\{\psi_i\}_{i=1}^{2K}$ is an explicitly constructed (in (3.3.5) of [4]) system of functions related to the space of eigenfunctions corresponding to the unstable eigenvalues of such linearized system. Here A is the Stokes operator defined by (6); H the space in (5); and $(\cdot\,,\,\cdot)_\omega$ is the scalar product in $(L^2(\omega))^d$. The present closed-loop feedback stabilization result has two main features, besides being finite-dimensional:

(1) it is more precise and less restrictive concerning the vectors y_0 and y_e than the open-loop version provided by the local exact controllability result established in [9], or the closed-loop stabilization in [2] (in that smallness of the distance between y_0 and y_e is measured in the $\mathcal{D}(A^{\frac{1}{4}})$-norm, i.e., the $(H^{\frac{1}{2}}(\Omega))^d$-norm, see the set $\mathcal{V}_\rho$ in (21) below, rather than in the $(H^2(\Omega))^d$-norm, as recalled above, where A is defined in (6).);

(2) it is independent of the Carleman inequality for the Stokes equation, which is necessary for the proof of local controllability.

There is a large literature on the stabilization problem of steady-state solutions to Navier-Stokes equations. Here we confine ourselves to mention only a few of the papers ([2], [7]) which are more related to this present work. We also refer to the recent paper of Fursikov [8] for a study of a boundary—rather than interior—problem for the N-S equations, which, however, does not pertain to the topic of feedback stabilization in the established sense, as in the present paper.

Notation. Here we shall use the standard notation for the spaces of summable functions and Sobolev spaces on Ω. In particular, $H^s(\Omega)$ is the Sobolev space of order s with the norm denoted by $\|\cdot\|_s$. The following notation will be also used:

$$\nabla \cdot y = \text{div } y, \ (y \cdot \nabla)y = y_i D_i y_j = y \cdot \nabla y_j,$$

$$j = 1, \ldots, d, \ D_i = \frac{\partial}{\partial x_i};$$

$$H = \{y \in (L^2(\Omega))^d; \ \nabla \cdot y = 0, \ y \cdot n = 0 \text{ on } \partial\Omega\} \quad [6, \text{p. } 7]; \tag{5}$$

$$H^{\perp} = \{y \in (L^2(\Omega))^d : y = \nabla p, \ p \in H^1(\Omega)\}, \ (L^2(\Omega))^d = H + H^{\perp},$$

$H^{\perp}$ being the orthogonal complement of H in $(L^2(\Omega))^d$ [13, p. 15] with summation convention to be used throughout the paper, presently in $i = 1, \ldots, d$, where n is the outward normal to the boundary $\partial\Omega$ of Ω. We shall denote by $P : (L^2(\Omega))^d \to H$ the orthogonal Leray projector [6, p. 9], and moreover [6, p. 31],

$$Ay = -P\Delta y, \quad \forall \ y \in \mathcal{D}(A) = (H^2(\Omega))^d \cap V, \qquad V = \mathcal{D}(A^{\frac{1}{2}}), \tag{6}$$

which is a self-adjoint positive definite operator in H with compact (resolvent) A^{-1} on H [6, p. 32]. Accordingly, the fractional powers A^s, $0 < s < 1$, are well-defined [6, p. 33]. We have $V = \mathcal{D}(A^{\frac{1}{2}})$ [6, p. 33]. Furthermore, we define $B : V \to V'$ by [6, p. 47, p. 54], [13, p. 162],

$$By = P[(y \cdot \nabla)y], \ (By, w) = b(y, y, w), \quad \forall \ y, w \in V, \tag{7}$$

where the trilinear form is defined by [6, p. 49], [13, p. 161]

$$b(y, z, w) = \int_{\Omega} y_i (D_i z_j) w_j dx = \int_{\Omega} \langle y \cdot \nabla z, w \rangle_{R^d} d\Omega,$$

$$y, w \in H, \ z \in V. \tag{8}$$

We shall denote by $(\cdot, \cdot)$ the scalar product in both H and $(L^2(\Omega))^d$. Similarly, we shall denote by the same symbol $|\cdot|$ the norm of both $(L^2(\Omega))^d$ and H, and by $\|\cdot\|$ the norm of the space V as defined in (3).

Preliminaries. In the notation introduced above, Eqn. (1) can be equivalently rewritten in abstract form as

$$\frac{dy}{dt} + \nu Ay + By = P(mu + f_e); \qquad y(0) = y_0 \in H, \tag{9}$$

since the procedure of applying P to (1) eliminates the pressure from the equations [6, p. 47], the orthogonal space $H^{\perp}$ to H being made up of $(L^2(\Omega)^d$-functions which are the gradients of $H^1(\Omega)$-functions by (5). Moreover, $y \in H$ for (1) implies $Py_t = y_t$.

2. The Main Results

Assumptions. (i) The boundary $\partial\Omega$ of Ω is a finite union of $d-1$ dimensional C^2-connected manifolds. Moreover, the boundary $\partial\omega$ of ω is of class C^2.

(ii) The steady-state solution (y_e, p_e) defined in (2) belongs to $((H^2(\Omega))^d \cap V) \times H^1(\Omega)$, where we recall from (6) that then $y_e \in \mathcal{D}(A)$. [For $d = 2, 3$, this property is guaranteed by [6, Theorem 7.3, p. 59] on y_e, for $f_e \in H$, followed by [6, Theorem 3.11, p. 30] on p_e, for sufficiently smooth $\partial\Omega$.]

Preliminaries. The translated problem. By the substitutions $y \to y_e + y$, $p \to p_e + p$, we are readily led via (1), (2) to the study of *null stabilization* of the equation

$$y_t - \nu\Delta y + (y\cdot\nabla)y + (y_e\cdot\nabla)y + (y\cdot\nabla)y_e = mu + \nabla p \quad \text{in } Q \tag{10a}$$

$$\nabla\cdot y = 0 \quad \text{in } Q \tag{10b}$$

$$y = 0 \quad \text{on } \Sigma \tag{10c}$$

$$y(x,0) = y^0(x) = y_0(x) - y_e(x). \tag{10d}$$

By use of (5) on H, (6), (7) on A and B, we see that (10), after application of P, can be rewritten abstractly as

$$\frac{dy}{dt} + \nu Ay + A_0 y + By = P(mu), \; t > 0; \quad y(0) = y^0 \tag{11}$$

(compare with (9) again $Py_t = y_t$, since $y \in H$ by (10)), where we have now introduced the operator $A_0 \in \mathcal{L}(V; H)$,

$$A_0 y = P((y_e\cdot\nabla)y + (y\cdot\nabla)y_e), \quad \mathcal{D}(A_0) = V = \mathcal{D}(A^{\frac{1}{2}}), \tag{12a}$$

or equivalently, recalling (7),

$$(A_0 y, z) = b(y_e, y, z) + b(y, y_e, z), \quad \forall\, y \in V,\; z \in H. \tag{12b}$$

The operator A_0 in (12) is well-defined $H \supset V = \mathcal{D}(A_0) \to H$. This follows from the estimate

$$|A_0 y| \le C_1 \|y_e\|_2 \|y\|, \qquad \forall\, y \in V = \mathcal{D}(A_0) = \mathcal{D}(A^{\frac{1}{2}}), \tag{13}$$

which is obtained directly by use of the definition (12b).

The linearized problem. Next, we consider the following linearized system of the translated model (10) or (11):

$$\frac{dy}{dt} + \nu Ay + A_0 y = P(mu), \; t > 0; \quad y(0) = y^0 \in H. \tag{14}$$

We have already noted below (6) that the operator $-\nu A$ ($\nu > 0$, the viscosity coefficient) is negative self-adjoint and has compact resolvent on H. Thus, $-\nu A$ generates an analytic (self-adjoint) C_0-semigroup on H. It then follows from here and from $\mathcal{D}(A_0) = V = \mathcal{D}(A^{\frac{1}{2}})$, as noted in (6) and in (13), that: *the perturbed operator*

$$\mathcal{A} = -(\nu A + A_0), \text{ with domain } \mathcal{D}(\mathcal{A}) = \mathcal{D}(A) = (H^2(\Omega))^d \cap V \quad (15a)$$

likewise has compact resolvent and generates an analytic C_0-semigroup on H. This is well-known. It follows from the above claim that the operator $\mathcal{A}$ has a finite number N of eigenvalues λ_j with Re $\lambda_j \geq 0$ (the unstable eigenvalues). The eigenvalues are repeated according to their algebraic multiplicity ℓ_j. Let $\{\varphi_j\}_{j=1}^N$ be a corresponding system of generalized eigenfunctions, $\varphi_j = \varphi_j^1 + i\varphi_j^2$, $j = 1, \dots, N$ of $\mathcal{A}$. (See [10, p. 41, 181].) More precisely, we shall denote by M the number of *distinct* unstable eigenvalues, so that $\ell_1 + \ell_2 + \cdots + \ell_M = N$. In order to state our first result, we finally need to introduce the following finite-dimensional real spaces X_N^α, $\alpha = 1, 2$ as well as the following natural number K:

$$X_N^\alpha = \text{span}\{\varphi_j^\alpha\}_{j=1}^N; \quad K = \max\{\ell_j;\ 1 \leq j \leq M\}. \quad (16)$$

Main results. Linearized problem (14)**.** We first state the following feedback stabilization result for the linearized system (14).

THEOREM 1 *Let $\epsilon > 0$ be arbitrary but fixed, and let $\gamma_0 = |Re\lambda_{N+1}| - \epsilon$. Then, for each λ, $0 \leq \lambda \leq \gamma_0$, there are functions $\{\psi_i\}_{i=1}^K \subset X_N^1$, $\{\psi_i\}_{i=K+1}^{2K} \subset X_N^2$ and a linear self-adjoint operator $R_N : \mathcal{D}(R_N) \subset H \to H$ such that for some constants $0 < a_1 < a_2 < \infty$ and $C_1 > 0$, we have:*

(i)

$$a_1|A^{\frac{1}{4}}y|^2 \leq (R_N y, y) \leq a_2|A^{\frac{1}{4}}y|^2, \quad \forall\, y \in \mathcal{D}(A^{\frac{1}{4}}), \quad (17a)$$

so that $\mathcal{D}(A^{\frac{1}{4}}) \subset \mathcal{D}(R_N^{\frac{1}{2}})$;

(ii)

$$|R_N y| \leq C_1\|y\|, \quad \forall\, y \in V = \mathcal{D}(A^{\frac{1}{2}}); \quad (17b)$$

(iii) R_N satisfies the following algebraic Riccati equation:

$$-((\mathcal{A} + \lambda)y, R_N y) + \frac{1}{2}\sum_{i=1}^{2K}(\psi_i, R_N y)_\omega^2 = \frac{1}{2}|A^{\frac{3}{4}}y|^2, \quad \forall\, y \in \mathcal{D}(A). \quad (18)$$

The vectors $\{\psi_i\}_{i=1}^{2K}$ are explicitly constructed in (3.3.5) of Lemma 4 of [4]. Moreover, with $2K \leq N$, the feedback controller,

$$u = -\sum_{i=1}^{2K}(R_N y, \psi_i)_\omega \psi_i \quad (19a)$$

once inserted in (14), *exponentially stabilizes the corresponding closed-loop system* (14). *The margin of stability for such closed loop system is* λ. *[See Remark 3.3.1 of [4] for the effective number of controls* $2K \leq N$.*] More specifically, this means that the solution of*

$$\frac{dy}{dt} + \nu Ay + A_0 y + P\left(m\sum_{i=1}^{2K}(R_N y, \psi_i)_\omega \psi_i\right) = 0, \; y(0) = y^0 \in \mathcal{D}(A^{\frac{1}{4}}) \tag{19b}$$

satisfies

$$|A^{\frac{1}{4}}y(t)| \leq C_\lambda e^{-\lambda t}|A^{\frac{1}{4}}y^0|, \; t \geq 0. \tag{19c}$$

Non-linear system (9). We next use the stabilizer in Theorem 1 to the linearized system (14) of the translated problem (10), or (11), to obtain the sought-after closed loop, local, feedback stabilization of the steady state solution y_e to the N-S equation (9).

THEOREM 2 *With reference to Theorem 1, the feedback controller*

$$u = -\sum_{i=1}^{2K}(R_N(y - y_e), \psi_i)_\omega \psi_i \tag{20}$$

[where the vectors ψ_i *are defined in (3.3.5) of Lemma 4 in [4]], once inserted in the N-S system* (9) *exponentially stabilizes the steady state solution* y_e *to* (1) *in a neighborhood*

$$\mathcal{V}_\rho = \{y_0 \in D(A^{\frac{1}{4}}); \; |A^{\frac{1}{4}}(y_0 - y_e)| < \rho\} \tag{21}$$

of y_e, *for suitable* $\rho > 0$. *More precisely, if* $\rho > 0$ *is sufficiently small, then for each* $y_0 \in \mathcal{V}_\rho$ *there exists a weak solution* $y \in L^\infty(0,T;H) \cap L^2(0,T;V)$, $\frac{dy}{dt} \in L^{\frac{4}{3}}(0,T;V')$ *for* $d = 3$, *and* $\frac{dy}{dt} \in L^2(0,T;V')$ *for* $d = 2$, $\forall\, T > 0$, *to the closed loop system*

$$\frac{dy}{dt} + \nu Ay + By + P\left(m\sum_{i=1}^{2K}(R_N(y - y_e), \psi_i)_\omega \psi_i\right) = Pf_e, \; t \geq 0, \; y(0) = y^0 \tag{22}$$

obtained from inserting the control (20) *in* (9), *such that the following two properties hold:*

(i)

$$\int_0^\infty e^{2\lambda t}|A^{\frac{3}{4}}(y(t) - y_e)|^2 dt \leq C_2|A^{\frac{1}{4}}(y_0 - y_e)|^2; \tag{23}$$

(ii)

$$|A^{\frac{1}{4}}(y(t) - y_e)| \leq C_3 e^{-\lambda t}|A^{\frac{1}{4}}(y_0 - y_e)|, \quad \forall\, t \geq 0. \tag{24}$$

We refer to [6, p. 71] for definition of weak solutions to equations of the form (22) and the asserted regularity. If $d = 2$ the solution to (22) is strong and unique [6, p. 83].

The pressure p. Theorem 2 implies the following result giving corresponding asymptotic properties of the pressure p.

THEOREM 3 *The solution y provided by Theorem 2 satisfies also the equation*

$$y_t - \nu\Delta y + (y \cdot \nabla)y + m\left(\sum_{i=1}^{2K}(R_N(y - y_e), \psi_i)_\omega \psi_i\right)$$

$$= \nabla p + f_e \text{ in } Q \equiv \Omega \times (0, \infty); \qquad (25)$$

$$\begin{aligned} \nabla \cdot y &\equiv 0 && \text{in } Q; \\ y &\equiv 0 && \text{on } \Sigma = \partial\Omega \times (0, \infty); \\ y(x,0) &= y_0(x) && \text{in } \Omega. \end{aligned}$$

Moreover, the following relations hold true for the pressure p:
(i) for $d = 2$, we have

$$\int_0^\infty t|p(t) - p_e|^2_{(H^1(\Omega))^d} dt \le C|A^{\frac{1}{4}}(y_0 - y_e)|^2[1 + |A^{\frac{1}{4}}(y_0 - y_e)|^2]; \quad (26)$$

(ii) for $d = 3$, we have

$$\int_0^\infty |p(t) - p_e|^2_{(L^2(\Omega))^d/R} dt \le C|A^{\frac{1}{4}}(y - y_e)|^2][1 + |A^{\frac{1}{4}}(y_0 - y_e)|^2]. \quad (27)$$

Part II: Boundary Control [3]

3. Introduction

Boundary controlled Navier-Stokes equations. We consider the controlled Navier-Stokes equations (see [6, p. 45], [13, p. 253] for the uncontrolled case $u \equiv 0$) with boundary control u in the Dirichlet B.C.:

$$\begin{aligned} y_t(x,t) - \nu_0\Delta y(x,t) + (y \cdot \nabla)y(x,t) &= f_e(x) \\ &\quad +\nabla p(x,t) \quad \text{in } G; && (28a) \\ \nabla \cdot y &= 0 \quad \text{in } G; && (28b) \\ y &= u \quad \text{on } \Sigma; && (28c) \\ y(x,0) &= y_0(x) \quad \text{in } \Omega. && (28d) \end{aligned}$$

Here, $G = \Omega \times (0, \infty); \Sigma = \partial\Omega \times (0, \infty)$ and Ω is an open smooth bounded domain of R^d, $d = 2, 3$; $u \in L^2(0, T; (L^2(\partial\Omega))^d)$ is the boundary control input; and $y = (y_1, y_2, \ldots, y_d)$ is the state (velocity) of the system. The constant $\nu_0 > 0$ is the viscosity coefficient. The functions $y_0, f_e \in (L^2(\Omega))^d$ are given, the latter being a body force, while p is the unknown pressure. Because of the divergence theorem: $\int_\Omega \nabla \cdot y \, d\Omega = \int_\Gamma y \cdot \nu \, d\Gamma$, $[\Gamma = \partial\Omega$, $\nu =$ unit outward normal to $\partial\Omega]$, we must require (at least) the integral boundary compatibility condition: $\int_\Gamma u \cdot \nu \, d\Gamma = 0$ on the control function u. Actually, a more stringent condition has to be imposed, in our final results: $u \cdot \nu \equiv 0$ on Σ, to sustain the pointwise boundary compatibility condition contained in the definition of the critical state space H in (28e) below. To summarize we shall then assume

$$\text{either } u \cdot \nu \equiv 0 \text{ on } \Sigma; \text{ or at least } \int_\Gamma u \cdot \nu \, d\Gamma \equiv 0, \text{ a.e. } t > 0, \tag{28e}$$

as it will be specified on a case-by-case basis.

Steady-state solutions and space V: sames as in (2), (3).

Goal. Our goal is to construct a boundary control u, subject to the boundary compatibility condition (c.c.) given by (28e) in the strong pointwise form $u \cdot \nu \equiv 0$ on $\partial\Omega$, and, moreover, in feedback form $u = u(y)$ via some linear operator $y \to u$, such that, once $u(y)$ is substituted in the translated problem (28), the resulting well-posed, closed-loop system (28a)–(28c) possesses the following desirable property: the steady-state solutions y_e defined in (2) are locally exponentially stable. In particular, motivated by our prior effort [4] to be described below, we seek to investigate if and when the feedback controller $u = u(y)$ can be chosen to be finite-dimensional, and, moreover, to act on an arbitrarily small portion (of positive measure) of the boundary $\Gamma = \partial\Omega$.

Orientation. Use of the Optimal Control Problem and Algebraic Riccati Theory. ($d = 3$) We emphasize here only the more demanding case of $d = 3$. A preliminary difficulty (for $d = 2, 3$) is the requirement in (28e) that the boundary control u must always be tangential at each point of the boundary. It is standard that this requirement is intrinsically built in the definition of the state space H (above in (5) Part I) of the velocity vectory y, which is critical to eliminate the second unknown of the N-S model, the pressure term ∇p (see the orthogonal complement $H^\perp$ in (5) above, Part I), by virtue of the Leray projection P. Evolution of the velocity must occur in H. Accordingly, we must then have that the boundary controls be pointwise tangential: $u \cdot \nu \equiv 0$ on Σ in (28e). Next, a second difficulty, this time for $d = 3$, is that the non-linearity of the N-S equation dictates and forces the requirement that stabilization must occur in the space $(H^{\frac{3}{2}+\epsilon}(\Omega))^3$, $\epsilon > 0$, see Eqn. (5.18a–b) of [3]. This is a high topological level, of which we shall have to say more below. A third source of difficulty consists in deciding how to inject 'dissipation' into the N-S model, in fact, as required, through *a boundary* tangential controller expressed in feedback form. Here, motivated by [4] and, in turn, by optimal control theory [12], in order to inject dissipation into the N-S system as to force local exponential boundary stabilization of its steady-state solutions, we choose the strategy of introducing an Optimal Control Problem (OCP) with a quadratic cost functional, over an infinite time-horizon, for the linearized N-S model subject to tangential Dirichlet-boundary control u, i.e., satisfying $u \cdot \nu \equiv 0$ on Σ. One then seeks to express the boundary feedback, closed-loop controller of the optimal solution of the OCP, in terms of the Riccati operator arising in the corresponding algebraic Riccati theory. As a result, the same Riccati-based boundary feedback optimal controller that is obtained in the linearized OCP is then selected and implemented also on the full N-S system. This controller in feedback form is both dissipative as well as 'robust' (with respect to a certain class of perturbations). For $d = 3$, however, the OCP must be resolved at the *high* $(H^{\frac{3}{2}+\epsilon}(\Omega))^3$-topological level, within the class of Dirichlet *boundary* controls in $L^2(0, \infty; (\partial\Omega)^3)$, which are further constrained to be *tangential to the boundary.*

Thus, the OCP faces two additional difficulties that set it apart and definitely outside the boundaries of established optimal control theory for parabolic systems with boundary controls: (1) the high degree of unboundedness of the *boundary* control operator, of order $(\frac{3}{4} + \epsilon)$ as expressed in terms of fractional powers of the basic free-dynamics generator; and (2) the high degree of unboundedness of the 'penalization' or 'observation' operator of order also $(\frac{3}{4} + \epsilon)$, as expressed in terms of fractional powers of the basic free-dynamics generator. This yields a 'combined index' of unboundedness *strictly greater than* $\frac{3}{2}$. By contrast, the established (and rich) optimal control theory of boundary control parabolic problems and corresponding algebraic Riccati theory requires a 'combined index' of unboundedness *strictly less than* 1 [12, Vol. 1, in particular, p. 501–503], which is the maximum limit handled by perturbation theory of analytic semigroups. To implement this program, however, one must first overcome, at the very outset, the preliminary stumbling obstacle of showing that the present highly non-standard OCP—with the aforementioned high level of combined unboundedness in control and observation operators and further restricted within the class of *tangential* boundary controllers—is, in fact, *non-empty.* This result is achieved in Theorem 3.5.1 of [3]in full generality (and in Proposition 3.7.1 of [3] under the assumption that the linearized operator is diagonalizable over the finite-dimensional unstable subspace). Thus, after this result, the study of the OCP may then begin. Because of the aforementioned intrinsic difficulties of the OCP with a combined index of unboundedness $> \frac{3}{2}$, one cannot (and cannot hope to) recover in full all desirable features of the corresponding algebraic Riccati theory which are available when the combined index of unboundedness in control and observation operators is *strictly less than* 1 ([12] and references therein). For instance, existence of a solution (Riccati operator) of the algebraic Riccati equation is here asserted only on the domain of the generator of the optimal feedback dynamics (Proposition 4.5.1 of [3]); not on the domain of the free-dynamics operator, as it would be required by, or at least desirable from, the viewpoint of the OCP. However, in our present treatment, the OCP is a means to extract dissipation and stability, not an end in itself. And indeed, the present study of the algebraic Riccati theory, with a combined index of unboundedness in control and observation operator *strictly above* $\frac{3}{2}$ (rather than *strictly less than* 1) does manage, at the end, to draw out the key sought-after features of interest—dissipativity and decay—for the resulting optimal solution in feedback form of the OCP for the linearized N-S equation. All this is accomplished in Section 4 of [3].

The subsequent step of the strategy is then to select and use the same Riccati-based, boundary feedback operator, which was found to describe the optimal solution of OCP of the *linearized* N-S equation, directly into the full N-S model. For $d = 3$, the heavy groundwork for the feedback stabilization of the linearized problem via optimal control theory makes then the resulting analysis of well-posedness (in Section 5 of [3]) and stabilization (in Section 6) of the N-S model more amenable than would otherwise be the case.

To this end, key use is made of the Algebraic Riccati Equation satisfied by the Riccati operator that describes the stabilizing control in closed-loop feedback form.

Literature. This paper [3]is a successor to [4], which instead considered the *interior* stabilization problem of the Navier-Stokes equations, that is, problem (1), Part I, with (i) non-slip boundary condition $y \equiv 0$ on $\Sigma \equiv \partial\Omega \times (0, \infty)$ in place of the boundary controlled condition (28b); and (ii) interior control $m(x)u(x,t)$ on the right-hand side of Eqn. (28a), where $m(x)$ is the characteristic function of an *arbitrary* open subset $\omega \subset \Omega$ of positive measure. In this case [4] (Part I) proves that (the linearized problem is exponentially stabilizable, hence that) the steady-state solutions y_e to the Navier-Stokes equations are locally exponentially stabilizable by a *finite-dimensional* feedback controller, in fact, *of minimal size*, see Part I. In addition, one may select the *finite-dimensional* feedback controller to be expressed in terms of a Riccati operator (solution of an algebraic Riccati equation, which arises in an optimization problem associated with the linearized equation). We shall need to invoke this interior stabilization problem (though not in its full strength) in Section 3.5 of [3].

The work in the literature which is most relevant to our present paper is that of A. Fursikov, see [8] (of which we become aware after completing [4]), which culminates a series of papers quoted therein. A statement of the main contribution of [8], as it pertains to the linearized problem (31) below, is contained in [8, Theorems 3.3 and 3.5, pp. 104–5]. Given the pair $\{\Omega, \Gamma_0\}$, where Γ_0 is a portion of the boundary $\partial\Omega$ of Ω, the approach of [8] starts with a special class, called $V^1(\Omega, \Gamma_0)$, of initial conditions y^0 for the linearized problem (31) defined on Ω. More specifically, $y^0 \in V^1(\Omega, \Gamma_0)$ means that the initial vector y^0 is the restriction $y^0 = Y^0|_\Omega$ on Ω of the vector Y^0 on the extended domain $G = \Omega \cup \omega$, where: (i) $y^0 \in (H^1(G))^d$, $d = 2, 3$; (ii) $\nabla \cdot Y^0 \equiv 0$ in G; (iii) $Y^0|_{\partial G} = 0$; (iv) the set ω is an estension of Ω across the pre-assigned portion Γ_0 of $\partial\Omega$. Then, [8, Theorem 3.3] establishes that such $y^0 \in V^1(\Omega, \Gamma_0)$ can, in turn, be re-extended from Ω to G as a new vector $\eta^0 \in (H^1(G))^d$, $\nabla \cdot \eta^0 \equiv 0$ in G, $\eta^0|_{\partial G} = 0$, possessing now the new critical property that, in addition, η^0 belongs to the *stable* subspace of the linearized operator $\tilde{\mathcal{A}}$. Here $\tilde{\mathcal{A}}$ is an extension of the original operator $\mathcal{A}$ in (6) Part I from Ω to G, obtained by a corresponding extension of the steady-state solution y_e from Ω to G, while preserving the required properties of being divergence-free across G and vanishing on ∂G. As a consequence of this extension, one obtains a function $\eta(t; \eta^0) = e^{\tilde{\mathcal{A}}t}\eta^0$, which is the solution of the corresponding linearized problem, except this time on G, and with homogeneous Dirichlet (non-slip) boundary condition on ∂G. Then, [8, Theorem 3.5] concludes (because of the obvious invariance of the stable subspace for the s.c. analytic semigroup $e^{\tilde{\mathcal{A}}t}$) that such $\eta(t; \eta^0)$ is exponentially decaying: $\|e^{\tilde{\mathcal{A}}t}\eta^0\| \leq Ce^{-\sigma t}\|\eta^0\|$, in the $(H^1(G))^d$-norm $\| \ \|$, with a controlled decay rate σ, $0 < \sigma < |\text{Re } \lambda_{N+1}|$. At this point, [8] takes the restrictions of $\eta(t; \eta^0)$ on Ω and Γ_0: $y(t) = \eta(t; \eta^0)|_\Omega$, $u(t) = \eta(t; \eta^0)|_{\Gamma_0}$, with $u(t) \equiv 0$ on $\Gamma \setminus \Gamma_0$, and defines such u as a "feedback control" of the corresponding solution y of problem (31) below on Ω, which then stabilizes such solution $y(t)$ of (31) below over Ω. A similar definition of "feedback control" is given in [8] with respect to the non-linear Navier-Stokes problem.

One should note, however, that the aforementioned controller for problem (31) below given in [8] is not a feedback controller in the standard sense. Instead, our main results (in [4] as well as) in the present paper construct genuine, authentic, and real feedback controls (Riccati-based, in fact, hence with some feature of 'robustness'), that use at time t only the state information on Ω at time t. The present paper, therefore, encounters a host of technical problems not present in [8]: from the need for the genuine feedback control u to satisfy the pointwise compatibility condition $u \cdot \nu \equiv 0$ on Σ; to the high topological level $(H^{\frac{3}{2}+\epsilon}(\Omega))^3$ at which stabilization must occur in our case, as dictated by the non-linearity for $d = 3$, see Eqn. (5.18a–b) of [3], versus the H^1-topology decay obtained in [8]; to the treatment of the Riccati theory for a corresponding optimal control problem with a combined 'index of unboundedness' in control and observation operators exceeding $\frac{3}{2}$ —thus $\frac{1}{2} + 2\epsilon$ beyond the (rich) theory of the literature [12], as explained in the *Orientation.*

Main contributions of the present paper. Qualitative summary of main results. A first qualitative description of the main results of the present paper follows next. First of all, the pre-set goal is achieved: *with no assumptions whatsoever (except mild assumptions on the domain), we prove here that the steady-state solutions to Navier-Stokes equations on $\Omega \subset \mathbb{R}^d$, $d = 2, 3$, are locally exponentially stabilizable by a closed-loop boundary feedback controller acting in the Dirichlet boundary conditions in the required topologies [Theorem 2.3 for $d = 3$ and Theorem 2.6 for $d = 2$ of [3]].* The feedback controller is expressed in terms of a Riccati operator (solution of a suitable algebraic Riccati equation): as such, via standard arguments (e.g., [4]) this feedback controller is 'robust' with respect to a certain class of exogeneous perturbations.

More precisely, the following main results are established in the present paper:

(i) For the general cases $d = 2, 3$, an infinite-dimensional, closed-loop boundary feedback, stabilizing controller is constructed, as acting (for general initial data) on the entire boundary $\partial\Omega$ for $d = 3$; or on an arbitrarily small portion of the boundary, for $d = 2$.

(ii) By contrast, for $d = 2$ and under a finite-dimensional spectral assumption FDSA = (3.6.2) of [3] (diagonalizability of the restriction of the linearized operator over the finite-dimensional unstable subspace), the feedback controller can be chosen to be *finite-dimensional*, with dimension related to properties of the unstable eigenvalues, and, moreover, still to act on an *arbitrarily small portion* of the boundary.

(iii) For $d = 3$, local exponential feedback stabilization of the steady-state solutions to Navier-Stokes equations is *not* possible with a *finite-*

dimensional boundary feedback controller (except for a meager set of special initial conditions).

(iv) The pathology noted in (iii) for $d = 3$ is due to the non-linearity (see Eqn. (5.18a–b) of [3]) which (by Sobolev embedding and multiplier theory for $d = 3$) forces the the requirement that solutions of the linearized problem be considered at the high regularity space $H^{\frac{3}{2}+\epsilon}(\Omega) \cap H$, $\epsilon > 0$, under initial conditions $y_0 \in H^{\frac{1}{2}+\epsilon}(\Omega) \cap H$ and $L^2(0,\infty;(L^2(\Gamma))^d)$-boundary controls u. In turn, this high regularity space $H^{\frac{3}{2}+\epsilon}(\Omega)$ causes the occurrence of the compatibility condition $y_0|_\Gamma = u(0)$ at $t = 0$ on the boundary to be satisfied. Thus, for $d = 3$, the constructed feedback controller must be infinite-dimensional in general.

(v) By contrast, the *linearized* problem for $d = 2, 3$ is exponentially stabilizable with a closed-loop boundary, *finite-dimensional* feedback-controller acting on an *arbitrarily small portion* of the boundary up to the topological level $(H^{\frac{3}{2}-\epsilon}(\Omega))^d$ and with initial conditions $y_0 \in (H^{\frac{1}{2}-\epsilon}(\Omega))^d \cap H$, under the same FDSA = (3.6.2) of [3].

Notation and preliminaries. Same as in Part I.

4. Main Results (Case $d = 3$)

The following assumptions will be in effect throughout the paper.

Assumptions. (i) The boundary $\partial\Omega$ of Ω is a finite union of $d-1$ dimensional C^2-connected manifolds.

(ii) The steady-state solution (y_e, p_e) defined in (2) Part I, belongs to $((H^2(\Omega))^d \cap V) \times H^1(\Omega)$. [For $d = 2, 3$, this property is guaranteed by [6, Theorem 7.3, p. 59] on y_e, for $f_e \in H$, followed by [6, Theorem 3.11, p. 30] on p_e, for sufficiently smooth $\partial\Omega$.]

Preliminaries. The translated non-linear N-S problem. By the substitutions $y \to y_e + y$, $p \to p_e + p$ and $u \to y_e|_\Gamma + u$ ($y_e|_\Gamma$ being the Dirichlet trace of y_e on $\Gamma \equiv \partial\Omega$), we are readily led via (28), (2) to study the boundary *null stabilization* of the equation

$$
\begin{aligned}
y_t - \nu_0 \Delta y + (y \cdot \nabla) y + (y_e \cdot \nabla) y & \\
+ (y \cdot \nabla) y_e &= \nabla p && \text{in } Q; && \text{(29a)}\\
\nabla \cdot y &= 0 && \text{in } Q; && \text{(29b)}\\
y &= u && \text{on } \Sigma; && \text{(29c)}\\
y(x,0) = y^0(x) &= y_0(x) - y_e(x) && \text{in } \Omega. && \text{(29d)}
\end{aligned}
$$

Abstract model of the N-S problem (29) **projected on** H. We shall see in Section 3.1 of [3] that, under the pointwise compatibility condition (c.c.) $u \cdot \nu = 0$ on Σ of (28e) (whereby then $Py_t = y_t$), application of the Leray projection P on (29a)–(29d) leads to a corresponding equation in H, without the pressure terms, whose abstract version can be written as

$$y_t - \mathcal{A}y + By = -\mathcal{A}Du, \qquad y(0) = y^0 \in H, \quad u \cdot \nu \equiv 0 \text{ on } \Sigma, \tag{30}$$

where the infinitesimal generator $\mathcal{A}$ and the non-linear operator B are defined in (15) and (7), respectively, of Part I. Moreover, the operator D: $(L^2(\Gamma))^d \to (H^{\frac{1}{2}}(\Omega))^d \cap H \in \mathcal{D}(A^{\frac{1}{4}-\epsilon})$ is defined in (3.1.3) of [3].

The translated linearized problem. PDE version. The translated linearized problem corresponding to (29) is then

$$y_t - \nu_0 \Delta y + (y_e \cdot \nabla)y + (y \cdot \nabla)y_e = \nabla p \quad \text{in } Q; \tag{31a}$$

$$\nabla \cdot y = 0 \quad \text{in } Q; \tag{31b}$$

$$y = u \quad \text{on } \Sigma; \tag{31c}$$

$$y(x,0) = y^0(x) \quad \text{in } \Omega. \tag{31d}$$

Abstract model of problem (31) **projected on** H**.** Its *abstract* version on H is then

$$y_t = \mathcal{A}y - \mathcal{A}Du \in [\mathcal{D}(\mathcal{A}^*)]', \qquad y(0) = y^0 \in H; \quad u \cdot \nu \equiv \text{ on } \Sigma. \tag{32}$$

Main results: Case $d = 3$. The linearized model. We begin with the translated linearized problem (31) or its projected version (32). For the first result—the main result on problem (32)—essentially no assumptions are required.

THEOREM 4 *With reference to the linearized problem* (32), *Part II, the following results hold true:*

(i) Let $d = 3$ and assume further that Ω is simply connected. Then, given any $y^0 \in W \equiv (H^{\frac{1}{2}+\epsilon}(\Omega))^3 \cap H$, $\epsilon > 0$ arbitrary, there exists an open-loop, infinite-dimensional boundary control $u \in L^2(0,\infty; (L^2(\Gamma))^3)$, $u \cdot \nu \equiv 0$ on Σ, such that the corresponding solution y of (32) *satisfies $y \in L^2(0,\infty; (H^{\frac{3}{2}+\epsilon}(\Omega))^3 \cap H) \cap H^{\frac{3}{4}+\frac{\epsilon}{2}}(0,\infty; H)$. Moreover, if y^0 vanishes on the portion Γ_0 of the boundary $\Gamma = \partial\Omega$, then u may be required to act only on the comlementary part $\Gamma_1 = \Gamma \setminus \Gamma_0$ of the boundary. In particular, if y^0 vanishes on all of Γ, then u may be required to have an arbitrarily small support Γ_1, meas$(\Gamma_1) > 0$. [This is Theorem 3.5.1 along with Remark 3.5.1, illustrated by Figures 3.5.1 and 3.5.2 in [3].]*

(ii) Let $d = 3$*. Then, the control* u *claimed in (i) cannot generally be finite-dimensional except for a meager set of special initial conditions. [This is Proposition 3.1.3 of [3].]*

Case $d = 3$**. Original N-S model** (28). We now report the main result of the present paper, which provides the sought-after closed-loop boundary feedback control for the original N-S equations (28) [or its projected version (30)], which *exponentially stabilizes the stationary solution* y_e *of* (28) *in a neighborhood of* y_e. The stabilizing feedback control that we shall find is 'robust,' as it is expressed in terms of a Riccati operator R, which arises in an associated corresponding Optimal Control Problem. To state our (local) stabilizing result, we need to introduce the set

$$\mathcal{V}_\rho \equiv \left\{ y_0 \in W \equiv (H^{\frac{1}{2}+\epsilon}(\Omega))^3 \cap H : |y_0 - y_e|_W < \rho \right\} \tag{33}$$

of initial conditions y_0 of (28), whose distance in the norm of W from a stationary solution y_e is less than $\rho > 0$. Here, $\epsilon > 0$ arbitrary is fixed once and for all.

THEOREM 5 (MAIN THEOREM) *Let* $d = 3$ *and assume further that* Ω *is simply connected. If* $\rho > 0$ *in (33) is sufficiently small, then: for each* $y_0 \in \mathcal{V}_\rho$*, there exists a unique fixed-point, mild, semigroup solution* y *of the following closed-loop problem:*

$$y_t(x,t) - \nu_0 \Delta y(x,t) + (y \cdot \nabla) y(x,t) = f_e(x) + \nabla p(x,t) \quad \text{in } G; \tag{34a}$$

$$\nabla \cdot y = 0 \quad \text{in } G; \tag{34b}$$

$$y = \nu_0 \frac{\partial}{\partial \nu} R(y - y_e) \quad \text{on } \Sigma; \tag{34c}$$

$$y(x,0) = y_0(x) \quad \text{in } \Omega. \tag{34d}$$

obtained from (28) *by replacing* u *with the boundary feedback control* $u = y_e + \nu_0 \frac{\partial}{\partial \nu} R(y - y_e)$ *having the following regularity and asymptotic properties:*

(i)

$$(y - y_e) \in C([0,\infty); W) \cap L^2(0,\infty; (H^{\frac{3}{2}+\epsilon}(\Omega))^3 \cap H) \tag{35}$$

continuously in $y_0 \in W \equiv (H^{\frac{1}{2}+\epsilon}(\Omega))^3 \cap H$*:*

$$|y(t) - y_e|_W^2 + \int_0^\infty |y(t) - y_e|^2_{(H^{\frac{3}{2}+\epsilon}(\Omega))^3 \cap H} dt \le C |y_0 - y_e|_W^2, \; t \ge 0. \tag{36}$$

[This follows from Theorem 5.1 and Corollary 5.5 of [3], via the translation $y \to y_e$, etc., performed above problem (29).*]*

(ii) there exist constants $M \geq 1$, $\omega > 0$ (independent of $\rho > 0$) such that such solution $y(t)$ satisfies

$$|y(t) - y_e|_W \leq Me^{-\omega t}|y_0 - y_e|_W, \quad t \geq 0. \tag{37}$$

[This follows from Theorem 6.1(i) of [3], via the translation $y \to y - y_e$, etc., performed above problem (29).*]*

Here R is a Riccati operator, in the sense that it [arises in the Optimal Control Problem of Section 4.1 and] satisfies the Algebraic Riccati Equation (4.5.1) of [3]. The operator R is positive self-adjoint on H and, moreover, $R \in \mathcal{L}(W; W')$ where W' is the dual of W with respect to H as a pivot space. In addition [Proposition 4.1.4 of [3]],

$$c|x|_W^2 \leq (Rx, x)_H \leq C|x|_W^2, \ 0 < c < C < \infty, \ \forall \ x \in W,$$

so that the $|R^{\frac{1}{2}}x|$-norm is equivalent to the W-norm. By a solution to Eqn. (5)*), we mean, of course, a weak solution (see, e.g., [6], [13]). (This part is Theorem 5.1 of [3].)*

References

[1] V. Barbu. Local internal controllability of the Navier-Stokes equations. *Advances in Differential Equations*, 6(12):1443–1462, 2001.

[2] V. Barbu. Feedback stabilization of Navier-Stokes equations. *ESAIM COCV*, 9:197–206, 2003.

[3] V. Barbu, I. Lasiecka, and R. Triggiani. Boundary stabilization of Navier-Stokes equations. 2003. Preprint.

[4] V. Barbu and R. Triggiani. Internal stabilization of Navier-Stokes equations with finite-dimensional control. *Indiana University Mathematics Journal*, 2004. To appear.

[5] V. Barbu and G. Wang. Feedback stabilization of the semilinear heat equations. *Abstract and Applied Analysis*, 12:697–714, 2003.

[6] P. Constantin and C. Foias. *Navier-Stokes Equations.* University of Chicago Press, Chicago, London, 1988.

[7] J. M. Coron. On the null asymptotic stabilization of the 2-d incompressible Euler equations in a simply connected domain. *SIAM J. Control Optimiz.*, 37:1874–1896, 1999.

[8] A. Fursikov. *Real Process Corresponding to the 3D Navier-Stokes System and its Feedback Stabilization from the Boundary*, volume 206 of *Translations of the AMS, Series 2.* American Mathematical Society, 2002. *Partial Differential Equations, Mark Vishik's Seminar*, pp. 95–123, M. S. Agranovich and M. A. Shubin, editors.

[9] O. A. Imanuvilov. Local controllability of Navier-Stokes equations. *ESAIM COCV*, 3:97–131, 1998. On local controllability of Navier-Stokes equations, *ESAIM COCV* 6 (2001), 49–97.

[10] T. Kato. *Perturbation Theory of Linear Operators.* Springer-Verlag, New York, Berlin, 1966.

[11] I. Lasiecka. Exponential stabilization of hyperbolic systems with nonlinear unbounded perturbations—a riccati operator approach. *Applicable Analysis*, 42:243–261, 1991.

[12] I. Lasiecka and R. Triggiani. *Control Theory for Partial Differential Equations: Continuous and Approximation Theories*, volume 1 of *Encyclopedia of Mathematics and its Applications.* Cambridge University Press, 2000. 644 pp.

[13] R. Teman. *Navier-Stokes Equations, Revised edition*, volume 2 of *Studies in Mathematics and its Applications.* North Holland, 1979.

[14] R. Triggiani. On the stabilizability problem in Banach space. *J. Math. Anal. Appl.*, 52:383–403, 1975.

[15] R. Triggiani. Boundary feedback stabilizability of parabolic equations. *Appl. Math. Optimiz.*, 6:201–220, 1980.

MATRIX ROUNDING AND RELATED PROBLEMS WITH APPLICATION TO DIGITAL HALFTONING

Naoki Katoh
Department of Architecture and Architectural Systems
Kyoto University, Japan
naoki@archi.kyoto-u.ac.jp

Abstract In this paper, we first study the problem of rounding a real-valued matrix into an integer-valued matrix to minimize an L_p-discrepancy measure between them. To define the L_p-discrepancy measure, we introduce a family $\mathcal{F}$ of regions (rigid submatrices) of the matrix, and consider a hypergraph defined by the family. The difficulty of the problem depends on the choice of the region family $\mathcal{F}$. We first investigate the rounding problem by using integer programming problems with convex piecewise-linear objective functions. Then, we propose "laminar family" for constructing a practical and well-solvable class of $\mathcal{F}$. Indeed, we show that the problem is solvable in polynomial time if $\mathcal{F}$ is the union of two laminar families. We shall present experimental results. We also give some nontrivial upper bounds for the L_p-discrepancy. We then foucs on the number of global roundings defined on a hypergraph $H_G = (V, \mathcal{P}_G)$ which corresponds to a set of shortest paths for a weighted graph $G = (V, E)$. For a given real assignment $\mathbf{a}$ on V satisfying $0 \le \mathbf{a}(v) \le 1$, a global rounding α with respect to H_G is a binary assignment satisfying that $|\sum_{v \in F} \mathbf{a}(v) - \alpha(v)| < 1$ for every $F \in \mathcal{P}_G$. We conjecture that there are at most $|V| + 1$ global roundings for H_G.

1. Introduction

Rounding is an important operation in numerical computation, and plays key roles in digitization of analogue data. Rounding of a real number a is basically a simple problem: We round it to either $\lfloor a \rfloor$ or $\lceil a \rceil$, and we usually choose the one nearer to a. However, we often encounter a data consisting of more than one real numbers instead of a singleton. If it has n numbers, we have 2^n choices for rounding. If the original data set has some feature, we need to choose a rounding so that the rounded result inherits as much of the feature as possible. The feature is described by using some combinatorial structure; we indeed consider a hypergraph $\mathcal{H}$ on the set. A typical input set is a multi-dimensional array of real numbers, and we consider a hypergraph whose hyperedges are its subarrays with contiguous indices. In this paper, we first focus on two-dimensional arrays; In other words, we consider rounding problems on matrices. We then consider the roounding problems on hypergraphs derived from a set of shortest paths of a weighted graph.

This article is a summary of our recent papers [4, 3].

1.1 Rounding Problem and Discrepancy Measure

Given an $N \times N$ matrix $A = (a_{ij})_{1 \le i,j \le N}$ of real numbers, its rounding is a matrix $B = (b_{ij})_{1 \le i,j \le N}$ of integral values such that b_{ij} is either $\lfloor a_{ij} \rfloor$ or $\lceil a_{ij} \rceil$ for each (i, j). There are 2^{N^2} possible roundings of a given A, and we would like to find an optimal rounding with respect to a given criterion. This is called the *matrix rounding problem.* Without loss of generality, we can assume that each entry of A is in the closed interval $[0, 1]$ and each entry is rounded to either 0 or 1.

In order to give a criterion to determine the quality of roundings, we define a distance in the space of all $[0, 1]$-valued $N \times N$ matrices. We introduce a family $\mathcal{F}$ of regions over the $N \times N$ integer grid

$$G_n = \{(1,1), (1,2) \dots, (N,N)\} = \{p_1, p_2, \dots, p_n\}, \quad n = N^2. \quad (1)$$

This means that each entry location of a matrix is denoted by a symbol p_i and that the order $(p_1, \dots, p_n)$ is arbitrary. Let $\mathcal{A} = \mathcal{A}(G_n)$ be the space of all $[0, 1]$-valued matrices with the index set G_n, and let $\mathcal{B} = \mathcal{B}(G_n)$ be its subset consisting of all $\{0, 1\}$-valued matrices. Let R be a region in $\mathcal{F}$ [1]. For an element $A \in \mathcal{A}$, let $A(R)$ be the sum of entries of A located in the region R, that is, $A(R) = \sum_{p_i \in R} a_{p_i}$. We define a distance $Dist_p^{\mathcal{F}}(A, A')$ between two elements A and A' in $\mathcal{A}$ for a positive integer p by

$$Dist_p^{\mathcal{F}}(A, A') = [\sum_{R \in \mathcal{F}} |A(R) - A'(R)|^p]^{1/p}.$$

The distance is called the L_p-distance with respect to $\mathcal{F}$. The L_∞ distance with respect to $\mathcal{F}$ is defined by

$$Dist_\infty^{\mathcal{F}}(A, A') = \lim_{p \to \infty} Dist_p^{\mathcal{F}}(A, A') = \max_{R \in \mathcal{F}} |A(R) - A'(R)|.$$

Using the notations above, we can formally define the matrix rounding problem:

L_p-Optimal Matrix Rounding Problem:
$\mathcal{P}(G_n, \mathcal{F}, p)$: Given a $[0,1]$-matrix $A \in \mathcal{A}$, a family $\mathcal{F}$ of subsets of G_n, and a positive integer p, find a $\{0,1\}$-matrix $B \in \mathcal{B}$ that minimizes

$$Dist_p^{\mathcal{F}}(A, B) = [\sum_{R \in \mathcal{F}} |A(R) - B(R)|^p]^{1/p}.$$

Also, we are interested in the following combinatorial problem:

L_p-Discrepancy Bound:
Given a $[0,1]$-matrix $A \in \mathcal{A}$, a family $\mathcal{F}$ of subsets of G_n, and a positive integer p, investigate upper and lower bounds of

$$\mathcal{D}(G_n, \mathcal{F}, p) = \sup_{A \in \mathcal{A}} \min_{B \in \mathcal{B}} Dist_p^{\mathcal{F}}(A, B).$$

The pair $(G_n, \mathcal{F})$ defines a hypergraph on G_n, and $\mathcal{D}(G_n, \mathcal{F}, \infty)$ is called the *inhomogeneous discrepancy* of the hypergraph. Abusing the notation, we call $\mathcal{D}(G_n, \mathcal{F}, p)$ the (inhomogeneous) L_p-discrepancy of the hypergraph, and also often call $Dist_p^{\mathcal{F}}(A, B)$ the L_p-discrepancy measure of (quality of) the output B with respect to $\mathcal{F}$.

1.2 Motivation and our Application

The most popular example of the family $\mathcal{F}$ is the set of all rectangular subregions in G_n, and the corresponding L_∞-discrepancy measure is utilized in many application areas such as Monte Carlo simulation and computational geometry. Unfortunately, if we consider the family of all rectangular subregions, the discrepancy bound (for the L_∞ measure) is known to be $\Omega(\log n)$ and $O(\log^3 n)$ [7]. It seems hard to find an optimal solution to minimize the discrepancy. In fact, it is NP-hard [2].

Therefore, we seek a family of regions for which low discrepancy rounding is useful in an important application and also can be computed in polynomial time. For the application, L_∞ rounding is not always suitable, and L_p-discrepancy (with $p = 1$ or 2) is preferable. For the purpose, we present a geometric structure of a family of regions reflecting the combinatorial discrepancy bound and computational difficulty of the matrix rounding problem.

In particular, we focus on the digital halftoning application of the matrix rounding problem, where we should consider smaller families of rectangular subregions as $\mathcal{F}$. More precisely, the input matrix represents a digital (gray) image, where a_{ij} represents the brightness level of the (i, j)-pixel in the $N \times N$ pixel grid. Typically, N is between 256 and 4096, and a_{ij} is an integral multiple of $1/256$: This means that we use 256 brightness levels. If we want to send an image using fax or print it out by a dot (or ink-jet) printer, brightness levels available are limited. Instead, we replace A by an integral matrix B so that each pixel uses only two brightness levels. Here, it is important that B looks similar to A; in other words, B should be a good approximation of A.

For each pixel (i, j), if the average brightness level of B in each of its neighborhoods is similar to that of A, we can expect that B is a good approximation of A. For this purpose, the set of all rectangles is not suitable (i.e., it is too large), and we may use a more compact family. Moreover, since human vision detects global features, the L_1 or L_2 measure should be better than the L_∞ measure to obtain a clear output image. This intuition is supported by our experimental results.

We should mention here that there exist known algorithms for digital halftoning. We briefly review two popular methods among them. The first one is *ordered Dither* which extends a simple thresholding in such a way that it uses different thresholds depending on the positions [6]. Namely, we prepare an $M \times M$ matrix of integers ranging from 1 to M^2. This matrix (*dither matrix*) is tiled periodically to cover the whole image. Each pixel a_{ij} in the image is compared with the corresponding threshold (the integer of dither matrix divided by M^2) to decide whether the output $b_{ij} = 0$ or 1. The second method is called *error diffusion* which propagates the quantization erros to unprocessed neighboring pixels according to a predeteremined way. More precisely, pixels are processed in a raster order, from left to right and from top to bottom. Each pixel is compared with a fixed threshold, 0.5 and round it up if it is greater than or equal to the threshold and round it down otherwise. The quantization error caused by the rounding is diffuesed over the pixels around it with fixed ratios. For example, if a pixel level is 0.7, it is rounded up to 1 and the error -0.3 is diffused to the unprocessed pixels nearby.

This method gives excellent image quality in many cases, but it tends to produce visible artifacts in an area of uniform intensity.

Those two existing methods do not have any theoretical background which guarantee the quality of the goodness of the output image. This is one of the motivations of our theoretical study.

1.3 Known Results on L_∞ Measure

For the L_∞ measure, the following beautiful combinatorial result is classically known:

THEOREM 1 [Baranyai[5]1974] *Given a real-valued matrix $A = (a_{ij})$ and a family $\mathcal{F}$ of regions consisting of all rows, all columns and the whole matrix, there exists an integer-valued matrix $B = (b_{ij})$ such that $|A(R) - B(R)| < 1$ holds for every $R \in \mathcal{F}$.*

The combinatorial structure and algorithmic aspects of roundings of (one-dimensional) sequences with respect to the L_∞-discrepancy measure are investigated in recent studies [2, 17].

The *incidence matrix* $\mathcal{C}(G_n, \mathcal{F}) = (\mathcal{C}_{ij})$ of the hypergraph $(G_n, \mathcal{F})$ is defined by $\mathcal{C}_{ij} = 1$ if the j-th element of G_n belongs to the i-th region R_i in $\mathcal{F}$ and 0 otherwise. [2] A hypergraph is called *unimodular* if its incidence matrix is totally unimodular, where a matrix $\mathcal{C}$ is *totally unimodular* if the determinant of each square submatrix of $\mathcal{C}$ is equal to $0, 1$, or -1.

The L_∞-discrepancy problem can be formulated as an integer programming problem, and the unimodularity implies that its relaxation has an integral solution. A classical theorem of Ghouila-Houri [9]implies that total unimodularity is a necessary and sufficient condition for the existence of a rounding with L_∞ discrepancy less than 1.

1.4 Organization of the Paper

We shall consider L_p-discrepancy measure instead of L_∞-discrepancy measure. Starting with one-dimensional array, we shall show several interesting unimodular families of $\mathcal{F}$ where we can find an optimal solution of $\mathcal{P}(G_n, \mathcal{F}, p)$. we consider the optimization problem. In fact, if the hypergraph is unimodular, the rounding minimizing the L_p-discrepancy can be computed in polynomial time by translating it to a separable convex programming problem and applying known general algorithms [10]. However, we want to define a class of region families for which we can compute the optimal solution more efficiently, as well as the class is useful in applications (in particular, the digital halftoning application). We consider the union of two laminar families (defined in Section 3), and show that the matrix rounding problem can be formulated into a minimum cost flow problem, and hence solved in polynomial time. We then show recent results on the L_p-discrepancy bound [4].

We implemented the algorithm using LEDA[11]. Some output pictures of the algorithm applying to the digital halftoning problem are included.

Finally, we briefly review the results concerning global roundings.

2. Mathematical Programming Formulations

2.1 One-Dimensional Case

We shall begin with a basic case for the rounding problem. We take a $[0, 1]$-valued one-dimensional array $A = (a_i)_{i=1,2,\ldots,n}$ as an input, and a binary array $B = (b_i)_{i=1,2,\ldots,n}$ as an output. A family $\mathcal{F}$ of regions we consdier here is a set of all subintervals of the entire interval $[1, n]$. Now the L_1-optimal rounding problem on $\mathcal{F}$ is described as follows:

$$(I1) : \text{minimize } \{\sum_{I_i \in \mathcal{F}} |\sum_{j \in I_i} a_j - \sum_{j \in I_i} b_j| \mid b_j = 0, 1, \quad j = 1, 2, \ldots, n\}.$$

If we introduce new variables $y_i = B(I_i)$ and constants $c_i = A(I_i)$ the problem can be replaced by

$$\begin{aligned} (I2): \quad & \text{minimize} \quad \sum_{I_i \in \mathcal{F}} |y_i - c_i| \\ & \text{subject to} \quad y_i = \sum_{j \in I_i} b_j, j = 1, 2, \ldots, m = |\mathcal{F}| \text{ and} \\ & \qquad\qquad b_j = 0, 1, \quad j = 1, 2, \ldots, n. \end{aligned}$$

The constarints concerning the variables y_i are represented in a form:

$$(-I, \mathcal{C}(G_n, \mathcal{F}))Y = 0$$

where I is an identity matrix and $Y = (y_1, \ldots, y_m, b_1, \ldots, b_n)$, $G_n = \{1, 2, \ldots, n\}$ and $\mathcal{C}(G_n, \mathcal{F})$ is the incidence matrix defined by $c_{ij} = 1$ if $b_j \in I_i$, 0, otherwise. Notice that the above problem can be viewed as a separable piecewise linear convex minimization problem under linear constraints. It is known that the incidence matrix $\mathcal{C}(G_n, \mathcal{F})$ is totally unimodular [18]. Thus, using the following theorem, the problem $(I2)$ can be solve in polynomial time.

THEOREM 2 [Hochbaum and Shanthikumar [10]] *A nonlinear separable convex optimization problem* $\min\{\sum_{i=1}^{n} f_i(x_i) \mid Ax \geq b\}$ *on linear constraints with a totally unimodular matrix* A *can be solved in polynomial time in* n.

2.2 Mathematical Programming Formulation of Matrix Rounding Problem

We will extend the formulation of the one-dimensional rounding problem to that for matrix rounding problems.

Introducing a new variable $y_i = B(R_i) = \sum_{(j,k) \in R_i} b_{jk}$ for each $R_i \in \mathcal{F}$, the problem $\mathcal{P}(G_n, \mathcal{F}, p)$ is described in the following form:

$$\begin{aligned} (P1): \quad & \text{minimize} \quad [\sum_{R_i \in \mathcal{F}} |y_i - A(R_i)|^p]^{1/p} \\ & \text{subject to} \quad y_i = \sum_{(j,k) \in R_i} b_{jk}, \ i = 1, \ldots, m = |\mathcal{F}| \\ & \qquad\qquad B \in \mathcal{B}(G_n). \end{aligned}$$

Notice that G_n denotes the one of (1) from now one. The objective function can be replaced with $\sum_{R_i \in \mathcal{F}} |y_i - c_i|^p$, where $c_i = A(R_i) = \sum_{(j,k)\in R_i} a_{jk}$ is a constant depending only on input values. Now, $|y_i - c_i|^p$ is a convex function independent of other y_js. The constraints $y_i = \sum_{(j,k)\in R_i} b_{jk}$, $i = 1, \ldots, m$ are represented by $(-I, \mathcal{C}(G_n, \mathcal{F}))Y = 0$ using the incidence matrix $\mathcal{C}(G_n, \mathcal{F})$ defined in Section 1.3 where $Y = (y_1, \ldots, y_m, b_{11}, \ldots, b_{nn})^T$ and I is an identity matrix.

Although the objective function is now a separable convex function, its nonlinearity gives difficulty to analyze the properties of the solution. Thus, we apply the idea of Hochbaum and Shanthikumar [10]to replace $|y_i - c_i|^p$ with a piecewise linear convex continuous function $f_i(y_i)$ which is equal to $|y_i - c_i|^p$ for each integral value of y_i in $[0, |R_i|]$. This is because we only need integral solutions, and if each b_{p_j} is integral, y_i must be a nonnegative integer less than or equal to $|R_i|$. Typically for $p = 1$, $f_i(y_i)$ is illustrated in Figure 1.

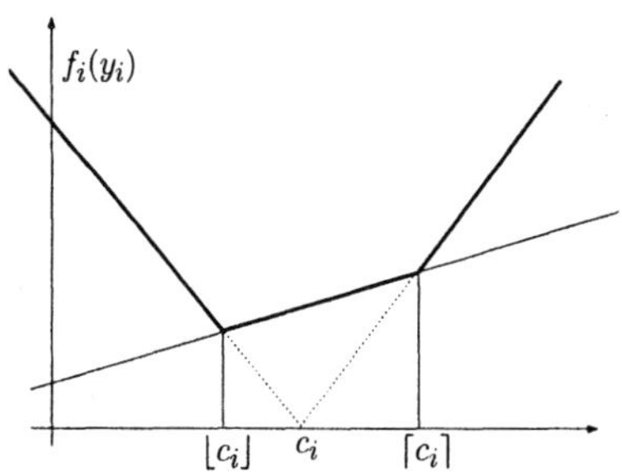

Figure 1. Conversion of the convex objective function $|y_i - c_i|$ into a piecewise linear convex function $f_i(y_i)$ with integral breakpoints (shown in bold lines).

Thus, we obtain the following problem (P2):

$$(P2) : \text{minimize} \, \{ \sum_{R_i \in \mathcal{F}} f_i(y_i) \mid y_i = \sum_{(j,k)\in R_i} b_{jk}, \, i = 1, \ldots, m, B \in \mathcal{B}(G_n) \}$$

Here $m = |\mathcal{F}|$. Thus, we can formulate the problem into an integer programming problem where the objective function is a separable piecewise-linear convex function.

Let (P3) be the continuous relaxation obtained from (P2) by replacing the integral condition of b_{ij} with the condition $0 \leq b_{ij} \leq 1$. Note that this is different from the continuous relaxation of (P1), since the objective function of (P2) is larger than that of (P1) at non-integral values. If the matrix is totally unimodular, (P3) has an integral optimal solution by Theorem 2. This is a key to derive discrepancy bounds and also algorithms. We thus have the following result.

COROLLARY 3 *The matrix rounding problem $\mathcal{P}(G_n, \mathcal{F}, p)$ is solved in polynomial time in n if its associated incidence matrix $C(G_n, \mathcal{F})$ is totally unimodular.*

3. Geometric Families of Regions Defining Unimodular Hypergraphs

In this section we consider interesting classes of families whose associated incidence matrices are totally unimodular. We call such a family a *unimodular family*, since the associated hypergraph is unimodular. A family $\mathcal{F} = \{R_1, R_2, \ldots, R_m\}$ is a *partition family* (or a partition) of G_n if $\bigcup_{i=1}^{m} R_i = G_n$ and $R_i \cap R_j = \emptyset$ for any $R_i \neq R_j$ in $\mathcal{F}$. A *k-partition family* is a family of regions on a matrix which is the union of k different partitions of G_n.

A family $\mathcal{F}$ of regions on a grid G_n is a *laminar family* if one of the following holds for any pair R_i and R_j in $\mathcal{F}$: (1) $R_i \cap R_j = \emptyset$, (2) $R_i \subset R_j$ and (3) $R_j \subset R_i$. The family is also called a laminar decomposition of the grid G_n. In general, a *k-laminar family* is a family of regions on a matrix which is the union of k different laminar families.

The following theorem is known [18].

THEOREM 4 *A 2-laminar family is unimodular.*

Direct applications of Theorem 4 lead to various unimodular families of regions. The family of regions defined in Baranyai's theorem is a 2-laminar family. Also, take any 2−partition family consisting of 2×2 regions on a matrix. For example, take all 2×2 regions with their upper left corners located in even points (where the sums of their row and column indices are even). The set of all those regions defines two partition families $\mathcal{F}_{even}$ and $\mathcal{F}_{odd}$ where $\mathcal{F}_{even}$ (resp. $\mathcal{F}_{odd}$) consists of all 2×2 squares with their upper left corners lying at even (resp. odd) rows (see Figure 2). This kind of families plays an important role in the following Section and also in our experiment. Notice that a 3−partition family is not unimodular in general.

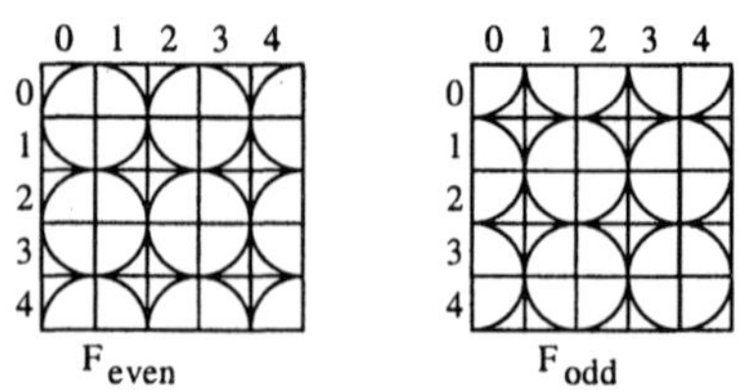

Figure 2. 2-partition family of 2×2 regions.

4. Algorithms for Computing the Optimal Rounding

The arguments so far guarantee polynomial-time solvability of our problem. However, we needed a more practical algorithm for our experiments that runs fast for large-scale problem instances. In this section we will show how to solve the matrix rounding problem for a 2-laminar family based on the minimum-cost flow algorithm.

Our main result is the following:

THEOREM 5 *Given a* $[0,1]$*-matrix* A *and a* 2*-laminar family* $\mathcal{F}$*, an optimal binary matrix* B *that minimizes the distance* $Dist_p^{\mathcal{F}}(A,B)$ *is computed in* $O(n^2 \log^2 n)$ *time, where* n *is the number of matrix elements.*

PROOF. We can transform the problem into that of finding a minimum-cost circulation flow in the network defined as follows.

Let $\mathcal{F}$ be a 2-laminar family given as the union of two laminar families $\mathcal{F}_1 = \{R_0, R_1, \ldots, R_m\}$ and $\mathcal{F}_2 = \{R'_0, R'_1, \ldots, R'_{m'}\}$ over the grid G_n, where R_0 and R'_0 are the entire region G_n. The network to be constructed consists of three parts. The first part is an in-tree T_1 derived from $\mathcal{F}_1$ whose root is R_0, the second one an out-tree T_2 from $\mathcal{F}_2$ whose root is R'_0, and the third part connects T_1 and T_2. The lattice structure implied by $\mathcal{F}_1$ naturally defines an in-tree T_1 such that the vertex set is the set of regions in $\mathcal{F}_1$ and there is a directed edge (R_i, R_j) if and only if $R_i \subsetneq R_j$ and there is no other region R_k such that $R_j \subsetneq R_k \subsetneq R_i$. Then, each region $R_i, i \geq 1$ has a unique outgoing edge, which is denoted by $e(R_i)$. We can similarly define T_2 for the laminar family $\mathcal{F}_2$, in which the edge direction is reversed in T_2, that is, each node $R'_i, i \geq 1$ has an unique incoming edge, which is denoted by $e(R'_i)$.

In addition, leaves of T_1 and T_2 are connected by edges corresponding to elements of G_n. Because of the definition of $\mathcal{F}_1$ and $\mathcal{F}_2$, each element (k,l) of G_n belongs to exactly one region in $\mathcal{F}_1$ which is a leaf in T_1 and to exactly one region in $\mathcal{F}_2$ which is a leaf in T_2. If (k,l) belongs to R_i and R'_j, then we have a directed edge $e(i,j)$ from R'_j to R_i. Finally, we draw an edge from R_0 to R'_0.

Now, we define capacity and cost coefficient of each edge. (The lower bound on the flow of each edge is defined to be 0.) The capacity of an edge $e(i,j)$ is determined simply as 1 because b_{ij} is to be rounded to 0 or 1. Other edges do not have capacity constraint.

The cost associated of an edge $e(R_i)$ ($e(R'_j)$, respectively) is $f_i(y_i)$ where y_i is defined in $(P1)$. As already mentioned, $f_i(y_i)$ is piecewise linear convex in y_i. We then apply the result of [16](see also Chapter

14 of the book by Ahuja et al. [1]) for the algorithm of minimum-cost circulation with separable convex costs.

From this result, we can find an optimal rounding in time $O(|E| \log U(|E| + |V| \log |V|))$ for a network with node set V and edge set E and the largest integral capacity U. In our case, $|V|, |E|$ and U are all $O(n)$, and thus we have $O(n^2 \log^2 n)$, where n is the number of matrix elements. Q.E.D.

5. Upper Bounds for the L_p-Discrepancy

In this section, we show the following theorem for the L_p-discrepancy of a unimodular family. We will omit the proof for the space limit.

THEOREM 6 [Asano, Katoh, Obokata and Tokuyama [4]] *If $\mathcal{F}$ is unimodular and $p \leq 3$, for any $A \in \mathcal{A}$ we have*

$$\min_{B \in \mathcal{B}} Dist_p^{\mathcal{F}}(A, B) \leq \frac{1}{2}|\mathcal{F}|^{1/p}.$$

It is easy to give an instance to show that the bound is tight: Consider Baranyai's problem on a matrix having $\frac{1}{2}$ entries in its diagonal position (other entries are zeros).

6. Application to Digital Halftoning

The quality of color printers has been drastically improved in recent years, mainly based on the development of fine control mechanism. On the other hand, there seems to be no great invention on the software side of the printing technology. What is required is a technique to convert a continuous-tone image into a binary image consisting of black and white dots so that the binary image looks very similar to the input image. From a theoretical standpoint, the problem is how to approximate an input $[0, 1]$-array by a binary array. Since this is one of the central techniques in computer vision and computer graphics, a great number of algorithms have been proposed (see, e.g., [12, 8, 6, 13, 15]). However, there have been very few studies toward the goal of achieving an optimal binary image under some reasonable criterion; maybe because the problem itself is very practically oriented. A desired output image is the one which looks similar to the input image to the human visual system. The most popular distortion criterion that is used in practice is perhaps Frequency Weighted Mean Square Error (FWMSE)[14]which is defined by

$$W(G, X) = \sum_{(i,j) \in G_n} [\sum_{k=-K}^{K} \sum_{l=-K}^{K} v_{|k||l|} a_{i+k,j+l} - \sum_{k=-K}^{K} \sum_{l=-K}^{K} v_{|k||l|} b_{i+k,j+l}]^2.$$

Here, $V = (v_{|k||l|}), k, l = 0, \ldots, K$ is an impulse response that approximates the characteristics of the human visual system and K is some small constant, say 3. Our discrepancy measure which has been discussed in this paper is a hopeful replacement; Indeed, the L_2-discrepancy measure can be regarded as a simplified version of the FWMSE criterion.

We have implemented the algorithm using LEDA [11]functions for finding minimum-cost flow, and applied to several test images to compare its results with the error diffusion algorithm which is most commonly used in practice. The data we used for our experiments are *Standard high precision picture data* created by the Institute of Image Electronics Engineers of Japan, which include four standard pictures called, "Bride," "Harbor," "Wool," and "Bottles." They are color pictures of 8 bits each in RGB. Their original picture size is 4096×3072. In our experiments we scaled them down to 1024×768 in order to shorten the running time of the program. Figures 3 and 4 show experimental results for "Wool" and "Wine" to compare our algorithm with error diffusion. Our algorithm has been implemented using a 2-laminar family defined by the two tiles (b) and (c) depicted in Figure 5.

Figure 3. Experimental results for "Wool". Output images by error diffusion algorithm (left) and the algorithm in this paper (right).

7. Global Roundings

Let $H = (V, \mathcal{F})$, where $\mathcal{F} \subset 2^V$, be a hypergraph on a set V of n nodes. Given a real valued function $\mathbf{a}$ on V, we say that an integer valued function α on V is a *global rounding* of $\mathbf{a}$ with respect to H, if $w_F(\alpha)$ is a rounding of $w_F(\mathbf{a})$ for each $F \in \mathcal{F}$, where $w_F(f)$ denotes $\sum_{v \in F} f(v)$. Without loss of generality we restrict our attention to the case where the ranges of $\mathbf{a}$ and α are $[0,1]$ and $\{0,1\}$ respectively.

This notion of global roundings on hypergraphs is closely related to that of $_\infty$-*discrepancy* of hypergraphs[7]. Given $\mathbf{a}$ and $\mathbf{b} \in [0,1]^V$, define

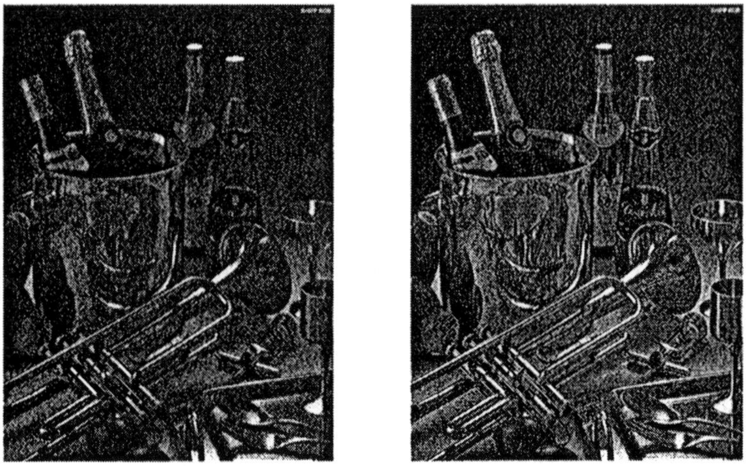

Figure 4. Experimental results for "Wine". Output images by error diffusion (left) and by the algorithm in this paper (right).

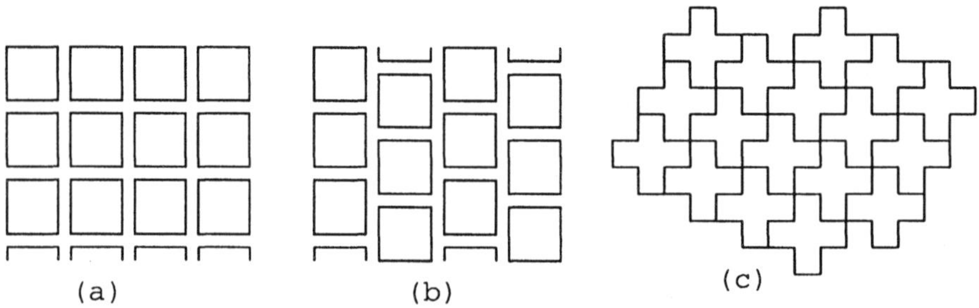

Figure 5. Three different partitions of the image plane (a) by 2×2 squares, (b) vertically shifted 2×2 squares, and (3) cross patterns consisting of 5 pixels.

the *discrepancy* $D_H(\mathbf{a}, \mathbf{b})$ between them on H by

$$D_H(\mathbf{a}, \mathbf{b}) = \max_{F \in \mathcal{F}} |w_F(\mathbf{a}) - w_F(\mathbf{b})|.$$

Our recent paper [3]investigated maximum number $\nu(H)$ of global roundings.

This direction of research is initiated by Sadakane *et al.*[17]where the authors discovered a somewhat surprising fact that $\nu(I_n) \leq n+1$ where I_n is a hypergraph on $V = \{1, 2, .., n\}$ with edge set $\{[i, j]; 1 \leq i \leq j \leq n\}$ consisting of all subintervals of V. We can also see that $\nu(H) \geq n+1$ for any hypergraph H: if we let $\mathbf{a}(v) = \epsilon$ for every v, where $\epsilon < 1/n$, then any binary assignment on V that assigns 1 to at most one vertex is a global rounding of H, and hence $\nu(H) \geq n+1$.

Given this discovery, it is natural to ask for which class of hypergraphs this property $\nu(H) = n + 1$ holds.

We showed [3]that $\nu(H) = n+1$ holds for a considerably wider class of hypergraphs. Given a connected G in which edges are possibly weighted by a positive value, we define a *shortest-path hypergraph* H_G generated by G as follows: a set F of vertices of G is an edge of H_G, if and only if F is the set of vertices of some shortest path in G with respect to the given edge weights. Note that we permit more than one shortest path between a pair of nodes if they have the same length. The following theorem is our main result [3]:

THEOREM 7 $\nu(H_G) = n + 1$ *holds for the shortest-path hypergraph* H_G, *if* G *is a tree, a cycle, a tree of cycles, an unweighted mesh, or an unweighted* k*-tree.*

We conjecture that the result holds for general connected graphs.

8. Concluding Remarks

We have considered the matrix rounding problem based on L_p-discrepancy measure. Although we have shown that the measure is useful in application to the digital halftoning application, the current algorithm is too slow if we want to require speed together with the high-quality requirement. It is desired to design a faster algorithm (even an approximation algorithm). Moreover, it is an interesting question to investigate what kind of region families give the best criterion for the halftoning application. Once we know such a region family, it is valuable to design an algorithm (heuristic algorithm if the problem for solving the optimal solution is intractable) for the criterion.

Acknowledgment

The authors would like to thank Tetsuo Asano, Koji, Obokata, Tomomi Matsui, Koji Nakano, Hiroshi Nagamochi, and Takeshi Tokuyama for their valuable comments and helpful discussions.

Notes

1. Strictly speaking, R can be any subset of G_n. Although we implicitly assume that R forms some connected portion on the grid G_n, the connectivity assumption is not used throughout the paper.
2. We implicitly assume a one-dimensional ordering of elements in G_n

References

[1] R. Ahuja, T. Magnanti, and J. Orlin. *Network Flows, Theory Algorithms and Applications.* Prentice Hall, 1993.

[2] T. Asano, T. Matsui, and T. Tokuyama. Optimal roundings of sequences and matrices. *Nordic Journal of Computing*, 7:241–256, 2000.

[3] Tetsuo Asano, Naoki Katoh, Hisao Tamaki, and Takeshi Tokuyama. The structure and number of global roundings of a graph. *Proc. of COCOON 2003, LNCS 2697*, pages 130–138, 2003.

[4] Tetsuo Asano, Naoki Katoha, Koji Obokata, and Takeshi Tokuyama. Matrix rounding under the lp-discrepancy measure and its application to digital halftoning. *SODA 2002 (also to appear in SIAM J. Comput.)*, pages 896–904, 2002.

[5] Z. Baranyai. On the factorization of the complete uniform hypergraphs. *Infinite and Finite Sets, (A. Hanaj, R. Rado and V. T. Sós, eds.), Colloq. Math. Soc. J'anos Bolyai*, 10:91–108, 1974.

[6] B. E. Bayer. An optimum method for two-level rendition of continuous-tone pictures. *Conference Record, IEEE International Conf. on Communications*, 1:26–11–26–15, 1973.

[7] J. Beck and V. T. Sós. *Discrepancy Theory, in* Handbook of Combinatorics, volume II. Elsevier, 1995.

[8] R. W. Floyd and L. Steinberg. An adaptive algorithm for spatial gray scale. *SID 75 Digest, Society for Information Display*, pages 36–37, 1975.

[9] A. Ghoulia-Houri. Characterisation des matrices totalement unimodulaires. *C.R. Acad/ Sci. Paris*, 254:1192–1194, 1962.

[10] D.S. Hochbaum and J.G. Shanthikumar. Nonlinear separable optimization is not much harder than linear optimization. *Journal of ACM*, 37:843–862, 1990.

[11] *http://www.algorithmic-solutions.com/as_html/products/products.html.* LEDA homepage. Algorithmic Solutions Software GmbH, 2003.

[12] D.E. Knuth. D.e. knuth. *ACM Trans. Graphics*, 6:245–273, 1987.

[13] J. O. Limb. Design of dither waveforms for quantized visual signals. *Bell Syst. Tech. J.*, 48-7:2555–2582, 1969.

[14] Q. Lin. Halftone image quality analysis based on a human vision model. *Proceedings of SPIE*, 1913:378–389, 1993.

[15] B. Lippel and M. Kurland. The effect of dither on luminance quantization of pictures. *IEEE Trans. Commun. Tech.*, COM-19:879–888, 1971.

[16] M. Minoux. Solving integer minimum cost flows with separable cost objective polynomially. *Mathematical Programming Study*, 46:237–239, 1986.

[17] K. Sadakane, N. Takki-Chebihi, and T. Tokuyama. Combinatorics and algorithms on low-discrepancy roundings of a real sequence. *Proc. 28th ICALP, Springer LNCS 2076*, pages 166–177, 2001.

[18] A. Schrijver. *Theory of Linear and Integer Programming.* Wiley-Interscience Series in Discrete Mathematics. John Wiley and Sons, 1986.

NONLINEAR PROGRAMMING: ALGORITHMS, SOFTWARE, AND APPLICATIONS

From Small to Very Large Scale Optimization

Klaus Schittkowski
University of Bayreuth
Department of Computer Science
D-95440 Bayreuth
klaus.schittkowski@uni-bayreuth.de

Christian Zillober
University of Bayreuth
Department of Computer Science
D-95440 Bayreuth
christian.zillober@uni-bayreuth.de

Abstract We introduce some methods for constrained nonlinear programming that are widely used in practice and that are known under the names SQP for sequential quadratic programming and SCP for sequential convex programming. In both cases, convex subproblems are formulated, in the first case a quadratic programming problem, in the second case a separable nonlinear program in inverse variables. The methods are outlined in a uniform way and the results of some comparative performance tests are listed. We especially show the suitability of sequential convex programming methods to solve some classes of very large scale nonlinear programs, where implicitly defined systems of equations seem to support the usage of inverse approximations. The areas of interest are structural mechanical optimization, i.e., topology optimization, and optimal control of partial differential equations after a full discretization. In addition, a few industrial applications and case studies are shown to illustrate practical situations under which the codes implemented by the authors are in use.

Keywords: sequential quadratic programming, SQP, sequential convex programming, SCP, comparative tests, industrial applications, topology optimization, optimal control of semilinear elliptic equations

Introduction

Over the last years, mathematical optimization became a powerful tool in many *real-life* application areas. Proceeding from a formal mathematical model simulating the behavior of a practical system, optimization algorithms are applied to minimize a so-called cost function subject to some constraints. A typical example is the minimization of the weight of a mechanical structure under given loads and under constraints for admissible stresses, displacements, or dynamic responses. Highly complex industrial and academic design problems can be solved today by means of nonlinear programming algorithms without any chance to get equally qualified results by traditional empirical approaches.

We consider the smooth, constrained optimization problem to minimize a scalar objective function $f(x)$ under nonlinear inequality constraints,

$$x \in I\!R^n : \begin{array}{l} \min f(x) \\ g(x) \leq 0 \end{array} \quad , \tag{1}$$

where x is an n-dimensional parameter vector. The vector-valued function $g(x)$ defines m inequality constraints, $g(x) = (g_1(x), \ldots, g_m(x))^T$. To simplify the notation, equality constraints and upper or lower bounds of variables are omitted. We assume that the feasible domain of (1) is non-empty and bounded, and that the functions $f(x)$ and $g(x)$ are smooth, i.e., twice continuously differentiable.

Since we suppose that problem (1) is nonconvex and nonlinear in general, the basic idea is to replace it by a sequence of *simpler* problems. *Simpler* means in this case that the structure of the subproblem is much easier to analyze and that the subproblem is uniquely solvable by an available *black box* technique. In particular, it is assumed that the applied numerical algorithm does not require any additional function or gradient evaluations of the original functions $f(x)$ and $g(x)$. An essential requirement is that we get a first order approximation of (1), i.e., that function and gradient values of the original problem and the subproblem coincide at a current iterate.

Sequential quadratic programming (SQP) methods are very well known and are considered as the standard general purpose algorithm for solving smooth nonlinear optimization problems at least under the following assumptions:

- The problem is not too big.

- Function and especially gradient values can be evaluated within sufficient precision.

- The problem is smooth and well-scaled.

In this case, the subproblems consist of strictly convex quadratic programming problems with inequality constraints obtained by linearizing the constraints and by approximating the Lagrangian function of (1) quadratically. SQP methods have their roots in unconstrained optimization, and can be considered as extensions of quasi-Newton methods taking constraints into account. The basic idea is to establish a quadratic approximation based on second order information with the goal to achieve a fast local convergence speed. Second order information about the Hessian of the Lagrangian is updated by a positive definite quasi-Newton matrix. The linearly constrained, strictly convex quadratic program must be solved in each iteration step by an available *black box* solver.

Despite of the success of SQP methods, another class of efficient optimization algorithms was proposed by engineers, where the motivation is found in mechanical structural optimization. The method is based on the observation that in some special cases, typical structural constraints become linear in the inverse variables. Although this special situation is rarely observed in practice, a suitable substitution of structural variables by inverse ones depending on the sign of the corresponding partial derivatives and subsequent linearization is expected to linearize constraints somehow.

More general convex approximations are introduced known under the name *moving asymptotes* (MMA). The goal is to construct convex and separable subproblems, for which efficient solvers are available based on interior point techniques. Thus, we denote this class of methods by SCP, an abbreviation for *sequential convex programming.* In other words, SQP methods are based on local second order approximations, whereas SCP methods are applying global first order convex approximations.

SCP methods can be used to solve very large scale optimization or VLSO problems, respectively, without any further adoptions of the fundamental structure. Typically they are applied to solve topology optimization problems in mechanical engineering. It seems that these methods are particularly efficient in situations where objective function and constraints depend also on *state variables*, i.e., variables defined implicitly by an internal system of equations.

The subsequent section contains a brief outline of SQP and SCP methods to illustrate their basic structure. More details are found in the references cited. The results of some comparative numerical tests of both approaches are shown for a set of 79 structural optimization problems and for a set of 306 standard academic test problems. Section 3 shows how SCP methods can be applied to solve VLSO problems. The two classes of test problems under consideration are from topology optimization and optimal control of semilinear elliptic partial differential equations. Although the presented examples proceed from relatively simple *academic* formulations, they show at least the capability of SCP methods for solving large optimization problems without assuming any special sparsity patterns.

To show that SQP and SCP methods are of practical interest and are routinely used in industry for design and production, we outline a few case studies in Section 4. Especially, we briefly introduce applications how to find the

- optimal design of horn radiators for satellite communication (Astrium),
- optimal design of surface acoustic wave filters (Epcos),
- on-line control of a tubular reactor (BASF),
- weight-reduction of a cruise ship (Meyer Werft).

We want to give an impression of the numerical complexity of the optimization models and the importance and efficiency of optimization codes in these cases.

1. Sequential Quadratic Versus Sequential Convex Programming Methods

1.1 A General Framework

Since we assume that our optimization problem (1) is nonconvex and nonlinear in general, the basic idea is to replace (1) by a sequence of *simpler* problems. Starting from an initial design vector $x_0 \in I\!R^n$ and an initial multiplier estimate $u_0 \in I\!R^m$, iterates $x_k \in I\!R^n$ and $u_k \in I\!R^m$ are computed successively by solving subproblems of the form

$$y \in I\!R^n : \quad \begin{array}{l} \min f^k(y) \\ g^k(y) \le 0 \end{array} \quad . \tag{2}$$

Let y_k be the optimal solution and v_k the corresponding Lagrangian multiplier of (2). A new iterate is computed by

$$\begin{aligned} x_{k+1} &= x_k + \alpha_k(y_k - x_k) \ , \\ u_{k+1} &= u_k + \alpha_k(v_k - u_k) \ , \end{aligned} \tag{3}$$

where α_k is a steplength parameter discussed subsequently.

Simpler means that the subproblem has a specific mathematical structure which is easier to analyze, and that it is efficiently solvable by an available *black box* technique, more or less independently of the underlying model structure. In particular, it is assumed that the numerical algorithm for solving (2) does not require any additional function or gradient evaluations of the original functions $f(x)$ and $g_j(x)$, $j = 1, \ldots, m$. Note that we are looking for a simultaneous approximation of an optimal solution $x^\star$ and of the corresponding multiplier vector $u^\star$.

Now we summarize the requirements to describe SQP and SCP algorithms in a uniform way:

1 (2) is strictly convex and smooth, i.e., functions $f^k(x)$ and $g_j^k(x)$ are twice continuously differentiable, $j = 1, \ldots, m$.

2 (2) is a first order approximation of (1) at x_k, i.e., $f(x_k) = f^k(x_k)$, $\nabla f(x_k) = \nabla f^k(x_k)$, $g(x_k) = g^k(x_k)$, and $\nabla g(x_k) = \nabla g^k(x_k)$.

3 The search direction $(y_k - x_k, v_k - u_k)$ is a descent direction for an augmented Lagrangian merit function introduced below.

4 The feasible domain of (2) is non-empty and bounded.

Strict convexity of (2) means that the objective function $f^k(x)$ is strictly convex and that the constraints $g_j^k(x)$ are convex functions for all iterates x_k and $j = 1, \ldots, m$. Since the feasible domain is supposed to be non-empty, (2) has a unique solution $y_k \in I\!R^n$ with Lagrangian multiplier $v_k \in I\!R^m$. A further important consequence is that if $y_k = x_k$, then x_k and v_k solve the general nonlinear programming problem (1) in the sense of a stationary solution.

A line search is introduced to stabilize the solution process, particularly helpful when starting from a bad initial guess. We are looking for an α_k, see (3), $0 < \alpha_k \leq 1$, so that a step along a merit function $\Psi_k(\alpha)$ from the current iterate to the new one becomes *acceptable*. The idea is to penalize the Lagrange function in the L_2 norm as soon as constraints are violated, by defining

$$\Phi_r(x,u) = f(x) + \sum_{j \in J} \left(u_j g_j(x) + \frac{1}{2} r_j g_j(x)^2 \right) - \frac{1}{2} \sum_{j \in K} u_j^2 / r_j \ . \tag{4}$$

Then we set

$$\Psi_k(\alpha) = \Phi_{r_k}\left(\begin{pmatrix} x_k \\ u_k \end{pmatrix} + \alpha \begin{pmatrix} y_k - x_k \\ v_k - u_k \end{pmatrix} \right) \ , \tag{5}$$

where $J = \{j : g_j(x) \geq -u_j/r_j\}$ and $K = \{1, \ldots, m\} \setminus J$ define the constraints considered as active or inactive, respectively.

The steplength parameter α_k is required in (3) to enforce global convergence of the optimization method, i.e., the approximation of a point satisfying the necessary Karush-Kuhn-Tucker optimality conditions when starting from arbitrary initial values, e.g., a user-provided $x_0 \in I\!R^n$ and $u_0 = 0$. The merit function defined by (4) is also called augmented Lagrange function, see for example Rockafellar [31]. The corresponding penalty parameter r_k at the k-th iterate that controls the degree of constraint violation, must be chosen carefully to guarantee a descent direction of the merit function, so that the line search is well-defined,

$$\Psi_k'(0) = \nabla\Phi_{r_k}(x_k, u_k)^T \begin{pmatrix} y_k - x_k \\ v_k - u_k \end{pmatrix} < 0 \ . \tag{6}$$

The line search consists of a successive reduction of α starting at 1, usually combined with a quadratic interpolation, until a sufficient decrease condition is obtained.

Standard techniques to approximate constraints, for example successive linearization, can lead to inconsistent constraints in (2). In these cases, it is possible to introduce an additional variable and to modify objective function and constraints, for example by

$$\begin{array}{rl} & \min f^k(y) + \rho_k y_{n+1}^2 \\ y \in I\!R^{n+1}: & g^k(y) - y_{n+1} \le 0 \quad , \\ & -y_{n+1} \le 0 \end{array} \tag{7}$$

in the simplest form. The penalty term ρ_k is added to the objective function to reduce the influence of the additional variable y_{n+1} as much as possible. The index k implies that this parameter also needs to be updated during the algorithm. It is obvious that (7) always possesses a feasible solution.

1.2 Sequential Quadratic Programming

Sequential quadratic programming or SQP methods belong to the most powerful nonlinear programming algorithms we know today for solving differentiable nonlinear programming problems of the form (1). The theoretical background is described in Stoer [44] in form of a review, and in Spellucci [43] in form of an extensive text book. From the more practical point of view, SQP methods are also introduced in the books of Papalambros and Wilde [28] or Edgar and Himmelblau [9]. Their excellent numerical performance is tested and compared with other methods in Schittkowski [33–34] and Hock, Schittkowski [16]. Since many years they belong to the most frequently used algorithms to solve practical optimization problems.

The basic idea is to formulate and solve a quadratic programming subproblem in each iteration, which is obtained by linearizing the constraints and approximating the Lagrange function of (1) quadratically. To formulate the subproblem, we proceed from given iterates $x_k \in I\!R^n$, an approximation of the solution, $u_k \in I\!R^m$, an approximation of the vector of multipliers, and $B_k \in I\!R^{n \times n}$, an approximation of the Hessian of the Lagrange function. Then we obtain subproblem (2) by defining

$$\begin{aligned} f^k(y) &= \tfrac{1}{2}\,(y - x_k)^T B_k (y - x_k) + \nabla f(x_k)^T (y - x_k) + f(x_k)\ , \\ g_j^k(y) &= \nabla g_j(x_k)^T (y - x_k) + g_j(x_k)\ ,\ \ j = 1, \ldots, m\ . \end{aligned} \tag{8}$$

It is immediately seen that the requirements of the previous section for (2) are satisfied. The key idea is to approximate also second order information to get a fast final convergence speed. The update of the matrix B_k can be performed by standard quasi-Newton techniques known from unconstrained optimization subject to the Lagrangian function of the nonlinear program. In most cases, the BFGS-method is applied, see Powell [30, 29], or Stoer [44], starting from the identity matrix or any other positive definite matrix B_0. A simple modification as for example proposed by Powell [30] guarantees positive definite matrices.

Among the most attractive features of sequential quadratic programming methods is the superlinear convergence speed in the neighborhood of a solution, i.e.,

$$\|x_{k+1} - x^\star\| < \gamma_k \|x_k - x^\star\| \tag{9}$$

with $\gamma_k \to 0$.

The motivation for the fast convergence speed of SQP methods is based on the following observation: an SQP method is identical to Newton's method to solve the necessary optimality conditions of (1), if B_k is the Hessian of the Lagrange function at x_k and u_k and if we start sufficiently close to a solution. The statement is easily derived in case of equality constraints only, but holds also for inequality restrictions.

There remain a few comments to summarize some interesting features of SQP methods:

- Linear constraints and bounds of variables remain satisfied.

- In case of n active constraints, the SQP method behaves like Newton's method for solving the corresponding system of equations, i.e., the local convergence speed is even quadratically.

- The algorithm is globally convergent and the local convergence speed is superlinear.

- A simple reformulation allows the efficient solution of constrained nonlinear least squares problems, see Schittkowski [37, 40].

- A large number of constraints can be treated by an active set strategy, see Schittkowski [38]. In particular, the computation of gradients for inactive restrictions can be omitted.

- There exists a large variety of different extensions to solve also large scale problems, see Gould and Toint [13] for a review.

1.3 Sequential Convex Programming

Sequential convex programming methods are developed mainly for mechanical structural optimization. The first approach of Fleury and Braibant [11] and Fleury [10] is known under the name convex linearization (CONLIN) and exploits the observation that in some special cases, typical structural constraints become linear in the inverse variables. Although this special situation is always found in case of statically determinate structures, it is rarely observed in practice. However, a suitable substitution by inverse variables depending on the sign of the corresponding partial derivatives and subsequent linearization is expected to linearize constraints somehow.

For the CONLIN method, Nguyen et al. [26] gave a convergence proof but only for the case that (1) consists of a concave objective function and constraints which is of minor practical interest. They showed also that a generalization to non-concave constraints is not possible. More general convex approximations are introduced by Svanberg [45] known under the name *method of moving asymptotes* (MMA). The goal is always to construct nonlinear convex and separable subproblems, for which efficient solvers are available. Using the flexibility of the asymptotes which influence the curvature of the approximations, it is possible to avoid the concavity assumption.

Given an iterate x_k, the basic idea is to linearize f and g_j at x_k subject to transformed variables $(U_i^k - x_i)^{-1}$ and $(x_i - L_i^k)^{-1}$ depending on the sign of the corresponding first partial derivative. U_i^k and L_i^k are reasonable bounds and are adapted by the algorithm after each successful step. Also several other transformations have been developed in the past.

By defining suitable index sets

$$I_k^+ = \{i : \frac{\partial}{\partial x_i} f(x_k) \geq 0\} \ , \quad I_k^- = \{i : \frac{\partial}{\partial x_i} f(x_k) < 0\}$$

for objective function and, in a similar way, I_{jk}^+ and I_{jk}^- for constraints, we get the corresponding approximating functions of subproblem (2) by

$$\begin{aligned} f^k(y) &= \alpha_0^k + \sum_{i \in I_k^+} \frac{\beta_{i,0}^k}{U_i^k - y_i} - \sum_{i \in I_k^-} \frac{\beta_{i,0}^k}{y_i - L_i^k} , \\ g_j^k(y) &= \alpha_j^k + \sum_{i \in I_{jk}^+} \frac{\beta_{i,j}^k}{U_i^k - y_i} - \sum_{i \in I_{jk}^-} \frac{\beta_{i,j}^k}{y_i - L_i^k} , \end{aligned} \tag{10}$$

$j = 1, \ldots, m$, where $y = (y_1, \ldots, y_n)^T$. The coefficients α_j^k and $\beta_{i,j}^k$, $j = 0, \ldots, m$ are chosen to satisfy the requirements of Section 2.1, i.e., that (2) is convex and a first order approximation of (1) at x_k. By an appropriate regularization of the objective function, strict convexity of $f^k(y)$ is guaranteed, see Zillober [50]. As shown there, the search direction $(y_k - x_k, v_k - u_k)$ is a descent direction for the augmented Lagrangian merit function (4,5), see also (6). The approximation scheme (10) can be applied only to inequality constraints. Additional equality constraints can be added, but are linearized as for SQP methods.

The choice of the asymptotes L_i^k and U_i^k, is crucial for the computational behavior of the method, in particular since additional lower and upper bounds are usually available. Additional safeguards ensure the compatibility of this procedure with the overall scheme and guarantee global convergence. A small positive constant is introduced to avoid that the difference between the asymptotes and the current iteration point becomes too small. However, these safeguards are rarely used in practice, see Zillober [50] for more details.

For the first SCP codes developed, the convex and separable subproblems are solved by a dual approach, where dense linear systems of equations with m rows and columns are solved, cf. Svanberg [45] or Fleury [10]. Recently, a predictor-corrector interior point method for the solution of the subproblems was proposed by Zillober [49]. The advantage is to formulate either $n \times n$ or $m \times m$ linear systems of equations leading to a more flexible treatment of large problems. The resulting algorithm is very efficient especially for large scale mechanical engineering problems, and given sparsity patterns of the original problem data can be exploited.

To summarize, the most important features of SCP methods are:

- Linear equality constraints and bounds of variables remain satisfied.

- The algorithm is globally convergent.

- As for SQP methods, a large number of constraints can be treated by an active set strategy, see Zillober [52, 51]. In particular, the computation of gradients for inactive restrictions can be omitted.
- Large scale problems can be handled by different variants of the solution procedure for the subproblem, see Zillober et al. [53], where sparsity of problem data can be exploited.

1.4 Comparative Results

Our numerical tests use all 306 academic and real-life test problems published in Hock and Schittkowski [16] and in Schittkowski [36]. Part of them are also available in the CUTE library, see Bongartz et al. [5]. The test problems possess also nonlinear equality constraints and additional lower and upper bounds for the variables. The two codes under consideration, NLPQLP of Schittkowski [39] and SCPIP of Zillober [51], are able to solve more general problems

$$x \in I\!R^n : \quad \begin{array}{ll} \min f(x) & \\ h(x) = 0 & , \\ g(x) \leq 0 & , \\ x_l \leq x \leq x_u & , \end{array} \qquad (11)$$

with additional smooth functions $h(x) = (h_1, \ldots, h_{m_e})^T$ for equality constraints and bounds $x_l < x_u$.

Since analytical derivatives are not available for all problems, we approximate them numerically by a five-point difference formula. The test examples are provided with exact solutions, either known from analytical evaluation or from the best numerical data found so far. Since the calculation times are very short, we count only function and gradient evaluations. This is a realistic assumption, since for the practical applications in mind calculation times for evaluating model functions dominate and the numerical efforts within an optimization code are negligible.

The result of a test run is considered as a successful return, if the relative error in the objective function is less than a given tolerance ϵ and if the maximum constraint violation is less than ϵ^2. We take into account that a code returns a solution with a better function value than the known one subject to the error tolerance of the allowed constraint violation. However, there is still the possibility that an algorithm terminates at a local solution different from the one known in advance. Thus, we call a test run a successful one, if the internal termination conditions are satisfied subject to a reasonably small tolerance, if the obtained solution is feasible, and if the objective function value is significantly larger

code	p_{succ}	n_f	n_{it}
NLPQLP	100 %	33	21
SCPIP	93 %	74	42

Table 1. Performance Results for Standard Test Problems

than the known one. For our numerical tests, we use $\epsilon = 0.01$, i.e., we require a final accuracy of one per cent, see Zillober et al. [53] for more details.

The code NLPQLP of Schittkowski [39] represents the most recent version of NLPQL which is frequently used in academic and commercial institutions, see Schittkowski [35]. Functions and gradients must be provided by reverse communication and the quadratic programming subproblems are solved by the primal-dual method of Goldfarb and Idnani [12] based on numerically stable orthogonal decompositions. The SQP algorithm is executed with termination accuracy 10^{-8} and the maximum number of iterations is 500. In the SCP implementation SCPIP of Zillober [52, 51], the convex subproblems are solved by the predictor-corrector interior point method described in Zillober [49]. Input variables and tolerances are chosen in a way such that the termination conditions for SCPIP and NLPQLP are comparable.

Table 1 shows the percentage of successful test runs, p_{succ}, the average number of function calls, n_f, and the average number of iterations, n_{it}. Function evaluations needed for gradient approximations, are not counted for n_f. Their average number is $4 \times n_f$. Many test problems are unconstrained or possess a highly nonlinear objective function preventing SCP from converging as fast as SQP methods. Moreover, bounds are often set far away from the optimal solution, leading to initial asymptotes too far away from the region of interest. Since SCP methods do not possess fast local convergence properties, SCPIP needs about twice as many iterations and function evaluations.

The situation is different in mechanical structural optimization, where the SCP methods have been invented. In the numerical study of Schittkowski et al. [41], 79 finite element formulations of academic and practical problems are collected based on the simulation package MBB-LAGRANGE, see Kneppe et al. [19]. The maximum number of variables is 144 and a maximum number of constraints 1020 without box-constraints. NLPQL, see Schittkowski [35], and MMA, former versions of NLPQLP and SCPIP, respectively, are among the 11 optimization algorithms under consideration. To give an impression on the behavior of SQP versus MMA, we repeat some results of Schittkowski et al. [41], see Table 2.

One of the main difficulties of a comparative performance study is that the optimization program solve only a certain subset of test problems successfully, which differs from code to code. Thus, mean values of a performance criterion are evaluated pairwise over the set of successfully solved test problems of two algorithms, and then compared in form of a matrix, see the priority theory of Saaty [32] and also Lootsma [21]. The decision whether the result of a test run is considered as a successful one or not, depends on a tolerance ϵ which is set to $\epsilon = 0.01$ and $\epsilon = 0.00001$, respectively.

		$\epsilon = 0.01$			$\epsilon = 0.00001$	
code	p_{succ}	n_f	n_{it}	p_{succ}	n_f	n_{it}
NLPQL	84 %	2.0	1.6	77 %	1.3	1.3
MMA	73 %	1.0	1.0	73 %	1.0	1.0

Table 2. Performance Results for Structural Optimization Test Problems

The figures of Table 2 represent the scaled relative performance data when comparing the codes among each other. We conclude for example that for $\epsilon = 0.01$, NLPQL requires about twice as many gradient evaluations or iterations, respectively, as MMA. When requiring a higher termination accuracy, however, NLPQL needs only 30 % more gradient calls. On the other hand, NLPQL is a bit more reliable than MMA.

2. Very Large Scale Optimization by Sequential Convex Programming

2.1 Topology Optimization

To give an impression about the capabilities of an SCP implementation for solving very large scale nonlinear programming problems, we consider now structural mechanical optimization, more precisely topology optimization. Given a predefined domain in the 2D/3D space with boundary conditions and external load, the intention is to distribute a percentage of the initial mass on the given domain such that a global measure takes a minimum, see Bendsøe [2] or Bendsøe and Sigmund [3] for a broader introduction. Assuming isotropic material, the so-called power law approach, see also Bendsøe [1] or Mlejnek [25], leads to a nonlinear program of the form

$$x \in I\!R^n,\ u \in I\!R^d: \quad \begin{array}{l} \min\ u^T p \\ V(x) \le aV_0\ , \\ K(x)u = p\ , \\ 0 < x_l \le x \le 1\ , \end{array} \tag{12}$$

where $x = (x_1, \ldots, x_n)^T$ denotes the relative material densities, artificially introduced variables. In the final solution, we consider a small value of x_i as zero or no mass, a larger value as one or full mass. Theoretically, one is only interested in 0-1 solutions, which are not guaranteed by the continuous approach applied. $u = (u_1, \ldots, u_d)^T$ is the displacement vector computed from the linear system of equations $K(x)u = p$ with a positive definite stiffness matrix $K(x)$ and an external load vector p. d denotes the number of degrees of freedom of the structure. We assume without loss of generality that there is only one load case. The goal is to minimize the so-called compliance or, in other words, to make the structure as stiff as possible.

It is essential to understand that the system of linear equations $K(x)u = p$ can be considered as the state equations of our optimization problem. In practical situations, finite element simulation software is available to set up the stiffness matrix and to solve the system $K(x)u = p$ internally. To indicate that u depends on the relative densities x, we use the notation $u(x)$.

The relative densities and the elementary stiffness matrices K_i define $K(x)$ by

$$K(x) = \sum_{i=1}^{n} x_i^q\, K_i\ .$$

$V(x)$ is the volume of the structure, usually a linear function of the design variables,

$$V(x) = \sum_{i=1}^{n} x_i V_i\ ,$$

where V_i is the volume of the i-th finite element. V_0 is the available volume, $V_0 = \sum_{i=1}^{n} V_i$, and a with $0 < a < 1$ the given fraction of the full volume to distribute the available mass. x_l is a vector of small positive numbers for avoiding singularities. The nonlinearity x_i^q in the state equation is found heuristically and usually applied in practice with $q = 3$. Its role is to penalize intermediate values between the lower bound and 1.

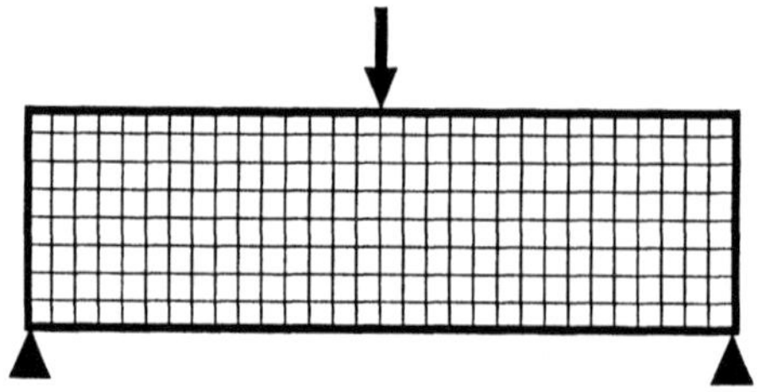

Figure 1. Design Region of Beam

The partial derivatives of the objective function of problem (12) are computed from

$$\frac{\partial}{\partial x_j}\left(u(x)^T p\right) = -q x_j^{q-1} u(x)^T K_j u(x) \tag{13}$$

for $j = 1, \ldots, n$, see Zillober et al. [53]. Since the elementary stiffness matrices K_j are very sparse, for example containing only non-zero entries on an $8 \times 8-$submatrix in case of the rectangular elements used in this section, the j-th partial derivative is computed very efficiently as soon as the displacement vector $u(x)$ is available.

The solution of topology optimization problems easily leads to very large scale, highly nonlinear programs. The probably most simple example is a beam, which is loaded in the middle and supported at the two lower vertices. The design region is a rectangular plate, see Figure 1, discretized by rectangular finite elements. For symmetry reasons, we consider only one half of the beam for our calculations. The number of horizontal grid lines is denoted by n_x, the number of vertical grid lines by n_y. The solution of topology optimization problems as outlined so far, produces a strange phenomenon, a checkerboard-type material distribution in certain regions. Thus, an additional filter is applied by which partial derivatives are modified by certain weights depending on a given radius r_f around the considered element, see for example Bendsøe and Sigmund [3]. In addition, Table 3 shows the total number of optimization variables, n, the number of iterations, n_{it}, the final objective function value, $f(x)$, and the gradient norm of Lagrangian function, $\|\nabla_x L(x, u)\|$. The results show that that SCP methods are able to solve dense nonlinear programs with more than 10^6 variables. More detailed computational results are presented in Zillober et al. [53].

n_x	n_y	n	n_{it}	$f(x)$	$\|\nabla_x L(x,u)\|$	r_f
600	400	240,000	22	52.63	1.3E-3	8
600	400	240,000	26	54.25	6.5E-4	0
1,050	700	735,000	38	54.39	4.6E-4	10
1,260	840	1,058,400	43	56.55	1.3E-5	0

Table 3. Numerical Results of SCPIP for the Half Beam

2.2 Optimal Control of Semilinear Elliptic Partial Differential Equations

The intention behind the numerical tests of this section is to show that SCP methods can be applied also to optimization problems which are completely different from the original mechanical engineering applications. We consider a series of test problems investigated by Maurer and Mittelmann [22–23] when studying necessary optimality conditions for optimal control of elliptic partial differential equations with state and control constraints. They differ by the type of control, boundary and interior control, the cost functional, the non-homogeneous part of the elliptic equation, and the boundary conditions. Results for a typical test problem are shown below. More detailed numerical results are presented in Zillober et al. [53], where the performance of the code SCPIP is compared with the best codes of the Maurer and Mittelmann study.

Proceeding from the two-dimensional unit square Ω and the boundary Γ, the optimal control problem is defined by

$$\begin{array}{ll} & \min\ \frac{1}{2}\int_\Omega (y(x)-\sin(2\pi x_1)\sin(2\pi x_2))^2 + u(x)^2\,dx \\ u\in L^\infty(\Omega), & \Delta y + e^y = 0\ ,\quad x\in\Omega\ , \\ y\in C^2(\Omega) \quad : & \partial_\nu y + y = 0\ ,\quad x\in\Gamma\ , \\ & y\le 0.371\ ,\quad -8\le u\le 9\ . \end{array} \tag{14}$$

u is the control function we want to compute subject to constant lower and upper bounds, and y denotes the state variable, i.e., the solution of the semilinear elliptic partial differential equation $\Delta y + e^y = 0$ subject to a Neumann boundary condition of the form $\partial_\nu y + y = 0$. $\partial_\nu y$ denotes the outward unit normal along the boundary Γ. The solution of the state equations depends on the control function u, and a state constraint for y is given in form of an upper constant bound. The cost function is of tracking type, see Ito and Kunisch [17] depending on the spatial variable x.

$N+1$	n	m	n_{it}	$f(x)$
100	19,998	10,197	20	0.0528
200	79,998	40,397	22	0.0530
300	179,998	90,597	16	0.0535
400	319,998	160,797	18	0.0534
500	499,998	250,997	16	0.0536
600	719,998	361,197	16	0.0537

Table 4. Numerical Results for a Semilinear Elliptic Control Problem

The elliptic control problem (14) is pointwise discretized subject to the control and state variables as proposed by Maurer and Mittelmann [22–23] based on a uniform grid of size N for discretizing Ω and a five-star-formula for the Laplace operator. Thus, we get a set of N^2 equality constraints

$$4y_{ij} - y_{i,j-1} - y_{i-1,j} - y_{i,j+1} - y_{i+1,j} + h^2 e^{y_{ij}} = 0 \ . \tag{15}$$

First derivatives in Neumann boundary conditions are approximated by forward or backward differences, respectively.

It is important to understand that SCP methods are not invented to solve equality constrained problems. Convex approximation cannot be applied to equality constraints, which are linearized internally, see Zillober [49]. Problem (14) is solved by the SCP code SCPIP with termination accuracy $\epsilon = 10^{-7}$ for the optimality condition and $\epsilon = 10^{-10}$ for maximum constraint violation. Starting values are $u_0 = 0$ and $y_0 = 0$ for all test runs. The total number of variables, n, and the number of equality constraints, m, are shown in Table 4 together with the number of SCPIP iterations, n_{it}, and the final objective function value, $f(u, y)$. The grid size N varies between 100 and 600.

3. Case Study: Horn Radiators for Satellite Communication

Corrugated horns are frequently used as reflector feed sources for large space antennae, for example for INTELSAT satellites. The goal is to achieve a given spatial energy distribution of the radio frequency (RF) waves, called the radiation or directional characteristic. The transmission quality of the information carried by the RF signals is strongly determined by the directional characteristics of the feeding horn as determined by its geometric structure.

The electromagnetic field theory is based on Maxwell's equations relating the electrical field E, the magnetic field H, the electrical displacement, and the magnetic induction to electrical charge density and current density, see Collin [8] or Silver [42]. Under some basic assumptions, particularly homogeneous and isotropic media, Maxwell's equations can be transformed into an equivalent system of two coupled equations. They have the form of a wave equation,

$$\nabla^2\Psi - c^2\frac{\partial^2}{\partial t^2}\Psi + f = 0$$

with displacement f enforcing the wave, and wave velocity c. Ψ is to be replaced either by E or H, respectively.

For circular horns with rotational symmetry, the usage of cylindrical coordinates (ρ, ϕ, z) is advantageous, especially since only waves propagating in z direction occur. Thus, a scalar wave equation in cylindrical coordinates can be derived from which general solution is obtained, see for example Collin [8] for more details.

By assuming that the surface of the wave guide has ideal conductivity, and that homogeneous Dirichlet boundary conditions $\Psi = 0$ for $\Psi = E$ and Neumann boundary conditions $\partial\Psi/\partial n = 0$ for $\Psi = H$ at the surface are applied, we get the eigenmodes or eigenwaves for the circular wave guide. Since they form a complete orthogonal system, electromagnetic field distribution in a circular wave guide can be expanded into an infinite series of eigenfunctions, and is completely described by the amplitudes of the modes. For the discussed problem, only the transversal eigenfunctions of the wave guides need to be considered and the eigenfunctions of the circular wave guide can be expressed analytically by trigonometric and Bessel functions.

In principle, the radiated far field pattern of a horn is determined by the field distribution of the waves emitted from the aperture. On the other hand, the aperture field distribution itself is uniquely determined by the excitation in the feeding wave guide and by the interior geometry of the horn. Therefore, assuming a given excitation, the far field is mainly influenced by the design of the interior geometry of the horn. Usually, the horn is excited by the TE_{11} mode, which is the fundamental, i.e., the first solution of the wave equation in cylindrical coordinates. In order to obtain a rotational symmetric distribution of the energy density of the field in the horn aperture, a quasi-periodical corrugated wall structure according to Figure 2 is assumed, see Johnson and Jasik [18].

To reduce the number of optimization parameters, the horn geometry is described by two envelope functions from which the actual geometric data for ridges and slots can be derived. Typically, a horn is subdivided into three sections, see Figure 3, consisting of an input section, a conical section, and an aperture section. For the input and the aperture

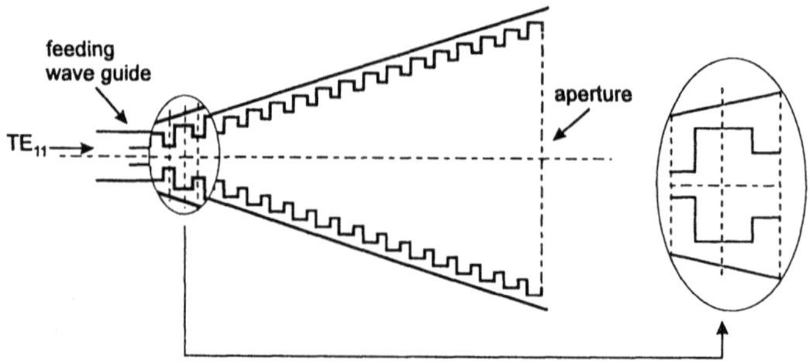

Figure 2. Cross Sectional View of a Circular Corrugated Horn

section, the interior and outer shape of slots and ridges is approximated by a second-order polynomial, while a linear function is used to describe the conical section. It is assumed that the envelope functions of ridges and slots are parallel in conical and aperture section. By this simple analytical approach, it is possible to approximate any reasonable geometry with sufficient accuracy by the design parameters.

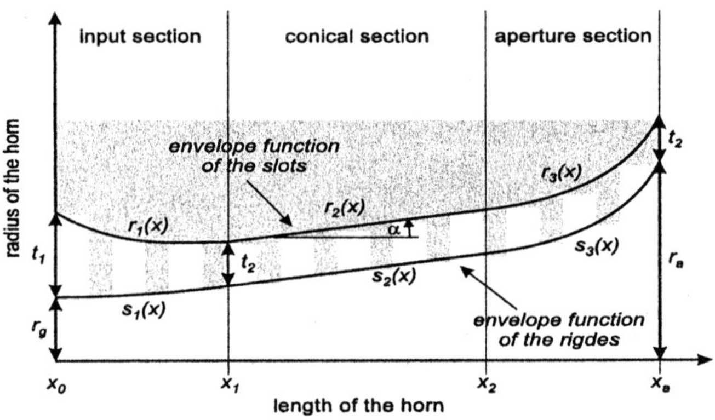

Figure 3. Envelope Functions of a Circular Corrugated Horn

A circular corrugated horn has a modular structure, where each module consists of a step transition between two circular wave guides with different diameters, see Figure 4. The amplitudes of waves, travelling towards and away from the break point, are coupled by a so-called scattering matrix. By combining all modules of the horn step by step, the corresponding scattering matrix describing the total transition of amplitudes from the entry point to the aperture can be computed by successive matrix operations, see Hartwanger et al. [15] or Mittra [24].

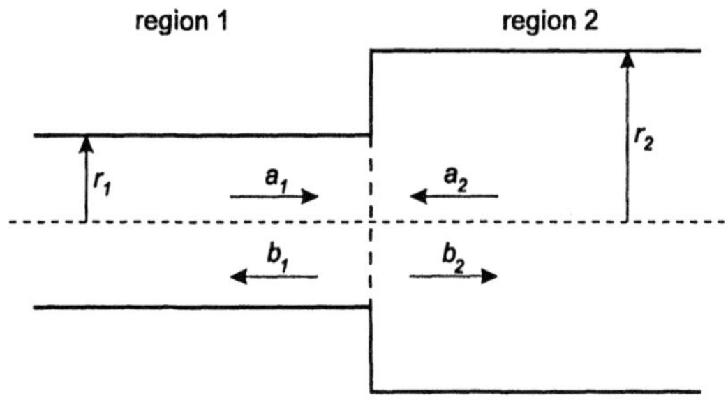

Figure 4. Cross Sectional View of One Module

From Maxwell's equations, it follows that the tangential electrical and magnetic field components must be continuous at the interface between two wave guides. This continuity condition is exploited to compute a relation between the mode amplitudes of the excident $b_{E,j}^k$, $b_{H,j}^k$ and incident $a_{E,j}^k$, $a_{H,j}^k$ waves in each wave guide of a module, see Figure 4, $k = 1, 2$. Then voltage and current coefficients $U_{H,j}^k$, $U_{E,j}^k$, $I_{H,j}^k$, and $I_{E,j}^k$ are defined by the amplitudes.

As mentioned before, the tangential fields must be continuous at the transition between two wave guides. Moreover, boundary conditions must be satisfied, $E_2 = 0$ for $r_1 \leq r \leq r_2$. Now only n_1 eigenwaves in region 1 and n_2 eigenwaves in region 2 are considered. The electric field in area 1 is expanded subject to the eigenfunctions in area 2 and the magnetic field in area 2 subject to the eigenfunctions in area 1. After some manipulations, in particular interchanging integrals and finite sums, the following relationship between voltage coefficients in region 1 and 2 can be formulated in matrix notation:

$$\begin{pmatrix} U_E^2 \\ U_H^2 \end{pmatrix} = \begin{pmatrix} X_{EE} & X_{HE} \\ X_{EH} & X_{HH} \end{pmatrix} \begin{pmatrix} U_E^1 \\ U_H^1 \end{pmatrix} . \tag{16}$$

Here U_E^k and U_H^k are vectors consisting of the coefficients $U_{E,j}^k$ and $U_{H,j}^k$ for $j = 1, \ldots, n_k$, respectively, $k = 1, 2$. The elements of the matrix X_{EE} are given by

$$X_{EE}^{ij} = \int_0^{r_2} \int_0^{2\pi} e_{E,i}^2(\rho, z, \phi)^T \; e_{E,j}^1(\rho, z, \phi) \; \rho \, d\phi \, d\rho \tag{17}$$

with tangential field vectors $e_{E,i}^k(\rho, z, \phi)$ for both regions $k = 1$ and $k = 2$. In the same way X_{HE}, X_{EH}, and X_{EE} are defined. Moreover, matrix equations for the current coefficients are available.

Next, the relationship between the mode amplitude vectors b_E^k and b_H^k of the excident waves $b_{E,j}^k$, $b_{H,j}^k$, and a_E^k and a_H^k of the incident waves $a_{E,j}^k$, $a_{H,j}^k$, $j = 1, \ldots, n_k$, $k = 1, 2$, are evaluated through a so-called scattering matrix. By combining all scattering matrices of successive modules, we compute the total scattering matrix relating the amplitudes at the feed input with those at the aperture,

$$\begin{pmatrix} b_1(p) \\ b_2(p) \end{pmatrix} = \begin{pmatrix} S_{11}^\star(p) & S_{12}^\star(p) \\ S_{21}^\star(p) & S_{22}^\star(p) \end{pmatrix} \begin{pmatrix} a_1 \\ a_2 \end{pmatrix} . \tag{18}$$

The vector a_1 describes the amplitudes of the modes exciting the horn, the TE_{11} mode in our case. Thus, a_1 is the $2n_1$-dimensional unity vector. The vector a_2 contains the amplitudes of the reflected modes at the horn aperture, known from the evaluation of the far field. Only a simple matrix $\times$ vector computation is performed to get the modes of reflected waves $b_1(p)$ and $b_2(p)$, once the scattering matrix is known.

The main goal of the optimization procedure is to find an interior geometry p of the horn so that the distances of $b_2(p)^j$ from given amplitudes $\overline{b}_2^j$ for $j = 1, \ldots, 2n_2$ become as small as possible. The first component of the vector $b_1(p)$ is a physically significant parameter, the so-called return loss, representing the power reflected at the throat of the horn. Obviously, this return loss should be minimized as well. The phase of the return loss and further components of $b_1(p)$ are not of interest.

From these considerations, the least squares optimization problem

$$p \in I\!R^n : \begin{array}{l} \min \sum_{j=1}^{2n_2} (b_2^j(p) - \overline{b}_2^j)^2 + \mu \, b_1^1(p)^2 \\ p_l \le p \le p_u \end{array} \tag{19}$$

is obtained. The upper index j denotes the j-th coefficient of the corresponding vector, μ a suitable weight, and p_l, p_u lower and upper bounds for the parameters to be optimized. Note also that complex numbers are evaluated throughout this section, leading to a separate evaluation of the regression function of (19) for the real and imaginary parts of $b_2^j(p)$.

The least squares problem is solved by a special variant of NLPQL called DFNLP, see Schittkowski [37], which retains typical features of a Gauss-Newton method after a certain transformation. For a typical test run under realistic assumptions, the radius of the feeding wave guide, and the radius of the aperture are kept constant, $r_g = 11.28\ mm$ and $r_a = 90.73\ mm$, where 37 ridges and slots are assumed. Parameter names, initial values p_0, and optimal solution values p_{opt} are listed in Table 5. The number of modes, needed to calculate the scattering matrix, is 70.

name	p_0^i	p_{opt}^i	*comment*
x_1	50.0	111.85	length of input section
x_{con}	50.0	0.00	length of conical section
x_o	50.0	47.00	length of output section
α	28.0	29.00	semi flare angle of conical section
q	0.25	0.20	quotient of slot and ridge width
t_1	12.5	11.97	depth of first slot in input section
t_2	7.2	7.82	depth of slots in conical section

Table 5. Initial and Optimal Parameter Values

Forward differences are used to evaluate numerical derivatives subject to a tolerance of 10^{-7}, and $\mu = 1$ was set for weighting the return loss. NLPQL needed 51 iterations to satisfy the stopping tolerance 10^{-7}.

4. Case Study: Design of Surface Acoustic Wave Filters

Computer-aided design optimization of electronic components is a powerful tool to reduce development costs on one hand and to improve the performance of the components on the other. A bandpass filter selects a band of frequencies out of the electro-magnetic spectrum. In this section, we consider surface-acoustic-wave (SAW) filters consisting of a piezo-electric substrate, where the surface is covered by metal structures. The incoming electrical signal is converted to a mechanical signal by this setup. The SAW filter acts as a transducer of electrical energy to mechanical energy and vice versa. The efficiency of the conversion depends strongly on the frequencies of the incoming signals and the geometry parameters of the metal structures, for example length, height, etc. On this basis, the characteristic properties of a filter are achieved.

Due to small physical sizes and unique electrical properties, SAW-bandpass filters raised tremendous interest in mobile phone applications. The large demand of the mobile phone industry is covered by large-scale, industrial mass-production of SAW-filters. For industrial applications, bandpass filters are designed in order to satisfy pre-defined electrical specifications. The *art of filter design* consists of defining the internal structure, or the geometry parameters, respectively, of a filter such that the specifications are satisfied. The electrical properties of the filters are simulated based on physical models. The simulation of a bandpass filter consists of the acoustic tracks, i.e., the areas on the piezo-electrical substrate on which the electrical energy is converted to mechanical vibrations and vice versa, and the electrical combinations of the different acoustic tracks. Typically, only the properties of the acoustic tracks are varied during the design process, and are defined by several physical parameters. Some of them are given in form of real numbers, some others in form of integer numbers. As soon as the filter properties fit to the demands, the mass production of the filter is started.

When observing the surface of a single-crystal, we see that any deviation of an ion from its equilibrium position provokes a restoring force and an electrical field due to the piezo-electric effect. Describing the deviations of ions at the surface in terms of a scalar potential, we conclude that the SAW is described by a scalar wave equation

$$\phi_{tt} = c^2 \Delta \phi \ . \tag{20}$$

The boundary conditions are given by the physical conditions at the surface and are non-trivial, since the surface is partly covered by a metal layer. In addition, the piezo-electric crystal is non-isotropic, and the velocity of the wave depends on its direction. For the numerical simulation, additional effects such as polarization charges in the metal layers have to be taken into account. Consequently, the fundamental wave equation is not solvable in a closed form.

For this reason, Tobolka [46] introduced the P-matrix model as an equivalent mathematical description of the SAW. One element is a simple base cell, which consists of two acoustic ports, and an additional electric port. The acoustic ports describe the incoming and outgoing acoustic signals, the electrical ports the electric voltage at this cell, see Figure 5. The quantities a_1, a_2, b_1 and b_2 denote the intensities of the acoustic waves. In terms of a description based on the wave equation, we have $a_1 \propto \phi$, u is the electrical voltage at the base cell, and i is the electrical current.

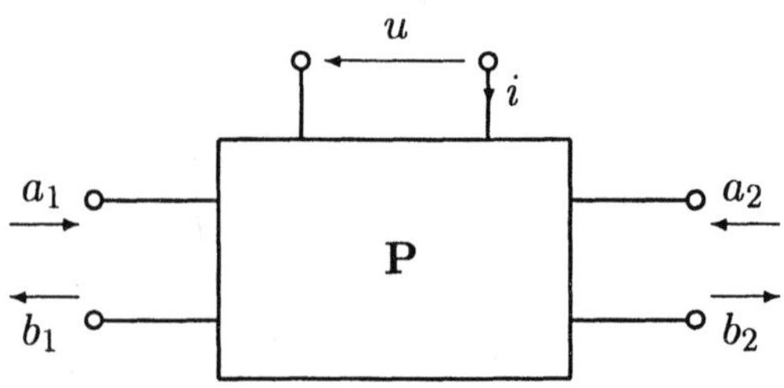

Figure 5. Base Cell of the P-Matrix Model with Two Acoustic and One Electric Port

The P-matrix model describes the interaction of the acoustic waves at the acoustic ports, with the electric port in linear form. Typically, a transformation is given in the form

$$\begin{pmatrix} b_1 \\ b_2 \\ i \end{pmatrix} = \mathbf{P} \begin{pmatrix} a_1 \\ a_2 \\ u \end{pmatrix} , \tag{21}$$

where for example

$$\mathbf{P} = \begin{pmatrix} 0 & 1 & E \\ 1 & 0 & -E^* \\ -2E & 2E^* & 2|E|^2 + i(-\mathcal{H}\{2|E|^2\} + \omega C) \end{pmatrix} . \tag{22}$$

$\mathcal{H}$ denotes the Hilbert transformation, C the static capacity between two fingers, E the excitation given by

$$E = -i\ 0.5\ \sqrt{\omega W K} \cdot \int_{tr} \sigma_e(x)\ \exp^{-ik|x|}\, dx \ ,$$

ω the frequency, W the aperture of the IDT, K a material constant, and σ_e the electric load distribution.

In general, the elements of P are the dimensionless acoustic reflection and transmission coefficients in the case of a short-circuited electrical port. The 2×2 upper diagonal submatrix is therefore the scattering matrix of the acoustic waves and describes the interaction of the incoming and outgoing waves. Other elements characterize the relation of the acoustic waves with the electric voltage, i.e., the piezo-electric effect of the substrate, or the admittance of the base cell, i.e., the the quotient of current to voltage and the reciprocal value of the impedance.

Proceeding from the P-matrix model, we calculate the scattering matrix S. This matrix is the basic physical unit that describes the electro-acoustic properties of the acoustic tracks, and finally the filter itself. The transmission coefficient T is one element of the scattering matrix, $T = S_{21}$.

Mobile phone manufacturers provide strict specifications towards the design of a bandpass filter. Typically, the transmission has to be above certain bounds in the pass band and below certain bounds in the stop band depending on the actual frequency. These specifications have to be achieved by designing the filter in a proper way. Depending on the exact requirements upon the filter to be designed, different optimization problems can be derived.

To formulate the optimization problem, let us assume that $x \in \mathbb{R}^n$ denotes the vector of continuous real design variables and $y \in Z^m$ the vector of the integer design variables. Z is the set of all integer values. By $T(f, x, y)$ we denote the transmission subject to frequency f and

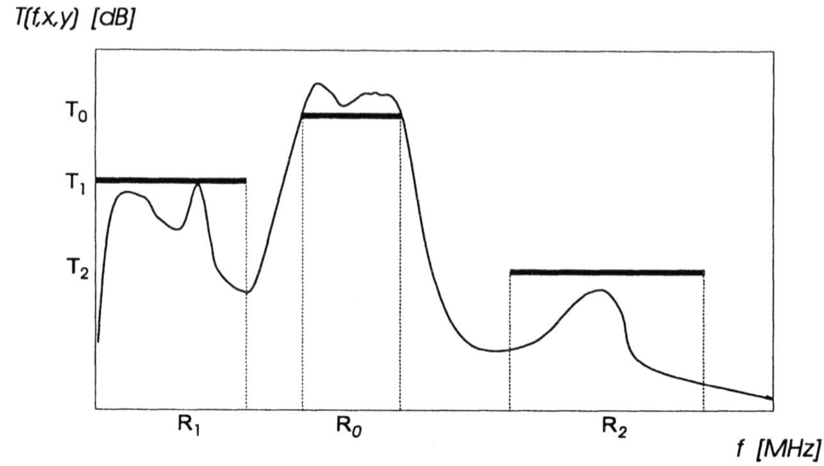

Figure 6. Design Goals of an SAW Filter

the optimization variables x and y. Some disjoint intervals R_0, ..., R_s define the design space within the frequency interval $f_l \leq f \leq f_u$. Our goal is to maximize the minimal distance of transmission $T(f,x,y)$ over the interval R_0, under lower bounds T_1, ..., T_s for the transmission in the remaining intervals R_1, ..., R_s. Moreover, it is required that the transmission is always above a certain bound in R_0, i.e., that $T(f,x,y) \geq T_0$ for all $f \in R_0$. The optimization problem is formulated as

$$\begin{array}{rl} & \max \min \{T(f,x,y) : f \in R_0\} \\ x \in I\!R^n, y \in Z^m : & T(f,x,y) \leq T_i \ \text{ for } f \in R_i \ , \ i = 1,\ldots,s \ , \\ & \underline{x} \leq x \leq \overline{x} \ , \ \underline{y} \leq y \leq \overline{y} \ . \end{array} \tag{23}$$

Here $\underline{x}, \overline{x} \in I\!R^n$ and $\underline{y}, \overline{y} \in Z^m$ are lower and upper bounds for the design variables.

To transform the infinite dimensional optimization problem into a finite dimensional one, we proceed from a given discretization of the frequency variable f by an equidistant grid in each interval. The corresponding index sets are called J_0, J_1, ..., J_s. Let l be the total number of all grid points. First we introduce the notation $T_j(x,y) = T(f_j,x,y)$, f_j suitable grid point, $j = 1$, ..., l. All indices are ordered sequentially so that $\{1,\ldots,l\} = J_0 \cup J_1 \cup \ldots \cup J_s$, i.e., $J_0 = \{1,\ldots,l_0\}$, $J_1 = \{l_0+1,\ldots,l_1\}$, ..., $J_s = \{l_{s-1}+1,\ldots,l\}$. Then the discretized optimization problem is

$$\begin{array}{rl} & \max \min \{T_j(x,y) : j \in J_0\} \\ x \in \mathbb{R}^n, y \in Z^m : & T_j(x,y) \le T_i \ \text{ for } j \in J_i \ , \ i = 1, \ldots, s \ , \\ & \underline{x} \le x \le \overline{x} \ , \ \ \underline{y} \le y \le \overline{y} \ . \end{array} \tag{24}$$

The existence of a feasible design is easily checked by performing the test $T_j(x,y) \ge T_0$ for all $j \in J_0$. Problem (24) is equivalent to a smooth nonlinear program after a simple standard transformation.

Integer variables are handled in a specific way, see van de Braak et al. [47]. Since continuous relaxation is not possible, a certain combination of a quadratic approximation and direct search algorithm in the integer space is developed. A function evaluation for a set of integer variables requires the complete solution of the continuous optimization problem (24) by the SQP code NLPQL.

Lower and upper bounds for the ten design variables under consideration are shown in Table 6 together with initial values and final ones obtained by the code NLPQL. Simulation is performed with respect to 154 frequency points leading to 174 constraints in the continuous model. Altogether 36 calls of NLPQL are made within four quadratic approximation cycles and one additional direct search iteration. The total number of simulations, i.e., the number of evaluations of the transmission energy $T_j(x,y)$ for all j, is 434 without the function calls needed for the gradient approximations. The purpose of the example is to show that the design goal is only achieved by taking also integer variables into account.

5. Case Study: Optimal Control of an Acetylene Reactor

The computation of optimal feed controls for chemical reactors, especially for tubular reactors, is a well-known technique, see Edgar and Himmelblau [9], Nishida et al. [27], and Buzzi-Ferraris et al. [6–7]. The mathematical model is given as a distributed parameter system consisting of a set of first-order partial differential equations in one space dimension. The chemical reactions and the temperature depend on the spatial variable, whereas the dynamical decrease of the cross-sectional area caused by coke deposition is time-dependent. In both cases, we know initial values either in the form of time-dependent feed control functions or a constant tube diameter. Alternative approaches to compute optimal reactor feed rates are discussed in Birk et al. [4], and Liepelt and Schittkowski [20].

variable	*lower bound*	*initial value*	*optimal value*	*upper bound*
y_1	7	19	12	19
y_2	12	24	14	25.0
y_3	100	130	124	150
x_1	5.0	11.58	9.589	15.0
x_2	50.0	50.0	92.39	150.0
x_3	10.5	11.39	11.25	11.5
x_4	10.0	10.61	10.62	11.0
x_5	0.3	0.3	0.3	0.5
x_6	0.95	1.033	1.031	1.05
x_7	0.95	1.031	1.023	1.05
x_8	0.95	1.012	1.015	1.02
x_9	0.985	1.001	0.998	1.03
x_{10}	1.0	1.0	1.000	1.03

Table 6. Bounds, Initial, and Optimal Values for Design Variables

We consider a chemical reactor producing acetylene (C_2H_2), reacting methane (CH_4) in natural gas with oxygen. This reaction requires less oxygen compared with complete combustion. The products are quickly quenched to keep the acetylene from being converted entirely to coke, see Wansbrough [48]. During the reaction process, a small part of the carbon is deposited in the reactor as coke. The quantity and its distribution in the reactor depend on the reaction equations. Since it is impossible to measure the cross-sectional area directly, we need a mathematical model that describes the functional dependence of the cross-sectional area upon other system parameters. If the deposition of coke reaches a certain limit, the reactor must be stopped and the tube cleaned.

There are six reactions to be taken into account. Reactions 1 through 5 are the main ones that produce acetylene, but also undesirable byproducts such as coke. Reaction 6 is included only to balance the hydrogen stoichiometry,

$$\begin{array}{ll} CH_4 \rightarrow \frac{1}{2}C_2H_2 + \frac{3}{2}H_2 \ , & CH_4 + O_2 \rightarrow CO + H_2O + H_2 \ , \\ CO + \frac{1}{2}O_2 \rightarrow CO_2 \ , & C_2H_2 \rightarrow 2C + H_2 \ , \\ H_2 + \frac{1}{2}O_2 \rightarrow H_2O \ , & C + \frac{n}{2}H_2 \rightarrow CH_n \ . \end{array} \qquad (25)$$

The reactions can be described by the following system of ordinary differential equations, where C_i denotes the molar concentration of the i-th component,

$$
\begin{aligned}
v(x,t)\,\frac{\partial}{\partial x}C_1(x,t) &= -r_1(x,t) - r_2(x,t) \ , \\
v(x,t)\,\frac{\partial}{\partial x}C_2(x,t) &= -r_2(x,t) - \tfrac{1}{2}r_3(x,t) - \tfrac{1}{2}r_5(x,t) \ , \\
v(x,t)\,\frac{\partial}{\partial x}C_3(x,t) &= \tfrac{1}{2}r_1(x,t) - r_4(x,t) \ , \\
v(x,t)\,\frac{\partial}{\partial x}C_4(x,t) &= r_3(x,t) \ , \\
v(x,t)\,\frac{\partial}{\partial x}C_5(x,t) &= \tfrac{3}{2}r_1(x,t) + r_2(x,t) + r_4(x,t) - r_5(x,t) \\
&\quad -n(1-\varepsilon)r_4(x,t) \ , \\
v(x,t)\,\frac{\partial}{\partial x}C_6(x,t) &= r_2(x,t) - r_3(x,t) \ , \\
v(x,t)\,\frac{\partial}{\partial x}C_7(x,t) &= r_2(x,t) + r_5(x,t) \ , \\
v(x,t)\,\frac{\partial}{\partial x}C_8(x,t) &= 2(1-\varepsilon)r_4(x,t)
\end{aligned}
\tag{26}
$$

with a reaction parameter ε. Since the acetylene reactor is controlled by the feeds of natural gas and oxygen, these are the only components with non-vanishing initial values. Initial molar concentrations are given by

$$
C_1^0(t) = \frac{\dot{m}_1(t)\rho_n}{M_1\dot{m}(t)} \ , \qquad C_2^0(t) = \frac{\dot{m}_2(t)\rho_n}{M_2\dot{m}(t)} \tag{27}
$$

with $\rho_n RT_0 = p_0$ assuming ideal gas law.

The velocity of the mixture in the reactor depends on the cross-sectional area $A(x,t)$, the total mass flow $\dot{m}(t)$ in the reactor, and the density $\rho(x,t)$, and is given by

$$
v(x,t) = \frac{\dot{m}(t)}{\rho(x,t)A(x,t)} \ , \tag{28}
$$

where the total mass flow $\dot{m}(t) = \dot{m}_1(t) + \dot{m}_2(t)$ is the sum of the two input flows. The density of the mixture is given by $\rho(x,t) = \sum_{j=1}^{8} C_j(x,t)M_j$, where M_j denotes the molar weight of the j-th component, and the temperature in the reactor can be described by the differential equation

$$
\frac{\partial}{\partial x}T(x,t) = \frac{1}{\rho(x,t)v(x,t)c_p(x,t)} \sum_{i=1}^{5} r_i(x,t)\Delta H_i \tag{29}
$$

with the initial condition $T(0,t) = T_0$. The incremental change of the temperature is determined by the rate of heat release for all reactions, which depends on the total heat capacity $c_p(x,t)$.

The eight material balance equations depend on the rates of the various reactions and on the velocity of the mixture in the reactor, because this speed determines the time that the components spent in the reactor. The reaction rates are expressed by

$$\begin{aligned}
r_1(x,t) &= k_1 \exp\Big(-\frac{E_1}{R}(1/T(x,t) - 1/T_r)\Big) C_1^{a_1}(x,t) \ , \\
r_2(x,t) &= k_2 \exp\Big(-\frac{E_2}{R}(1/T(x,t) - 1/T_r)\Big) C_1(x,t) C_2^{a_2}(x,t) \ , \\
r_3(x,t) &= k_3 \exp\Big(-\frac{E_3}{R}(1/T(x,t) - 1/T_r)\Big) C_6(x,t) C_2^{0.5}(x,t) \ , \\
r_4(x,t) &= k_4 \exp\Big(-\frac{E_4}{R}(1/T(x,t) - 1/T_r)\Big) C_3^{a_4}(x,t) \ , \\
r_5(x,t) &= k_5 \exp\Big(-\frac{E_5}{R}(1/T(x,t) - 1/T_r)\Big) C_5(x,t) C_2^{0.5}(x,t) \ ,
\end{aligned} \tag{30}$$

with five reaction constants k_1, ..., k_5, five activation energies E_1, ... E_5, and three reaction orders a_1, a_2, and a_4. For the smaller and less important reactions, the stoichiometric order can be used as an estimate for the reaction order. For the other reactions, these parameters have to be derived from the real reactor that is going to be examined. The average temperature T_r is used to scale the exponential functions and R denotes the gas constant.

If we neglect the deposition of coke, the underlying differential equation is stationary and does not depend on the time. But a decrease of the cross-sectional area $A(x,t)$ increases the velocity $v(x,t)$ of the mixture in the reactor, which influences the incremental change of the concentrations $C(x,t)$ and the temperature $T(x,t)$, see (26), (28), and (29). The coke deposition is modelled by the time-dependent differential equation

$$\frac{\partial}{\partial t} A(x,t) = -\beta r_4(x,t) \tag{31}$$

with initial condition $A(x,0) = A_0$ and reaction parameter β. A typical contour plot of the cross sectional area over time and spatial variable is shown in Figure 7.

The optimal control problem consists of maximizing

$$J(s) = \int_0^{T_{max}} \left(\sum_{i=1}^{8} P_i(t)\dot{m}_j(t) - \sum_{i=1}^{2} P_i(t)s_i(t) \right) dt \tag{32}$$

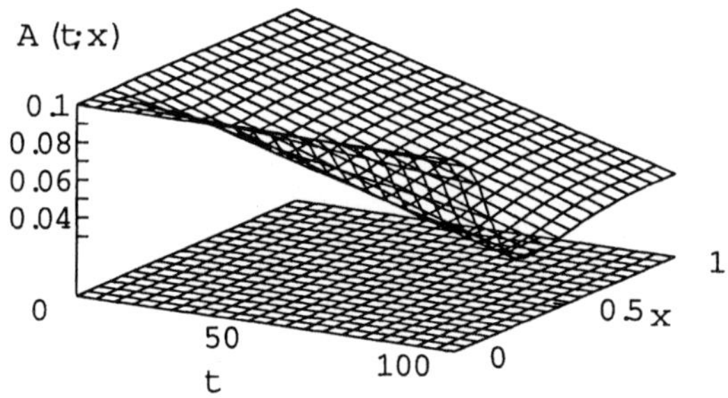

Figure 7. Cross Sectional Area

over all piecewise linear functions within a given range subject to the constraints

$$\begin{array}{lllll} A_{min} & \leq & A(x,t) & & , \\ 0 & \leq & T(x,t) & \leq & T_{max} \ , \\ \dot{m}_{k,min}(t) & \leq & \dot{m}_k(L,t) & \leq & \dot{m}_{k,max}(t) \ , \quad k=1,\ldots,8 \ , \end{array} \tag{33}$$

where $P_i(t)$ is the price of the i-th component. Maximum temperature is $T_{max} = 1.200$ and the tube length is 1, i.e., $x_L = 0$ and $x_R = 1$. More details and also the numerical data are presented in Birk et al. [4], where in addition also the optimal positions of maintenance intervals are to be computed.

The partial differential equation is discretizing subject to the time variable t, since the cross sectional area $A(x,t)$ is monotone decreasing without any steep fronts. Thus, it is possible to perform a few Euler steps for integrating (31) leading to a system of ordinary differential equations in x. We use 5 equidistant time values for approximating the two control variables by piecewise linear functions. The number of lines is 15 and a 5-point difference formula is applied to discretize spatial derivatives. The resulting ODE is integrated by RADAU5 of Hairer and Wanner [14]with a relative and absolute error tolerance of 10^{-6}. Thus we get 10 optimization variables and 510 nonlinear constraints by discretizing (26) at equidistant time and spatial values.

The SQP method NLPQL needs a relatively large number of 227 iterations to reach a solution. Obviously, the starting values are very poor as indicated by the values in the last column of the subsequent screen display. Moreover, objective function and constraints are badly scaled, see first and second column, and the stopping criterion KT for optimality, see last column, is decreased from 10^7 to 10^{-6}.

IT	F	SCV	NA	I	ALPHA	KT
1	-.13741137D+05	.16D+04	540	0	.00D+00	.17D+07
2	-.23624315D+05	.54D+01	12	1	.10D+01	.88D+06
3	-.23938874D+05	.95D-02	12	1	.10D+01	.71D+05
4	-.24234123D+05	.35D-02	9	1	.10D+01	.90D+04
5	-.24339670D+05	.93D-04	8	1	.10D+01	.17D+02
.	.	.	.	.	.	.
.	.	.	.	.	.	.
.	.	.	.	.	.	.
224	-.28064309D+05	.00D+00	0	1	.10D+01	.38D+00
225	-.28064499D+05	.00D+00	0	1	.10D+01	.14D-02
226	-.28064500D+05	.00D+00	0	1	.10D+01	.42D-04
227	-.28064500D+05	.00D+00	0	1	.10D+01	.49D-06

But the final convergence speed is very fast indicating that at least a local solution is approximated. The optimal control function for O_2 is shown in Figure 8, whereas the control variable for CH_4 attains its upper bound and is not displayed.

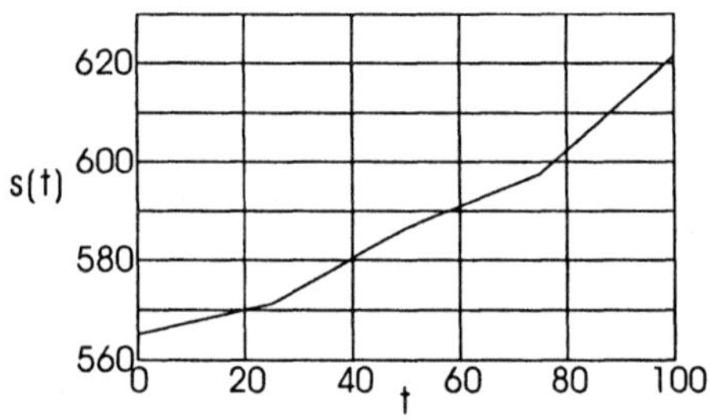

Figure 8. Optimal Input Control of O_2

6. Case Study: Weight Reduction of a Cruise Ship

In this section, we describe an example arising in ship building industry with the goal to minimize the total weight of the mechanical structure. The optimization problem is part of the overall design process of a new cruise ship and the results are essential to analyze further substructures by engineers. More details about the project can be found in Zillober and Vogel [54]. The problem is to find the optimal thickness distribution of beams and shell elements of one of the worlds largest cruise ships called *Radiance of the Seas* with a total length of about 300 meters with respect to minimal weight subject to thousands of stress constraints.

The discretized finite element model of the ship consists of 32946 shell and 33637 beam elements with 46986 nodes in total subject to two different load cases called *sagging* and *hogging*, see Figure 9. To formulate a practically relevant optimization problem, certain elements are linked to get finally $n = 415$ design variables x_i with different shell thicknesses, $i = 1, \ldots, n$. For each region, one element is selected for which the component stresses $(\sigma^x)^s$, $(\sigma^y)^s$, $(\sigma^{xy})^s$ for load case *sagging* and $(\sigma^x)^h$, $(\sigma^y)^h$, $(\sigma^{xy})^h$ for load case *hogging* are constrained. These quantities are calculated at the centroid of the middle layer of the element by the finite element analysis system ANSYS.

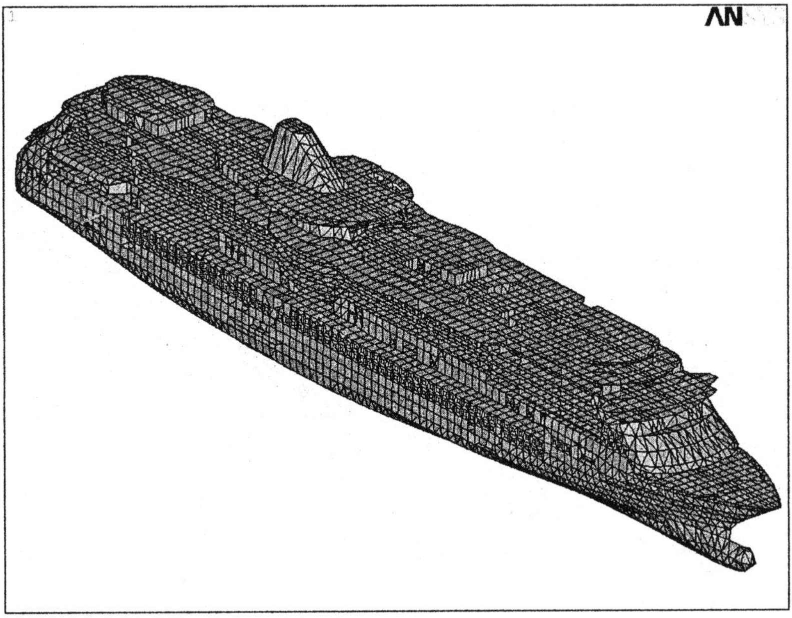

Figure 9. Finite Element Structure of the Ship to be Optimized

The optimization problem can be formulated as

$$
x \in \mathbb{R}^{415}: \quad
\begin{array}{lll}
\min & \text{Volume}(x), & \\
& \sigma^x_{\min} \leq (\sigma^x_j(x))^s \leq \sigma^x_{\max}, & j = 1, ..., n, \\
& \sigma^y_{\min} \leq (\sigma^y_j(x))^s \leq \sigma^y_{\max}, & j = 1, ..., n, \\
& \sigma^{xy}_{\min} \leq (\sigma^{xy}_j(x))^s \leq \sigma^{xy}_{\max}, & j = 1, ..., n, \\
& \sigma^x_{\min} \leq (\sigma^x_j(x))^h \leq \sigma^x_{\max}, & j = 1, ..., n, \\
& \sigma^y_{\min} \leq (\sigma^y_j(x))^h \leq \sigma^y_{\max}, & j = 1, ..., n, \\
& \sigma^{xy}_{\min} \leq (\sigma^{xy}_j(x))^h \leq \sigma^{xy}_{\max}, & j = 1, ..., n, \\
& 5 \leq x_i \leq 30 & i = 1, ..., n \ .
\end{array}
\qquad (34)
$$

Thus, the optimization problem consists of 415 optimization variables and 4980 constraints. It should be noted that the mechanical model does not take into account manufacturing constraints, such as fatigue life, buckling, etc. In other words, the weight reduction must be considered only as one part of the overall design process. Due to the high computational cost of the finite element simulations to be performed, the maximum number of iterations is set to 20. The total calculation time is about 30 hours for an SGI-ORIGIN workstation with 2 GB main memory allocated. SCPIP of Zillober [51] terminated at a thickness distribution which corresponds to an acceptable weight reduction compared to the initial weight of the ship. The objective function history is shown in Figure 10.

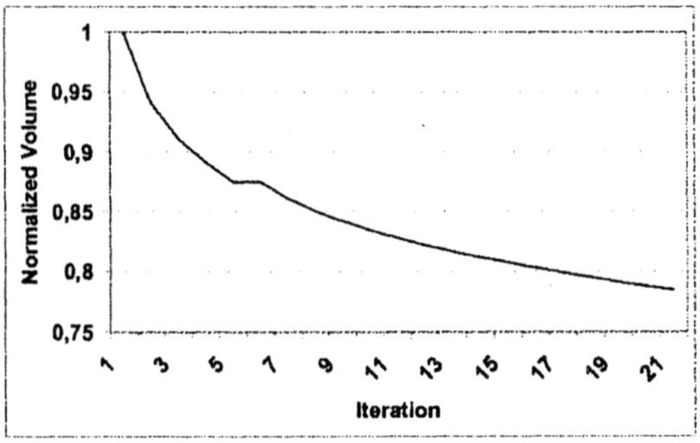

Figure 10. Iteration History of Structural Weight

Acknowledgments

The authors wish to thank Meyer Werft Papenburg for granting permission to publish Figure 9.

References

[1] M.P. Bendsøe (1989). Optimal shape design as a material distribution problem. *Structural Optimization*, 1:193–202, 1989.

[2] M.P. Bendsøe. *Optimization of Structural Topology, Shape and Material.* Springer, Heidelberg, 1995.

[3] M.P. Bendsøe and O. Sigmund O. *Topology Optimization — Theory, Methods and Applications.* Springer, Heidelberg, 2003.

[4] J. Birk, M. Liepelt, K. Schittkowski, and F. Vogel. Computation of optimal feed rates and operation intervals for tubular reactors. *Journal of Process Control*, 9:325–336, 1999.

[5] I. Bongartz, A.R. Conn, N.I.M. Gould, and P.L. Toint. CUTE: Constrained and unconstrained testing environment. *ACM Trans. Math. Software*, 21:123–160, 1995.

[6] G. Buzzi-Ferraris, G. Facchi, P. Forzetti, and E. Troncani. Control optimization of tubular catalytic decay. *Industrial Engineering in Chemistry*, 23:126–131, 1984.

[7] G. Buzzi-Ferraris, M. Morbidelli, P. Forzetti, and S. Carra. Deactivation of catalyst - mathematical models for the control and optimization of reactors. *International Chemical Engineering*, 24:441–451, 1984.

[8] R.E. Collin. *Field Theory of Guided Waves*. IEEE Press, New York, 1991.

[9] T.F. Edgar and D.M. Himmelblau. *Optimization of Chemical Processes*. McGraw-Hill, New York, 1988.

[10] C. Fleury. An efficient dual optimizer based on convex approximation concepts. *Structural Optimization*, 1:81–89, 1989.

[11] C. Fleury and V. Braibant. Structural optimization – a new dual method using mixed variables. *International Journal for Numerical Methods in Engineering*, 23:409–428, 1986.

[12] D. Goldfarb and A. Idnani. A numerically stable method for solving strictly convex quadratic programs. *Mathematical Programming*, 27:1–33, 1983.

[13] N.I.M. Gould and P.L. Toint. SQP methods for large-scale nonlinear programming. In *System Modelling and Optimization: Methods, Theory and Applications*. Kluwer, 2000.

[14] E. Hairer and G. Wanner. *Solving Ordinary Differential Equations II. Stiff and Differential-Algebraic Problems*. Springer, Berlin, Heidelberg, New York, 1991.

[15] C. Hartwanger, K. Schittkowski, and H. Wolf. Computer aided optimal design of horn radiators for satellite communication. *Engineering Optimization*, 33:221–244, 2000.

[16] W. Hock and K. Schittkowski. A comparative performance evaluation of 27 nonlinear programming codes. *Computing*, 30:335–358, 1983.

[17] K. Ito and K. Kunisch. Augmented Lagrangian-SQP methods for nonlinear optimal control problems of tracking type. *SIAM Journal on Optimization*, 6:96–125, 1996.

[18] R.C. Johnson and H. Jasik. *Antenna Engineering*. McGraw Hill, New York, 1984.

[19] G. Kneppe, J. Krammer, and E. Winkler. Structural optimization of large scale problems using MBB-LAGRANGE. Report MBB-S-PUB-305, Messerschmitt-Bölkow-Blohm, D-81663 Munich, 1987.

[20] M. Liepelt and K. Schittkowski. Optimal control of distributed systems with break points. In M. Grötschel, S.O. Krumke, and J. Rambau, editors, *Online Optimization of Large Scale Systems*, pages 271–294. Springer, Berlin, 2000.

[21] F.A. Lootsma. Fuzzy performance evaluation of nonlinear optimization methods. *Journal of Information and Optimization Sciences*, 10:15–44, 1981.

[22] H. Maurer and H.D. Mittelmann. Optimization techniques for solving elliptic control problems with control and state constraints: Part I. Boundary control. *Computational Optimization and Applications*, 16:29–55, 2000.

[23] H. Maurer and H.D. Mittelmann. Optimization techniques for solving elliptic control problems with control and state constraints: Part ii. distributed control. *Computational Optimization and Applications*, 18:141–160, 2001.

[24] R. Mittra. *Computer Techniques for Electromagnetics*. Pergamon Press, Oxford, 1973.

[25] H.P. Mlejnek. Some aspects of the genesis of structures. *Structural Optimization*, 5:64–69, 1992.

[26] V.H. Nguyen, J.J. Strodiot, and C. Fleury. A mathematical convergence analysis for the convex linearization method for engineering design optimization. *Engineering Optimization*, 11:195–216, 1987.

[27] N. Nishida, A. Ichikawa, and E. Tazaki. Optimal design and control in a class of distributed parameter systems under uncertainty. *AIChE Journal*, 18:561–568, 1972.

[28] P.Y. Papalambros and D.J. Wilde. *Principles of Optimal Design*. Cambridge University Press, Cambridge, 2000.

[29] M.J.D. Powell. The convergence of variable metric methods for nonlinearly constrained optimization calculations. In *Nonlinear Programming 3*. Academic Press, 1978.

[30] M.J.D. Powell. A fast algorithm for nonlinearly constraint optimization calculations. In *Numerical Analysis, G.A. Watson ed., Lecture Notes in Mathematics*, volume 630. Springer, 1978.

[31] R.T. Rockafellar. Augmented Lagrange multiplier functions and duality in nonconvex programming. *Journal on Control*, 12:268–285, 1974.

[32] T.L. Saaty. *The Analytic Hierarchy Process, Planning, Priority Setting, Resource Allocation*. McGraw Hill, New York, 1980.

[33] K. Schittkowski. *Nonlinear Programming Codes*, volume 183 of *Lecture Notes in Economics and Mathematical Systems*. Springer, Berlin, Heidelberg, New York, 1980.

[34] K. Schittkowski. Theory, implementation and test of a nonlinear programming algorithm. In *Optimization Methods in Structural Design*. N. Olhoff eds., Wissenschaftsverlag, 1983.

[35] K. Schittkowski. NLPQL: A Fortran subroutine solving constrained programming problems. *Annals of Operations Research*, 5:485–500, 1985.

[36] K. Schittkowski. *More Test Examples for Nonlinear Programming*, volume 187 of *Lecture Notes in Economics and Mathematical Systems*. Springer, Berlin, Heidelberg, New York, 1987.

[37] K. Schittkowski. Solving nonlinear least squares problems by a general purpose sqp-method. In *Trends in Mathematical Optimization, K.-H. Hoffmann, J.-B. Hiriart-Urruty, C. Lemarechal, J. Zowe eds., International Series of Numerical Mathematics*, volume 184, pages 295–309. Birkhäuser, 1988.

[38] K. Schittkowski. Solving nonlinear programming problems with very many constraints. *Optimization*, 25:179–196, 1992.

[39] K. Schittkowski. NLPQLP: A new fortran implementation of a sequential quadratic programming algorithm for parallel computing. Research report, Department of Mathematics, University of Bayreuth, D-95440 Bayreuth, 2001.

[40] K. Schittkowski. *Numerical Data Fitting in Dynamical Systems.* Kluwer, Dordrecht, 2002.

[41] K. Schittkowski, C. Zillober, and R. Zotemantel. Numerical comparison of nonlinear programming algorithms for structural optimization. *Structural Optimization*, 7:1–28, 1994.

[42] S. Silver. *Microwave Antenna Theory and Design.* McGraw Hill, 1949.

[43] P. Spellucci. *Numerische Verfahren der nichtlinearen Optimierung.* Birkhäuser, 1993.

[44] J. Stoer. Foundations of recursive quadratic programming methods for solving nonlinear programs. In *Computational Mathematical Programming, K. Schittkowski, ed., NATO ASI Series, Series F: Computer and Systems Sciences*, volume 15. Springer, 1985.

[45] K. Svanberg. The method of moving asymptotes – a new method for structural optimization. *International Journal for Numerical Methods in Engineering*, 24:359–373, 1987.

[46] G. Tobolka. Mixed matrix representation of SAW transducers. *Proceedings of the IEEE Ultrasonics Symposium*, 26:426–428, 1979.

[47] G. van de Braak, M. J. Bünner, and K. Schittkowski. Optimal design of electronic components by mixed-integer nonlinear programming. *To appear: Engineering Optimization*, 2003.

[48] R.W. Wansbrough. Modeling chemical reactors. *Chemical Engineering*, 5:95–102, 1985.

[49] C. Zillober. A combined convex approximation – interior point approach for large scale nonlinear programming. *Optimization and Engineering*, 2:51–73, 2001.

[50] C. Zillober. Global convergence of a nonlinear programming method using convex approximations. *Numerical Algorithms*, 27:256–289, 2001.

[51] C. Zillober. SCPIP – an efficient software tool for the solution of structural optimization problems. *Structural and Multidisciplinary Optimization*, 24:362–371, 2002.

[52] C. Zillober. Software manual for SCPIP 2.3. Research report, Department of Mathematics, University of Bayreuth, D-95440 Bayreuth, 2002.

[53] C. Zillober, K. Schittkowski, and K. Moritzen. Very large scale optimization by sequential convex programming. *Optimization Methods and Software*, 18:103–121, 2004.

[54] C. Zillober and F. Vogel. Solving large scale structural optimization problems. In *Proceedings of the 2nd ASMO UK/ISSMO Conference on Engineering Design Optimization*, pages 273–280. J. Sienz ed., University of Swansea, Wales, 2000.

STOCHASTIC MODELING AND OPTIMIZATION OF COMPLEX INFRASTRUCTURE SYSTEMS

P. Thoft-Christensen
Department of Building Technology and Structural Engineering
Aalborg University, Aalborg, Denmark
ptc@bt.aau.dk

Abstract In this paper it is shown that recent progress in stochastic modeling and optimization in combination with advanced computer systems has now made it possible to improve the design and the maintenance strategies for infrastructure systems. The paper concentrates on highway networks and single large bridges. United States has perhaps the largest highway networks in the world with more than 6 million kilometers of roadway and more than 0.5 million highway bridges; see [2]. About 40% of these bridges are considered deficient and more than $50 billion is estimated needed to correct the deficiencies; see [12]. The percentage of substandard bridges deemed to require urgent actions in other countries such as France (15%) and UK (20%) is also high; see [3].

Keywords: Stochastic modeling, Infrastructure systems, Bridge management systems, Suspension bridges.

Introduction

Obtaining and maintaining advanced infrastructure systems plays an important role in modern societies. Developed countries have in general well established infrastructure systems but most non-developed countries are characterized by having bad or no effective infrastructure systems. Therefore, in the transition from a non-developed country to a well developed country construction of effective infrastructure systems plays an important role. However, it is a fact that construction of new infrastructure systems requires great investments so a careful planning of all details in the system is essential for the effectiveness of the system from an operational but also economical point of view.

Obtaining the resources needed to establish infrastructure systems is only the first step. The next step and perhaps the most expensive step is to maintain the systems. It is recognized in most developed countries that good maintenance of infrastructure systems is in the long run the most economical way to keep the infrastructure in a satisfactory state. Effective maintenance requires however more resources than available in most countries. Therefore, careful planning of maintenance strategies is essential for all types of infrastructures.

1. Formulation of the Cost Optimization Problem

An infrastructure system consists of a number of structures. The objective is to minimize the cost of maintaining such a group of structures in the service life of the infrastructure. Estimation of the service life costs is a very uncertain so that a stochastic modeling is clearly needed. This can be expressed mathematically as

$$\min E[C] = \min \left(E[C_M] + E[C_U] + E[C_F]\right) \tag{1}$$

where

$E[C]$ is the expected total cost in the service life of the infrastructure
$E[C_M]$ is the expected maintenance cost in the service life of the infrastructure
$E[C_U]$ is the expected user costs e.g. traffic disruption costs due to works or restrictions on the structure
$E[C_F]$ is the expected costs due to failure of structures in the infrastructure.

For a *single* structure i in the infrastructure the expected cost can be written

$$\begin{aligned} E[C_i] &= E[C_{Mi}] + E[C_{Ui}] + E[C_{Fi}] \\ &= \sum_{t=1}^{T} \Big\{ (1+\gamma)^{-1} \left[E[C_{Mi}(t)] P(M_{it}) \right. \\ &\left. + E[C_{Ui}(t)] P(U_{it}) + E[C_{Fi}(t)] P(F_i(t)) \right] \Big\} \end{aligned} \tag{2}$$

where

γ is the discount rate (factor) e.g. 6%
$E[C_i]$ is the expected total cost for structure i
$E[C_{Mi}(t)]$ is the expected maintenance cost for structure i in year t
$E[C_{Ui}(t)]$ is the expected user costs for structure i in year t
$E[C_{Fi}(t)]$ is the expected failure cost for structure i in year t
$P(M_{it})$ is the probability of the event "maintenance is necessary" for structure i in year t
$P(D_{it})$ is the probability of the event "maintenance is necessary" for structure i in year t
$P(F_{it})$ is the probability of the event "maintenance is necessary" for structure i in year t
T is the remaining service life or reference period (in years).

Let the number of structures in the considered infrastructure be m. The expected total cost for the group can then be written

$$\begin{aligned} E[C] &= \sum_{i=1}^{m} \Big\{ E(C_{Mi}) + E(C_{Ui}) + E(C_{Fi}) \Big\} \\ &= \sum_{i=1}^{m} \sum_{t=1}^{T} \Big\{ (1+\gamma)^{-1} \left[E[C_{Mi}(t)] P(M_{it}) \right. \\ &\left. + E[C_{Ui}(t)] P(U_{it}) + E[C_{Fi}(t)] P(F_i(t)) \right] \Big\} \end{aligned} \tag{3}$$

2. Bridge Networks

Future advanced bridge management systems will be based on simple models for predicting the residual strength of structural elements. Improved stochastic modeling of the deterioration is needed to be able to formulate optimal strategies for inspection and maintenance of deteriorated bridges. However, such strategies will only be useful if they are also combined with expert knowledge. It is not possible to formulate all expert experience in mathematical terms. Therefore, it is believed that future management systems will be expert systems or at least knowledge-based systems; see [15].

Methods and computer programs for determining rational inspection and maintenance strategies for concrete bridges must be developed. The optimal decision should be based on the expected benefits and total cost of inspection, repair, maintenance and complete or partial failure of the bridge. Further, the reliability has to be acceptable during the expected lifetime.

The first major research on combining stochastic modeling, expert systems and optimal strategies for maintenance of reinforced concretes structures was sponsored by EU in 1990 to 1993. The research project is entitled *"Assessment of Performance and Optimal Strategies for inspection and Maintenance of Concrete Structures Using Reliability Based Expert systems"*. The results are presented in several reports and papers; see e.g. [15] and [5]. The methodology used in the project is analytic with traditional numerical analysis and rather advanced stochastic modeling.

Monte Carlo simulation has been used in decades to analyze complex engineering structures in many areas, e.g. in nuclear engineering. In modeling reliability profiles for reinforced concrete bridges Monte Carlo simulation seems to be used for the first time in December 1995 in the Highways Agency project *"Revision of the Bridge Assessment Rules based on Whole Life Performance: Concrete"* (1995-1996, Contract: DPU 9/3/44). The project is strongly inspired by the above-mentioned EU-project. The methodology used is presented in detail in the final project report, see [24].

In the Highways Agency project *"Optimum Maintenance Strategies for Different Bridge Types"* (1998-2000, Contract: 3/179), the simulation approach was extended in 1998, see [17] and [18] to include stochastic modeling of rehabilitation distributions and preventive and essential maintenance for reinforced concrete bridges. A similar approach is used in the project on steel/concrete composite bridges, see [6].

In a recent project *"Preventive Maintenance Strategies for Bridge Groups* (2001-2003, Contact 3/344 (A+B)), the simulation technique is extended further to modeling of condition profiles, and the interaction between reliability profiles and condition profiles for bridges, and the whole life costs. The simulation results are detailed presented in [7] and [23], [22].

3. Estimation of Service Life of Infrastructures

In this paper service life assessment of infrastructures is discussed based on stochastic models and with special emphasis on deterioration of reinforced structures due to reinforcement corrosion.

The service life $T^{(1)}_{service}$ for a reinforced concrete structure has been the subject of discussion between engineers for several decades. Several authors; see e.g. [16]; have defined the service life as the initiation time for corrosion T_{corr} of the reinforcement.

The service life $T^{(1)}_{service}$ has later been modified so that the time Δt_{crack} from corrosion initiation to corrosion crack initiation in the concrete is included; see [19]. The service life is then defined by $T^{(2)}_{service} = T_{crack} = T_{corr} + \Delta t_{crack}$. A stochastic model for Δt_{crack} may be developed on the basis of existing deterministic theories for crack initiation; see [10].

The service life may further be modified so that the time $\Delta t_{crack\ width}$ from corrosion crack initiation to formation of a certain (critical) crack width is included; see [20]. By this modeling it is possible to estimate the reliability of a given structure on the basis of measurements of the crack widths on the surface of the concrete structure.

Corrosion initiation period refers to the time during which the passivation of steel is destroyed and the reinforcement starts corroding actively. If Ficks law of diffusion can represent the rate of chloride penetration into concrete, then it can be shown that the time T_{corr} to initiation of reinforcement corrosion is

$$T^{(1)}_{service} = T_{corr} = \frac{d^2}{4D}\left((\mathrm{erf}^{-1}\left(\frac{C_{cr} - C_0}{C_i - C_0}\right)\right)^{-2} \qquad (4)$$

where d is the concrete cover, D is the diffusion coefficient, C_{cr} is the critical chloride concentration at the site of the corrosion, C_0 is the equilibrium chloride concentration on the concrete surface, C_i is the initial chloride concentration in the concrete, erf is the error function.

After corrosion initiation the rust products will initially fill the porous zone around the steel/concrete surface. As a result of this, tensile stresses are initiated in the concrete. With increasing corrosion the tensile stresses will reach a critical value and cracks will be developed. During this process the volume of the corrosion products at initial cracking of the concrete W_{cr} it will occupy three volumes, namely the porous zone W_{porous}, the expansion of the concrete due to rust pressure W_{expan}, and the space of the corroded steel W_{steel}. With this modeling and some minor simplifications it can then be shown that the time from corrosion imitation to crack initiation is; see [10]

$$\begin{aligned} \Delta t_{crack} &= \frac{1}{2 \times 0.383 \times 10^{-3} D_{bar} i_{corr}} \\ &\times \left(\frac{\rho_{steel}}{\rho_{steel} - 0.58\rho_{ust}} (W_{porous} - W_{expan}) \right)^2 \end{aligned} \quad (5)$$

where D_{bar} is the diameter of the reinforcement bar, i_{corr} is the annual mean corrosion rate, ρ_{steel} is the density of the steel, and ρ_{rust} is the density of the rust products.

After formation of the initial crack the rebar cross-section is further reduced due to the continued corrosion, and the width of the crack is increased. Experiments (see e.g. [1]) show that the function between the reduction of the rebar diameter ΔD_{bar} and the corresponding increase in crack width Δw_{crack} in a given time interval Δt measured on the surface of the concrete specimen can be approximated by a linear function

$$W_{crack} = \gamma \Delta D_{bar} \quad (6)$$

where the factor γ is of the order 1.5 to 5. This linearization has been confirmed by FEM analyses; see [21]. Let the critical crack width be $W_{critical}$ corresponding to the service life $T^{(3)}_{service}$. By setting $T^{(3)}_{service} = W_{critical}$ the following expression is obtained for $T^{(3)}_{service}$

$$T^{(3)}_{service} = \frac{W_{critical}(T_{crack})}{\gamma c_{corr} i_{corr}} + T_{crack} \quad (7)$$

$W_{crack}(T_{crack}) \approx 0$ is the initial crack width at the time T_{crack}. Using Monte Carlo simulation, the distribution functions of $T^{(1)}_{service}$, $T^{(2)}_{service}$ and $T^{(3)}_{service}$ can then for a given structure be estimated for any value of the critical crack width when stochastic distributions are known for all parameters.

4. Stochastic Modeling of Maintenance Strategies

After a structural assessment of the reliability of a reinforced concrete bridge deck at the time T_0 the problem is to decide if the bridge deck should be repaired and, if so, how and when it should be repaired. Solution of this optimization problem requires that all future inspections and repairs are taken into account. After each structural assessment the total expected benefits minus expected repair and failure costs in the residual lifetime of the bridge are maximized considering only the repair events in the residual service life of the bridge.

In order to simplify the decision modeling it is assumed that N_R repairs of the same type are performed in the residual service life $T_{service}$ of the bridge. The first repair is performed at the time T_{R_1}, and the remaining repairs are performed at equidistant times at the time interval $t_R = (T_{service} - T_{R_1})/N_R$. This decision model can be used in an adaptive way if the model is updated after an assessment (or repair) and a new optimal repair decision is made with regard to t_R. Therefore, it is mainly the time T_{R_1} of the first repair after an assessment, which is of importance. In order to decide which repair type is optimal after a structural assessment; the following optimization problem is considered for each repair technique, see [15]:

$$\begin{aligned} \max_{T_R, N_R} W(T_R, N_R) &= B(T_R, N_R) - C_R(T_R, N_R) - C_F(T_R, N_R) \\ \text{s.t.} \quad & \beta^U(T_{service}, T_R, N_R) \geq \beta^{\min} \\ & \text{or/and } T_{service}(T_R, N_R) \geq T_{service}^{\min} \end{aligned} \tag{8}$$

where the optimization variables are the expected number of repairs N_R in the residual service life and the time T_R of the first repair. W are the total expected benefits minus costs in the residual lifetime of the bridge. B is the benefit. C_R is the repair cost capitalized to the time $t = 0$ in the residual service life of the bridge. C_F are the expected failure costs capitalized to the time in the residual service life of the bridge. $T_{service}$ is the expected service life of the bridge. β^U is the updated reliability index. $\beta^{\min}$ is the minimum reliability index for the bridge (related to a critical element or to the total system). $T_{service}^{\min}$ is the minimum acceptable service life.

The benefits B play a significant role and are modelled by

$$B(T_R, N_R) = \sum_{i=[T_0]+1}^{[T_{service}]} B_i (1+r)^{T_0 - T_{ref}} \frac{1}{(1+r)^{T_i - T_0}} \tag{9}$$

where $[T]$ signifies the integer part of T measured in years and B_i are the benefits in year i (time interval $[T_{i-1}, T_i]$). T_i is the time from the construction of the bridge. The ith term in (9) represents the benefits from T_{i-1} to T_i. The benefits in year i are modelled by $B_i = k_0 V(T_i)$ where k_0 is a factor modeling the average benefits for one vehicle passing the bridge.

The expected repair costs C_R capitalized to the time $t = 0$ are modelled by

$$C_R(T_R, N_R) = \sum_{i=1}^{N_R} \left(1 - P_F^U(T_{R_i})\right) C_{R_0}(T_{R_i}) \frac{1}{(1+r)^{T_{R_i} - T_0}} \tag{10}$$

$P_F^U(T_R)$ is the updated probability of failure in the time interval $]T_0, T_R]$. The updating is based on a no failure event and the available inspection data at the time T_0. The factor $\left(1 - P_F^U(T_{R_i})\right)$ models the probability that the bridge has not failed at the time of repair. r is the discount rate. $C_{R_0}(T_{R_i})$ is the cost of repair.

The capitalized expected costs C_F due to failure are determined by

$$C_F(T_R, N_R) = \sum_{i=1}^{N_R+1} C_F(T_{R_i}) \left(P_F^U(T_{R_i}) - P_F^U(T_{i-1})\right) \frac{1}{(1+r)^{T_{R_i}}} \tag{11}$$

where $T_{R_0} = T_0$ is the time of the structural assessment and $T_{R_{N_R+1}} = T_{service}$ is the expected service life. The ith term in (11) represents the expected failure costs in the time interval $]T_{R_{i-1}}, T_{R_i}]$. $C_F(T)$ is the cost of failure at the time T.

5. Design of Long Bridges

Several short span (< 500 m) suspension bridges collapsed due to the wind. The famous and relatively long (854 m) Tacoma Narrows Bridge failed in 1940. In recent years much longer bridges have been constructed. The longest suspension bridge today is the Akashi Kaikyo Bridge in Japan (main span 1991 m) and the second longest is the Great Belt East Bridge in Denmark (main span 1624 m). Future designs with improved girder forms, lightweight cables, and control devices may be up to 3000-5000 m long. For such extremely long bridges, girder stability to wind action may be a serious problem, especially when the girder depth-to-width ratio is small compared with existing long bridges.

The main dynamic problem with long suspension bridges is the aeroelastic phenomenon called flutter. Flutter oscillation of a bridge girder is a stability problem and the oscillations are perpendicular to the direction of the wind and occur when the bridge is exposed to wind velocity above a critical value called the flutter wind velocity U_{cr}. U_{cr} decreases with decreasing stiffness and damping. Flutter is therefore a serious problem for bridges with a relatively low stiffness such as long bridges. Installation of passive and active control devices may be a solution to the girder stability problem.

Application of flaps to active control of flutter of long suspension bridges has been proposed in [11] to ensure the aerodynamic stability of slender bridge girders by attaching actively controlled flaps along the girders. The Ph.D. thesis [8] deals with wind tunnel experiments with a sectional model of a girder where the control flaps are installed as integrated parts of the leading and trailing edges of the girder. Several configurations of the flaps have been tested in a wind tunnel at Instituto Technico in Lisbon, Portugal. An analysis of a full span suspension bridge is performed in the Ph.D. thesis [9]. For the used configuration of the flaps it is shown that the flutter wind velocity U_{cr} can be increased by 50% compared with a girder with no flaps.

By assuming potential flow theory, it has been shown for thin airfoils in incompressible flow that the motion-induced vertical load $L_{ae}(x,t)$ and the motion-induced moment $M_{ae}(x,t)$ on the airfoil are linear in the theoretical displacement and the torsional angle and their first and second derivatives, where x is the coordinate in the direction of the bridge and t is the time, see [14]. Let y and z be the coordinates in the direction across the bridge and in the vertical direction. A similar formulation for bridges is introduced in [13]. The aeroelastic forces L^{deck} and M^{deck} per unit span and for small rotations can then be written, see [4]:

$$\begin{aligned} L^{deck}_{ae}(x,t) &= \frac{\rho U^2 B}{2}\Big[K H^\star_1(K)\frac{\dot{\nu}_z}{U} + K H^\star_2(K)\frac{B\dot{r}_x}{U} \\ &\quad + K^2 H^\star_3(K) r_x + K^2 H^\star_4(K)\frac{\nu_u}{B}\Big] \end{aligned} \tag{12}$$

$$\begin{aligned} M^{deck}_{ae}(x,t) &= \frac{\rho U^2 B^2}{2}\Big[K A^\star_1(K)\frac{\dot{\nu}_z}{U} + K A^\star_2(K)\frac{B\dot{r}_x}{U} \\ &\quad + K^2 A^\star_3(K) r_x + K^2 A^\star_4(K)\frac{\nu_z}{B}\Big] \end{aligned} \tag{13}$$

where $K = B\omega/U$ is the non-dimensional reduced frequency, B is the girder width, U is the mean wind velocity, ω is the bridge oscillating frequency (rad.) at the wind velocity U, and ρ is air density. $H^\star_i(K)$ and $A^\star_i(K)$ $(i = 1,2,3,4)$ are non-dimensional aerodynamic derivatives which can be estimated by wind tunnel experiments. The quantities $r_x\ \dot{\nu}_z/U$ and $B\dot{r}_x/U$ are non-dimensional, effective angles of attack. Two types of actively controlled flaps are shown in figure 1.

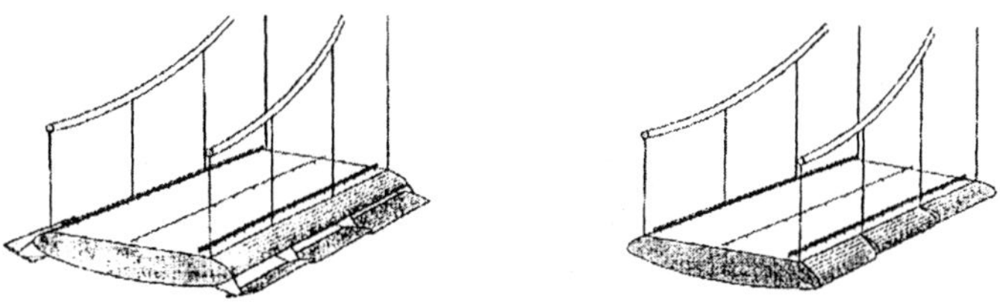

Figure 1. Sections with flaps on pylons and integrated in the section.

By assuming that the angle of a leading flap has no effect on the air circulation it can be shown that the loads due to movement of a leading flap on a thin airfoil are also linear in the angle of the leading flap and in the first and second derivatives. The motion-induced wind loads due to movement of the flaps can therefore be described by additional aerodynamic derivatives. The total motion-induced wind loads per unit

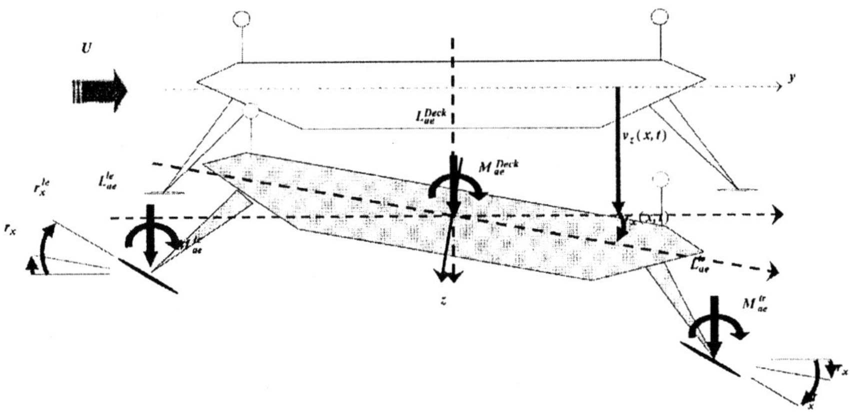

Figure 2. Motion-induced wind loads on the girder and on the flaps.

span on the girder and the flaps are, see figure 2

$$L_z^{total} = L_z^{deck} + L_z^{tr}(\nu_z, r_x^{tr}) + L_z^{le}(\nu_z, r_x^{le}) \tag{14}$$

$$\begin{aligned} M_x^{total} &= M_x^{deck} + M_x^{tr}(\nu_z, r_x^{tr}) + M_x^{le}(\nu_z, r_x^{le}) \\ &+ \left(L_z^{tr}(\nu, r_x^{tr}) - L_z^{le}(-\nu, r_z^{le})\right) \frac{B}{2} \end{aligned} \tag{15}$$

where $\nu_z(x,t)$ and $r_x(x,t)$ are the vertical motion and the rotation of the girder at position x along the bridge girder at the time t. $r_x^{le}(x,t)$ and $r_x^{tr}(x,t)$ are the rotations of the leading and the trailing flaps. Figure

3 shows the calculated flutter velocity U_{cr} for different combinations of flap rotations. α is the rotation of the girder, α_l and α_t are the rotations of the leading and the trailing flaps, φ_l and φ_t are the phase angles between the leading flap, the trailing flap and the girder, respectively.

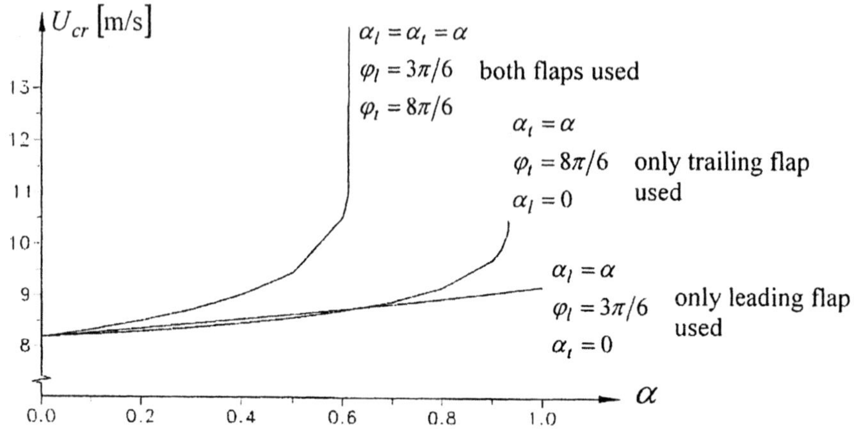

Figure 3. The theoretical effect on the flutter wind velocity of using flaps.

Figures 4 and 5 show the torsional movement of the model when the flaps are not regulated (configuration 0) and when they are regulated (configuration 2). The wind speed is 6.1 m/s. The conclusion is that configuration 2 is very efficient for controlling the torsional motion of the model. During the first second the torsional motion is reduced from 2.7° to 1.1°, i.e. by 62%.

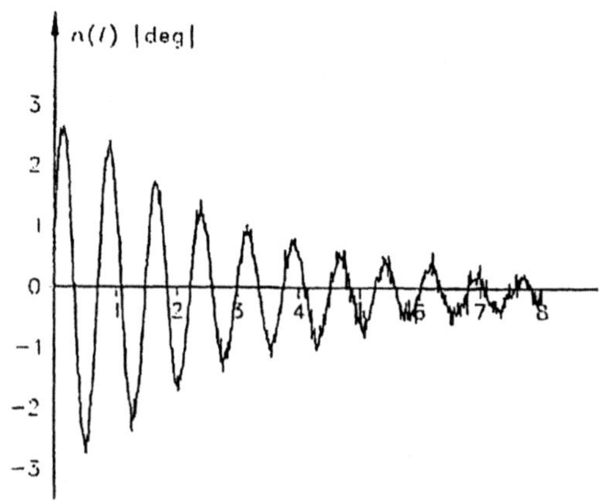

Figure 4. Torsional motion for flap configuration 0 and wind speed 6.1 m/s.

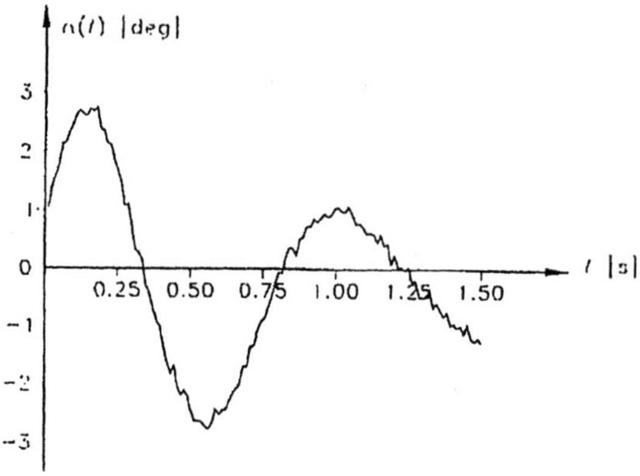

Figure 5. Torsional motion for flap configuration 2 and wind speed 6.1 m/s.

Let $\phi_i(x)$ and $\psi_j(x)$ be the vertical and the torsional mode shapes of the bridge in mode i and mode j which are assumed to be coupled at flutter. Then the governing modal equations for the two-mode flutter conditions are

$$M_z\left(\ddot{z}(t) + 2\omega_z\zeta_z\dot{z}(t) + \omega_z^2 z(t)\right) = F_z^{tot}(t) \tag{16}$$

$$M_x\left(\ddot{\alpha}(t) + 2\omega_\alpha\zeta_\alpha\dot{\alpha}(t) + \omega_\alpha^2 \alpha(t)\right) = F_x^{tot}(t) \tag{17}$$

where $z(t)$ and $\alpha(t)$ are the vertical and the torsional modal coordinates. ω_z, ζ_z, ω_α and ζ_α are the natural frequencies and the damping ratios of the vertical and torsional modes. M_z and M_x are the vertical and the torsional modal masses. At the coupled motion, the vertical and the torsional modal responses are both assumed to be proportional to $e^{i\omega t}$, when the critical wind velocity is acting on the bridge, i.e. $z(t) = z_0e^{i\omega t}$ and $\alpha(t) = \alpha_0 e^{i\omega t}$. When this is introduced into the above equations the following matrix equation can be derived

$$\mathbf{A}\begin{bmatrix} cz/B \\ \alpha \end{bmatrix} = \begin{bmatrix} 0 \\ 0 \end{bmatrix} \tag{18}$$

where the system matrix $\mathbf{A}$ depends on the natural mode shapes and frequencies, the damping ratios, the derivatives and the wind velocity. This matrix equation has non-trivial solutions when

$$\text{Det}(\mathbf{A}) = \text{Re Det}(\mathbf{A}) + i\ \text{Im Det}(\mathbf{A}) = 0 \tag{19}$$

resulting in the following two flutter conditions for a bridge with separate flaps, [9]:

$$\begin{aligned}\mathrm{Re(Det)} &= \tfrac{\omega^4}{\omega_z^4}\Big(1 + \tfrac{M3}{J\omega^2\Psi} + \tfrac{L4}{m\omega^2\Phi} + \tfrac{1}{mJ\omega^4\Psi\Phi} \\ &\times\left[-\omega^2 L1M2 + L4M3 - M4L3 + \omega^2 M1L2\right]\Big) \\ &+\tfrac{\omega^3}{\omega_z^3}\left(2\zeta_z\tfrac{M2}{J\omega\Psi} + 2\zeta_\alpha\tfrac{\omega_\alpha}{\omega_z}\tfrac{L1}{m\omega\Phi}\right) \\ &+\tfrac{\omega^2}{\omega_z^2}\left(-1 - \tfrac{\omega_\alpha^2}{\omega_z^2} - 4\tfrac{\omega_\alpha}{\omega_z}\zeta_z\zeta_\alpha - \tfrac{M3}{J\omega^2\Psi} - \tfrac{\omega_\alpha^2}{\omega_z^2}\tfrac{L4}{m\omega^2\Phi}\right) \\ &+\tfrac{\omega_\alpha^2}{\omega_z^2} = 0\end{aligned} \tag{20}$$

$$\begin{aligned}\mathrm{Im(Det)} &= \tfrac{\omega^3}{\omega_z^3}\Big(\tfrac{M2}{J\omega\Psi} + \tfrac{L1}{m\omega\Phi} + \tfrac{1}{m\omega^3\Psi J\Phi} \\ &\times\left[L1M3 + L4M2 + M1L3 - M4L2\right]\Big) \\ &+\tfrac{\omega^2}{\omega_z^2}\left(-2\zeta_z - 2\zeta_\alpha\tfrac{\omega_\alpha}{\omega_z} - 2\zeta_\alpha\tfrac{\omega_\alpha}{\omega_z}\tfrac{L4}{m\omega^2\Phi} - 2\zeta_z\tfrac{M3}{J\omega^2\Psi}\right) \\ &+\tfrac{\omega}{\omega_z}\left(-\tfrac{M2}{J\omega\Psi} - \tfrac{\omega_\alpha^2}{\omega_z^2}\tfrac{L1}{m\omega\Phi}\right) + 2\zeta_z\tfrac{\omega_\alpha^2}{\omega_z^2} + 2\zeta_\alpha\tfrac{\omega_\alpha}{\omega_z} = 0\end{aligned} \tag{21}$$

where m is the girder mass per unit span. Φ, Ξ and Ψ are the modal integrals of the girder given by:

$$\Psi = \int_0^L \phi_1^2(x)dx \ , \ \Xi = \int_0^L \phi_1(x)\psi_1(x)dx \ , \ \Psi = \int_0^L \psi_1^2(x)dx \tag{22}$$

and where $L1$ to $L4$ and $M1$ to $M4$ contain the modal integrals of the flaps Φ_f, Ξ_f and Ψ_f, the sum of flutter derivatives referred to the girder and the flaps (see [9] for full expressions). Finally, note that the flutter mode can be a coupling of more than two modes. In that case, an additional mode gives an additional equation. The determinant condition (19) is still valid, but the calculation of the solution is rather complicated analytically. The obtained critical wind velocity U_{cr} and the critical frequency ω_{cr} will not be varied by more than 5%, if several similar mode shapes with close frequencies are taken into account in the flutter computation, see [9].

6. Conclusions

It is shown in the paper that recent developed methodologies in stochastic and optimization may be used to solve complex problems related to infrastructure systems. A general formulation of the cost optimization problem is presented with special emphasis on bridge networks. Finally, the difficult (from a formulation and mathematical point of view) problem of estimating the flutter wind velocity for a suspension bridge is solved numerically to show how advanced research can take part in solving infrastructure problems.

References

[1] C. Andrade, C. Alonso, and F.J. Molina. Cover cracking as a function of bar corrosion: Part 1-experimental test. *Materials and Structures*, 26:453–464, 1993.

[2] S.B. Chase. The bridge maintenance programme of the united states federal highway administration, 1999. In: Management of Highway Structures (Editor: P.C. Das). Thomas Telford, pp. 14-23.

[3] P.C. Das. Prioritization of bridge maintenance needs, 1999. In: Case Studies in Optimal Design and Maintenance Planning of Civil Infrastructure Systems (Editor D.M. Frangopol), ASCE, pp. 26-44.

[4] Simiu E. and R.H. Scanlan. *Wind Effects on Structures: Fundamentals and Applications to Design. Third Edition.* John Wiley and Sons, 1996.

[5] P. Thoft-Christensen (Editor). Assessment of performance and optimal strategies for inspection and maintenance of concrete structures using reliability based expert systems, 2002. Report. CSRconsult ApS, Aalborg, Denmark.

[6] D.M. Frangopol. Optimum maintenance strategies for different bridge types, vol. 1, steel/concrete composite bridges, 2000. Final Report, HA-project 3/179.

[7] D.M. Frangopol. Preventive maintenance strategies for bridge groups - analysis - vol.1, 2003. Final Report, HA-project 3/344(B).

[8] H.I. Hansen. *Active Vibration Control of Long Suspension Bridges.* PhD thesis, Aalborg University, Department of Building Technology and Structural Engineering, Aalborg, Denmark, 1998.

[9] T. Huynh. *Suspension Bridge Aerodynamics and Active Vibration Control.* PhD thesis, Aalborg University, Department of Building Technology and Structural Engineering, Aalborg, Denmark, 2000.

[10] Y. Liu and R.E. Weyers. Modelling of the time to corrosion cracking in chloride contaminated reinforced concrete structures. *ACI Materials Journal*, 95:675–681, 1998.

[11] K.H. Ostenfeld and A. Larsen. Bridge engineering and aerodynamics, 1992. In: Aerodynamics of Large Bridges (editor A. Larsen), Proc. First Int. Symp. Aerodynamics of Large Bridges, Copenhagen, Denmark.

[12] J.E. Roberts. Bridge management for the 21st century, 2001. In: Maintaining the Safety of Deteriorating Civil Infrastructures (Editors A. Miyamoto and D.M. Frangopol), Ube, Yamaguchi, Japan, pp. 1-13.

[13] R.H. Scanlan and J.J.Tomko. Airfoil and bridge deck flutter derivatives. *J.Eng.Mech.Div., ASCE*, Paper 8601:1717–1737, 1971.

[14] T. Theodorsen. General theory of aerodynamic instability and the mechanism of flutter, 1935. NACA Report No. 496.

[15] P. Thoft-Christensen. Advanced bridge management systems. *Structural Engineering Review*, 7:151–163, 1995.

[16] P. Thoft-Christensen. Estimation of the service lifetime of concrete bridges, 1997. In: Proc. ASCE Structures Congress XV, Portland, Oregon, USA.

[17] P. Thoft-Christensen. Estimation of reliability distributions for reinforced concrete overbridges, 1998. HA-project 3/179, Working Document CSR-WD01.

[18] P. Thoft-Christensen. Optimum maintenance strategies for different bridge types, vol. 2, concrete bridges, 2000. Final Report, HA-project 3/179.

[19] P. Thoft-Christensen. Stochastic modelling of the crack initiation time for reinforced concrete structures, 2000. In: Proc. ASCE 2000 Structures Congress, Philadelphia.

[20] P. Thoft-Christensen. What happens with reinforced concrete structures when the reinforcement corrodes?, 2001. Keynote Speech at the 2nd International Workshop on Life-Cycle Cost Analysis and Design of Civil Infrastructure Systems, Ube, Yamaguchi, Japan. Proceedings: Maintaining the Safety of Deteriorating Civil Infrastructures, 2001, pp. 35-46.

[21] P. Thoft-Christensen. Modelling corrosion cracks, 2003. Presented at the IFIP TC-7, Conference, Sophia Antipolis, France.

[22] P. Thoft-Christensen. Preventive maintenance strategies for bridge groups- analysis - vol.2, 2003. Final Report, HA-project 3/344(B).

[23] P. Thoft-Christensen and C. Frier. Estimation of preventive maintenance costs using simulation, 2003. Report CSR-08, Aalborg, Denmark.

[24] P. Thoft-Christensen and F.M. Jensen. Revision of the bridge assessment rules based on whole life performance: Concrete, 1996. Final Report, HA-project DPU 9/3/44.

FEEDBACK ROBUST CONTROL FOR A PARABOLIC VARIATIONAL INEQUALITY

Vyacheslav Maksimov*
Institute of Mathematics and Mechanics
Ural Branch, Russian Academy of Sciences
Ekaterinburg, Russia
maksimov@imm.uran.ru

Abstract A problem of robust control of a parabolic variational inequality in the case of distributed control actions and disturbances is under consideration. The goal of the paper consists in the description and mathematical substantiation of the the method of feedback control in the formalization originated from works by N.N. Krasovskii [3], [2]. The paper continues investigations [5]–[4].

Keywords: robust control, parabolic variational inequality

1. Introduction

In the present work, the problem of robust control of distributed parameter systems is discussed. The essence of the problem under consideration may be formulated in the following way. A motion of a dynamical system Σ proceeds on a given time interval $T = [t_0, \vartheta]$. System's trajectory $x(t) = x(t; u(\cdot), v(\cdot))$, $t \in T$, depends on a time-varying unknown input $v = v(t) \in Q$, and a control $u = u(t) \in P$, where P and Q are given sets. Phase states of the system Σ are inaccurately measured at time moments $\tau_i \in \Delta = \{\tau_i\}_{i=0}^m$, $\tau_0 = t_0$, $\tau_m = \vartheta$, $\tau_{i+1} = \tau_i + \delta$. It is required to organize a process of control of the system Σ by the feedback principle in such a way that it is possible to preserve given properties of system's trajectory under the action of any admissible input $v = v(\cdot)$.

*This work was supported in part by the Russian Foundation for Basic Research (grant # 04-01-00059), Program on Basic Research of the Presidium of the Russian Acad. Sci. (project "Control of mechanical systems"), Program of supporting leading scientific schools of Russia (project 1846.2003.1) and Ural-Siberian Interdisciplinary Project.

The quality of a trajectory constructed is estimated either by the distance from a given (prescribed, standard) trajectory $x_*(t)$ or by some quality functional (a payoff). The problem under discussion is treated as the problem of constructing a control $u = u(t)$ providing (under the action of any possible but unknown disturbance $v = v(t)$) retention of a trajectory $x(t) = x(t; u(\cdot), v(\cdot))$ nearby $x_*(t)$ (in the first case) or minimizing maximally possible values of a payoff (in the second case). This is a meaningful formulation of the control problem under consideration.

2. Statement of the Problem

Let a system Σ be described by the parabolic variational inequality

$$\langle \dot{x}(t) + Ax(t), x(t) - z\rangle + \varphi(x(t)) - \varphi(z) \leq (f(t, u, v), x(t) - z) \quad (1)$$

$$\text{for a. a.} \quad t \in T \quad \text{and all} \quad z \in V, \quad x(t_0) = x_0.$$

Here $H = L_2(\Omega)$, V is a real Hilbert space, V is a dense subspace of H and $V \subset H \subset V^*$ algebraically and topologically, $(\cdot, \cdot)$ stands for the inner product in H, $\langle \cdot, \cdot \rangle$ stands for the duality relation between V and V^*, $A : V \to V^*$ is a linear continuous ($A \in \mathcal{L}(V; V^*)$) and symmetrical operator satisfying (for some $\omega > 0$ and real α_0) the coercitivity condition

$$(Ax, x) + \alpha_0 |x|_H^2 \geq \omega |x|_V^2 \quad \forall y \in V,$$

$|\cdot|_H$ and $|\cdot|_V$ stand for the norm in H and V, respectively, and $\varphi : V \to \overline{R} = \{r \in R : -\infty < r \leq +\infty\}$ is a lower semicontinuous convex function. Furthermore, without loss of generality, we assume that $\varphi(x) \geq 0 \quad \forall x \in V$. Let $x(t_0) = x_0 \in D(\varphi)$, where $D(\varphi) = \{x \in V : \varphi(x) < +\infty\}$. Let U_p and U_d be uniformly convex Banach spaces, $f : T \times U_p \times U_d \to H$ be a Lipschitz function, P and Q be given bounded and closed sets from space of controls U_p and disturbances U_d, respectively. It is known that under such conditions for any $\{u(\cdot), v(\cdot)\} \in L_2(T; U_p) \times L_2(T; U_d)$ there exists a unique solution $x(\cdot) = x(\cdot; t_0, x_0, u(\cdot), v(\cdot))$ of the inequality (1) with the following properties [1]: $x(\cdot) = x(\cdot; t_0, x_0, u(\cdot), v(\cdot)) \in W_*(T) = W^{1,2}(T; H) \cap L_2(T; V)$, $x(t) \in D(\varphi_\alpha)$ $\forall t \in T$, $t \to \varphi_\alpha(x(t)) \in AC(T)$. Here the function $\varphi_\alpha(y) : H \to \overline{R}$ is defined by

$$\varphi_\alpha(y) = \begin{cases} 1/2\langle Ay, y\rangle + \alpha_0/2|y|_H^2 + \varphi(y), & \text{if} \quad y \in D(\varphi) \\ +\infty, & \text{otherwise,} \end{cases}$$

$W^{1,2}(T; H) = \{w(\cdot) \in L_2(T; H) : w_t(\cdot) \in L_2(T; H)\}$, the derivative $w_t(\cdot)$ is understood in the sense of distributions, $AC(T)$ is the set of absolutely continuous functions $x(t) : T \to R$.

Let us give some definitions. Furthermore, we denote by $u_{a,b}(\cdot)$ a function $u(t)$, $t \in [a,b]$, considered as a whole. The symbol $P_{a,b}(\cdot)$ stands for restriction of a set $P_T(\cdot)$ onto the segment $[a,b] \subset T$. Any strongly measurable functions $u(\cdot) : T \to P$ and $v(\cdot) : T \to Q$ are called an open-loop control and a disturbance, respectively. The sets of all open-loop controls and disturbances are denoted by the symbols $P_T(\cdot)$ and $Q_T(\cdot)$. Elements of the product $T \times D(\varphi)$ are called positions. A unique solution of the inequality (1) with the properties $x(t_*) = x_*$, $x(\cdot) = x(\cdot; t_*, x_*, u_{t_*,\vartheta}(\cdot), v_{t_*,\vartheta}(\cdot)) \in W^{1,2}([t_*, \vartheta]; H) \cap L_2([t_*, \vartheta]; V)$, $x(t) \in D(\varphi_\alpha)$ $\forall t \in [t_*, \vartheta]$, $t \to \varphi_\alpha(x(t)) \in AC([t_*, \vartheta])$ is called a motion of the system (1) starting from a position $(t_*, x_*) \in T \times D(\varphi_\alpha)$ and corresponding to a control $u_{t,\vartheta}(\cdot) \in P_{t,\vartheta}(\cdot)$ and a disturbance $v_{t,\vartheta}(\cdot) \in Q_{t,\vartheta}(\cdot)$. A partition of T is any finite net $\Delta = \{\tau_i\}_{i=0}^m$, where $\tau_0 = t_0$, $\tau_m = \vartheta$, $\tau_{i+1} = \tau_i + \delta$, $\delta = \delta(\Delta)$ is a diameter of Δ. Any possible function (multifunction) $\mathcal{U} : T \times H \to P$ is said to be a feedback strategy. Feedback strategies correct controls at discrete time moments given by some partition of the interval T. A solution $x(\cdot)$ of the inequality (1) starting from an initial state (t_*, x_*) and corresponding to a piecewise constant control $u^h(\cdot)$ (formed by the feedback principle

$$u^h(t) = u_i \in \mathcal{U}(\tau_i, \xi_i), \quad t \in [\tau_i, \tau_{i+1}), \ i \in [0 : m-1], \ |\xi_i - x(\tau_i)|_H \le h)$$

and to a disturbance $v_{t_*,\vartheta}(\cdot) \in Q_{t_*,\vartheta}(\cdot)$ is called an (h, Δ)-motion $x_\Delta^h(\cdot; t_*, x_*, \mathcal{U}, v_{t_*,\vartheta}(\cdot))$ generated by a positional strategy $\mathcal{U}$ on a partition Δ. Thus, when we write $x_\Delta^h(\cdot)$, we mean a solution of the inequality (1) constructed by the feedback principle. The set of all (h, Δ)-motions is denoted by $X_h(t_*, x_*, \mathcal{U}, \Delta)$. It is clear that the set $X_h(t_*, x_*, \mathcal{U}, \Delta)$ is not empty for $(t_*, x_*) \in T \times D(\varphi_\alpha)$.

Thus, the problem under consideration may be formulated in the following way. Let the inequality (1) be considered on the given time interval T. Its solution $x(\cdot) = x(\cdot; t_0, x_0, u_T(\cdot), v_T(\cdot))$ depends on some control $u_T \in P_T(\cdot)$ and disturbance $v_T(\cdot) \in Q_T(\cdot)$. Let us fix a uniform net $\Delta = \{\tau_i\}_{i=0}^m$, $\tau_0 = t_0$, $\tau_m = \vartheta$, with a diameter $\delta = \delta(\Delta) = \tau_i - \tau_{i+1}$. Phase states $x(\tau_i)$ are inaccurately measured at the moments τ_i. Results of measurements $\xi_i \in H$ satisfy the inequalities

$$|\xi_i - x(\tau_i)|_H \le h, \quad i \in [0 : m-1]. \tag{2}$$

Here h is a value of the level of informational noise. Some prescribed trajectory $x_*(t)$, $t \in T$, and a number $\varepsilon > 0$ are given. It is required to construct an algorithm of feedback control of inequality (1) providing fulfillment of the following condition. Whatever the unknown disturbance $v_T(\cdot) \in Q_T(\cdot)$ may be, the deviation of the phase state $x(t)$ from the prescribed trajectory $x_*(t)$ at all moments $t \in T$ should not exceed the value of ε provided the values of h and δ are sufficiently small.

So, the problem (Problem 1) consists in construction of a positional strategy $\mathcal{U} : T \times H \to P$ with the following properties: whatever the value $\varepsilon > 0$ may be, one can indicate (explicitly) numbers $h_* > 0$ and $\delta_* > 0$ such that the inequalities

$$\rho(x_\Delta^h(\cdot), x_*(\cdot)) \leq \varepsilon \qquad \forall x_\Delta^h(\cdot) \in X_h(t_0, x_0, \mathcal{U}, \Delta), \tag{3}$$

are fulfilled uniformly with respect to all measurements ξ_i with the properties (2) if $h \leq h_*$ and the partition diameter $\delta = \delta(\Delta) \leq \delta_*$. Here the symbol $\rho(x(\cdot), y(\cdot))$ denotes the distance from $x(\cdot)$ to $y(\cdot)$ in the uniform metric, i.e.,

$$\rho(x(\cdot), y(\cdot)) = \sup_{t \in T} |x(t) - y(t)|_H.$$

Along with Problem 1, we consider also another problem. Let the following quality criterion of the process be given:

$$I(x_T(\cdot), u_T(\cdot), v_T(\cdot)) = \sigma(x(\vartheta)) + \int_{t_0}^{\vartheta} \chi(t, x(t), u(t), v(t))\, dt,$$

where $\sigma : H \to R$ and $\chi : T \times H \times U_p \times U_d \to R$ are given functions satisfying the local Lipschitz conditions. Introduce the following ordinary differential equation

$$\dot{p}(t) = \chi(t, x(t), u(t), v(t)). \tag{4}$$

A pair $\{x(\cdot), p(\cdot)\}$, where $x(\cdot)$ is a solution of the inequality (1) starting from an initial state (t_*, x_*) and $p(\cdot)$ is a solution of the equation (4) starting from an initial state (t_*, p_*) corresponding to a piecewise constant control $u^h(\cdot)$ (formed by the feedback principle

$$u^h(t) = u_i \in \mathcal{U}^e(\tau_i, \xi_i, \psi_i), \quad t \in [\tau_i, \tau_{i+1}), \quad i \in [0 : m-1], \tag{5}$$

$$|\psi_i - p(\tau_i)| \leq h, \quad |\xi_i - x(\tau_i)|_H \leq h)$$

and to a disturbance $v_{t_*,\vartheta}(\cdot) \in Q_{t_*,\vartheta}(\cdot)$ is called an (h, Δ, χ)-motion $z_\Delta^h(\cdot) = \{x_\Delta^h(\cdot; t_*, x_*, \mathcal{U}^e, v_{t_*,\vartheta}(\cdot)), p_\Delta^h(\cdot; t_*, p_*, \mathcal{U}^e, v_{t_*,\vartheta}(\cdot))\}$ generated by a positional strategy $\mathcal{U}^e : T \times H \times R \to P$ on a partition Δ. The set of all (h, Δ, χ)-motions is denoted by $Z_h^\chi(t_*, x_*, p_*, \mathcal{U}^e, \Delta)$. It is clear that the set $Z_h^\varphi(t_*, x_*, p_*, \mathcal{U}^e, \Delta)$ is not empty for $(t_*, x_*, p_*) \in T \times D(\varphi) \times R$.

Problem 2 consists in the following. A prescribed value of the criterion, number I_*, is fixed. It is necessary to construct a positional strategy $\mathcal{U}^e : T \times H \times R \to P$ with the following properties: whatever the value $\varepsilon > 0$ and disturbance $v_T(\cdot) \in Q_T(\cdot)$ may be, one can indicate (explicitly) numbers $h_* > 0$ and $\delta_* > 0$ such that the inequalities

$$|I(x^h_{\Delta T}(\cdot), u^h_T(\cdot), v_T(\cdot)) - I_*| \leq \varepsilon$$

are fulfilled uniformly with respect to all measurements ξ_i with the properties (2) and ψ_i, $|\psi_i - p^h_\Delta(\tau_i)| \leq h$, if $h \leq h_*$ and the partition diameter $\delta = \delta(\Delta) \leq \delta_*$. Here $\{x^h_\Delta(\cdot), p^h_\Delta(\cdot)\} \in Z^\chi_h(t_0, x_0, 0, \mathcal{U}^e, \Delta)$, $x^h_\Delta(\cdot) = x(\cdot; t_0, x_0, \mathcal{U}^e(\cdot), v_T(\cdot))$, $p^h_\Delta(\cdot) = p(\cdot; t_0, 0, \mathcal{U}^e(\cdot), v_T(\cdot))$, the control $u^h(\cdot)$ is defined by (5).

As one can see from the statement of Problem 2, to solve this problem at moments τ_i, it is necessary to know (perhaps, with an error) the realization $p^h_\Delta(\tau_i)$. To obtain this information, one should know the realization of disturbance $v(t)$, $t \in [t_0, \tau_i]$. If the function χ does not depend on v, i.e., $\chi = \chi(t, x, u)$, then the information on the disturbance v is not required. In this case the quality criterion does not depend on $v(\cdot)$:

$$I = I(x_T(\cdot), u_T(\cdot)) = \sigma(x(\vartheta)) + \int_{t_0}^{\vartheta} \chi(t, x(t), u(t))\, dt.$$

Therefore, in definition of an (h, Δ, χ)-motion one can assume that $p^h_\Delta(\cdot) = p^h_\Delta(\cdot; t_*, p_*, \mathcal{U}^e)$ is a solution of the equation

$$\dot{p}^h_\Delta(t) = \chi(\tau_i, \xi_i, u_i), \quad t \in [\tau_i, \tau_{i+1}), \quad t \geq t_*,$$

with the initial condition $p^h_\Delta(t_*) = p_*$.

3. The Algorithm for Solving Problem 1

Let us indicate the algorithm for solving Problem 1. Introduce sets

$$f_u(t, v) = \bigcup_{u \in P} f(t, u, v), \quad H(t) = \bigcap_{v \in Q} f_u(t, v),$$

$$H(\cdot) = \{u(\cdot) \in L_2(T; H) : u(t) \in H(t) \quad \text{for a. a. } t \in T\}.$$

Let the following condition be fulfilled.

Condition 1.

a) sets $H(t)$ for all $t \in T$ are nonempty;

b) there exists a control $u_*(\cdot) \in H(\cdot)$ such that $x_*(\cdot) = x(\cdot; t_0, x_0, u_*(\cdot))$ where the symbol $x(\cdot; t_0, x_0, u_*(\cdot))$ denotes a solution of the variational inequality

$$\langle \dot{x}(t) + Ax(t), x(t) - z \rangle + \varphi(x(t)) - \varphi(z) \leq (u_*(t), x(t) - z)$$

$$\text{for a. a.} \quad t \in T \quad \text{and all} \quad z \in V, \quad x(t_0) = x_0;$$

c) the saddle point condition is fulfilled:

$$\inf_{u \in P} \sup_{v \in Q} (s, f(t,u,v)) = \sup_{v \in Q} \inf_{u \in P} (s, f(t,u,v)) \quad \text{for any } t \in T, \; s \in H.$$

Let us give two examples of functions f satisfying condition 1c).

1 A function f does not depend on t and is linear with respect to u and v, i.e., $f(t,u,v) = Bu - Cv$, $B \in \mathcal{L}(U_p; H)$, $C \in \mathcal{L}(U_d; H)$. This case was under discussion in [4].

2 Let a control u and a disturbance v be elements of finite-dimensional Euclidean spaces, i.e., $u \in U_p = R^n$, $v \in U_d = R^m$. A mapping f is given according to the following rule: $f(t,u,v)(\eta) = F(t,\eta,u,v)$, where $F(\cdot) : T \times \Omega \times R^n \times R^m \to R$ possesses the Carathéodory property: a) for all $\eta \in \Omega$ the function $F_\eta(t,u,v) = F(t,\eta,u,v)$ satisfies the Lipschitz property; b) for all $(t,u,v) \in T \times P \times Q$ the function $F_{t,u,v}(\eta) = F(t,\eta,u,v)$ is Lebesgue measurable.

Condition 1c) is fulfilled if either a control u and a disturbance v are separated, i.e., $F(t,\eta,u,v) = F(t,\eta,u) + F(t,\eta,v)$, or the function F is of the following structure: $F(t,\eta,u,v) = \sum_{j=1}^{N} \omega_j(\eta) F_j(t,u,v)$, where $\omega_j \in L_2(\Omega)$, $j \in [1:N]$, and the vector function

$$f_*(t,u,v) = \{F_1(t,u,v), F_2(t,u,v), \ldots, F_N(t,u,v)\}$$

satisfies the saddle point condition in a "small game" [3]:

$$\min_{u \in P} \max_{v \in Q} s' f_*(t,u,v) = \max_{v \in Q} \min_{u \in P} s' f_*(t,u,v) \quad \text{for any } t \in T, \quad s \in R^N.$$

Here the symbol "$'$" ("prime") means transposition.

Let us describe the procedure of forming an (h,Δ)-motion $x^h_\Delta(t; t_0, x_0, \mathcal{U}, v_{t_0,t}(\cdot))$ corresponding to a fixed partition Δ and a strategy $\mathcal{U}$ of the form:

$$\mathcal{U}(t,x) = \{u^e \in P : \sup_{v \in Q} (x - x_*(t), f(t,u^e,v)) \leq \tag{6}$$

$$\inf_{u \in P} \sup_{v \in Q} (x - x_*(t), f(t,u,v)) + h\}.$$

Before the start of the work of the algorithm, we fix a value $h \in (0,1)$ and a partition $\Delta = \{\tau_i\}_{i=0}^m$ with a diameter $\delta = \delta(\Delta)$. Then we organize the process of control of the system (1) according to the feedback principle in such a way that the motion $x^h_\Delta(\cdot) = x(\cdot; t_0, x_0, \mathcal{U}, v_T(\cdot))$ remains in a sufficiently small neighborhood of $x_*(\cdot)$ at all moments $t \in T$ for sufficiently small h and δ under the action of any disturbance $v_T(\cdot) \in Q_T(\cdot)$, i.e., the inequality (3) is valid. The work of the algorithm is divided into $(m-1)$ identical steps. In the interval $[t_0, \tau_1)$ we assume

$$u^h(t) = u_0 \in \mathcal{U}(t_0, x_0) = P, \quad t \in [t_0, \tau_1).$$

Under the action of this control as well as of an unknown disturbance $v_{t_0,\tau_1}(\cdot)$ some (h, Δ)-motion $\{x^h_\Delta(\cdot; t_0, x_0, \mathcal{U}, v_{t_0,\tau_1}(\cdot))\}_{t_0,\tau_1}$ is realized. At the moment $t = \tau_1$ we determine u_1 from the condition

$$u_1 \in \mathcal{U}(\tau_1, \xi_1), \qquad |\xi_1 - x^h_\Delta(\tau_1)|_H \leq h,$$

i.e., $u^h(t) = u_1$ for $t \in [\tau_1, \tau_2)$. Then we calculate the realization of the (h, Δ)-motion $\{x^h_\Delta(\cdot; \tau_1, x^h_\Delta(\tau_1), \mathcal{U}, v_{\tau_1,\tau_2}(\cdot))\}_{\tau_1,\tau_2}$. Let the (h, Δ)-motion $x^h_\Delta(\cdot)$ be defined in the interval $[t_0, \tau_i]$. At the moment $t = \tau_i$ we assume

$$u_i \in \mathcal{U}(\tau_i, \xi_i), \qquad |\xi_i - x^h_\Delta(\tau_i)|_H \leq h,$$

i.e., $u^h(t) = u_i$ for $t \in [\tau_i, \tau_{i+1})$. As the result of the action of this control and of an unknown disturbance $v_{\tau_i,\tau_{i+1}}(\cdot)$ the (h, Δ)-motion of the system (1) $\{x^h_\Delta(\cdot; \tau_i, x^h_\Delta(\tau_i), \mathcal{U}, v_{\tau_i,\tau_{i+1}}(\cdot))\}_{\tau_i,\tau_{i+1}}$ is realized in the interval $[\tau_i, \tau_{i+1}]$. The indicated above procedure of forming the (h, Δ)-motion stops at the moment ϑ.

THEOREM 1 *The strategy $\mathcal{U}(t, x)$ of the form (6) solves the Problem 1.*

Proof. Let a partition $\Delta = \{\tau_i\}_{i=0}^m$ of the interval T with a diameter $\delta(\Delta) = \delta$ and a value of the level of informational noise h be fixed. Let us estimate the evolution of the function

$$\varepsilon(t; x^h_\Delta(\cdot), x_*(\cdot)) = 1/2|x^h_\Delta(t) - x_*(t)|^2_H + \omega \int_{t_0}^{t} |x^h_\Delta(\tau) - x_*(\tau)|^2_V \, d\tau$$

for $t \in T$. Introduce the functional $l(y(\cdot)) : W_*(T) \to R$,

$$l(y(\cdot)) = |y(\cdot)|_{C(T;H)} + |\dot{y}(\cdot)|_{L_2(T;H)} + |y(\cdot)|_{L_2(T;V)}.$$

Let $\alpha_* = \alpha_0$, if $\alpha_0 \geq 0$, and $\alpha_* = 0$, if $\alpha_0 < 0$. One can prove in a standard way [1]that there exists a number $K_* = K_*(\alpha_0, \omega)$ such that for any $x_0 \in D(\varphi)$, $u_T(\cdot) \in P_T(\cdot)$, $v_T(\cdot) \in Q_T(\cdot)$, $x(\cdot) = x(\cdot; t_0, x_0, u_T(\cdot), v_T(\cdot))$ the inequality

$$l(x(\cdot)) \leq K_*(1 + \alpha_*|x_0|_H + \varphi_\alpha^{1/2}(x_0) + |u(\cdot)|_{L_2(T;U_p)} + |v(\cdot)|_{L_2(T;U_d)}) \quad (7)$$

is true. It is easily seen that for a. a. $t \in [\tau_i, \tau_{i+1})$, $i \geq 1$, the inequality

$$\frac{d}{dt}\varepsilon(t; x^h_\Delta(\cdot), x_*(\cdot)) \leq \tag{8}$$

$$(f(t, u_i, v(t)) - u_*(t), x^h_\Delta(t) - x_*(t)) + \alpha_0 |x^h_\Delta(t) - x_*(t)|^2_H$$

holds. Here

$$u_i \in \mathcal{U}(\tau_i, \xi_i), \qquad |\xi_i - x^h_\Delta(\tau_i)|_H \leq h, \tag{9}$$

(ξ_i is an inaccurate measurement of phase state $x^h_\Delta(\tau_i)$), $v_{\tau_i, \tau_{i+1}}(\cdot)$ is an unknown realization of disturbance, the strategy $\mathcal{U}(t, x)$ is determined from (6). It follows from (7)–(9) that

$$\frac{d}{dt}\varepsilon(t, x^h_\Delta(\cdot), x_*(\cdot)) \leq (f(t, u_i, v(t)) - u_*(t), s_i)_H + \tag{10}$$

$$\alpha_0 |x^h_\Delta(t) - x_*(t)|^2_H +$$

$$k_1\Big(h + \int\limits_{\tau_i}^{t} \{|\dot{x}^h_\Delta(\tau)|_H + |\dot{x}_*(\tau)|_H\}\, d\tau\Big), \quad t \in \delta_i = [\tau_i, \tau_{i+1}), \quad s_i = \xi_i - x_*(\tau_i).$$

Let us define vectors v^e_i from the conditions

$$\inf_{u \in P}(s_i, f(\tau_i, u, v^e_i)) \geq \sup_{v \in Q} \inf_{u \in P}(s_i, f(\tau_i, u, v)) - h. \tag{11}$$

It is obvious that

$$u_*(t) \in H(t) \subset \bigcup_{u \in P} f(t, u, v^e_i), \quad \text{for a. a.} \quad t \in [\tau_i, \tau_{i+1}).$$

By virtue of condition 1a) there exists a control $u^{(1)}(t) \in P$, $t \in \delta_i$, such that

$$f(t, u^{(1)}(t), v^e_i) = u_*(t) \quad \text{for a. a.} \quad t \in [\tau_i, \tau_{i+1}]. \tag{12}$$

Using condition 1c) and (11), we deduce that

$$(s_i, f(\tau_i, u_i, v(t)) \leq \sup_{v \in Q}(s_i, f(t, u_i, v)) + L(t - \tau_i) \leq$$

$$\inf_{u \in P} \sup_{v \in Q}(s_i, f(t, u, v)) + h + L(t - \tau_i) =$$

$$\sup_{v \in Q} \inf_{u \in P}(s_i, f(t, u, v)) + h + L(t - \tau_i) \leq \tag{13}$$

$$\inf_{u \in P}(s_i, f(\tau_i, u, v^e_i)) + 2\{h + L(t - \tau_i)\} \leq (s_i, f(t, u^{(1)}, v^e_i)) + 3h + 2L(t - \tau_i).$$

Here L is a Lipschitz constant of the function $f(\cdot)$. In this case it follows from (12), (13) that

$$(s_i^*, f(t, u_i, v(t)) - u_*(t)) \leq 4h + 2L(t - \tau_i) \tag{14}$$

We derive from the inequalities (10), (14)

$$\varepsilon(t; x_\Delta^h(\cdot), x_*(\cdot)) \leq \varepsilon(\tau_i; x_\Delta^h(\cdot), x_*(\cdot)) + k_1\delta\Big(h + \delta + \int\limits_{\tau_i}^{\tau_{i+1}} \{|\dot{x}_\Delta^h(\tau)|_H + \tag{15}$$

$$|\dot{x}_*(\tau)|_H\} \, d\tau\Big) + \alpha_0 |x_\Delta^h(t) - x_*(t)|_H^2, \quad t \in \delta_i.$$

Since

$$\varepsilon(t_0; x_\Delta^h(\cdot), x_*(\cdot)) = 0, \quad \varepsilon(\tau_1; x_\Delta^h(\cdot), x_*(\cdot)) \leq k_2(h + \delta^{1/2}), \tag{16}$$

by (7), (15), and (16) we have for $t \in T$

$$\varepsilon(t; x_\Delta^h(\cdot), x_*(\cdot)) \leq k_2(h + \delta^{1/2}) + k_1\delta\Big(h(\vartheta - t_0)/\delta +$$

$$\int\limits_{t_0}^{t} \{|\dot{x}_\Delta^h(\tau)|_H + |\dot{x}_*(\tau)|_H\} \, d\tau\Big) + \alpha_0 \int\limits_{t_0}^{t} |x_\Delta^h(\tau) - x_*(\tau)|_H^2 \, d\tau.$$

Here constants k_1 and k_2 do not depend on h, δ and can be explicitly written. The conclusion of the theorem follows from Gronwall's lemma. **The theorem is proved.**

Remark 1. The trajectory $x_*(\cdot)$ plays the role of a "stable path", which is famous in the theory of differential games. By virtue of Theorem 1, the strategy $\mathcal{U}$ of form (6) guarantees that the solution of inequality (1) follows the trajectory $x_*(\cdot)$ (the strategy "leads" $x(\cdot)$ along $x_*(\cdot)$) irrespective of the unknown effective perturbation. As to the problem of choosing the stable path itself, it is a typical problem of a program control of a parabolic variational inequality and was studied by many authors.

4. The Algorithm for Solving Problem 2

Let us indicate the algorithm for solving Problem 2. Denote

$$\Phi(t, x, u, v) = \{f(t, u, v), \chi(t, x, u, v)\}$$

$$\Phi_u(t, x, v) = \bigcup_{u \in P} \Phi(t, x, u, v), \quad H_*(t; x) = \bigcap_{v \in Q} \Phi_u(t, x, v),$$

$$H_*(\cdot;x) = \{u(\cdot) \in L_2(T; H \times R) : u(t) \in H_*(t;x) \quad \text{for a. a. } t \in T\}.$$

Let the following condition be fulfilled.

Condition 2.

a)there exists a control $u^*(\cdot) = \{u_1(\cdot), u_2(\cdot)\} \in L_2(T; H \times R)$, $u^*(t) \in H_*(t; x_*(t))$ for a. a. $t \in T$, such that $I_* = \sigma(x_*(\vartheta)) + z_*(\vartheta)$, where $x_*(\cdot) = x(\cdot; t_0, x_0, u_1(\cdot))$, the symbol $x(\cdot; t_0, x_0, u_1(\cdot))$ denotes a solution of the variational inequality

$$\langle \dot{x}(t) + Ax(t), x(t) - z\rangle + \varphi(x(t)) - \varphi(z) \leq (u_1(t), x(t) - z) \tag{17}$$

$$\text{for a. a.} \quad t \in T \quad \text{and all} \quad z \in V, \quad x(t_0) = x_0,$$

and the symbol $z_*(\cdot)$ stands for a solution of the ordinary differential equation

$$\dot{z}(t) = u_2(t), \quad t \in T, \quad z(t_0) = 0;$$

b) the saddle point condition is fulfilled:

$$\inf_{u \in P} \sup_{v \in Q} \{(s, f(t, u, v)) + r\chi(t, x, u, v)\} =$$

$$\sup_{v \in Q} \inf_{u \in P} \{(s, f(t, u, v)) + r\chi(t, x, u, v)\} \quad \text{for any } t \in T, \ x, s \in H, \ r \in R.$$

Let us describe the procedure of forming an (h, Δ, χ)-motion $z^h_\Delta(\cdot) = \{x^h_\Delta(\cdot), p^h_\Delta(\cdot)\}$ corresponding to a fixed partition Δ and a strategy $\mathcal{U}^e$ of the form:

$$\mathcal{U}^e(t, x, p) = \tag{18}$$

$$\{u^e \in P : \sup_{v \in Q}\{(x - x_*(t), f(t, u^e, v)) + (p - z_*(t))\chi(t, x, u^e, v)\} \leq$$

$$\inf_{u \in P} \sup_{v \in Q}\{(x - x_*(t), f(t, u, v)) + (p - z_*(t))\chi(t, x, u, v)\} + h\}.$$

The algorithm for solving Problem 2 is analogous to the algorithm for solving Problem 1. Before the start of algorithm's work, we fix a value $h \in (0, 1)$ and a partition $\Delta = \{\tau_i\}_{i=0}^m$, with a diameter $\delta = \delta(\Delta)$. The work of the algorithm is subdivided into $(m - 1)$ identical steps. In the interval $[t_0, \tau_1)$ we assume

$$u^h(t) = u_0 \in \mathcal{U}^e(t_0, x, p) = P.$$

Under the action of this control as well as of an unknown disturbance $v_{t_0,\tau_1}(\cdot)$, some (h, Δ, χ)-motion $\{z^h_\Delta(\cdot)\}_{t_0,\tau_1} = \{x^h_\Delta(\cdot; t_0, x_0, \mathcal{U}^e, v_{t_0,\tau_1}(\cdot)), p^h_\Delta(\cdot, t_0, 0, \mathcal{U}^e, v_{t_0,\tau_t}(\cdot)\}_{t_0,\tau_1}$ is realized. At the moment $t = \tau_1$ we determine u_1 from the condition

$$u_1 \in \mathcal{U}^e(\tau_1, \xi_1, \psi_1), \quad |\xi_1 - x^h_\Delta(\tau_1)|_H \leq h, \quad |\psi_1 - p^h_\Delta(\tau_1)| \leq h,$$

i.e., $u^h(t) = u_1$ for $t \in [\tau_1, \tau_2)$. Then we calculate the realization of the (h, Δ, φ)-motion $\{z^h_\Delta(\cdot)\}_{\tau_1,\tau_2} = \{x^h_\Delta(\cdot;\tau_1, x^h_\Delta(\tau_1), \mathcal{U}, v_{\tau_1,\tau_2}(\cdot)),\ p^h_\Delta(\cdot;\tau_1, p^h_\Delta(\tau_1), \mathcal{U}, v_{\tau_1,\tau_2}(\cdot))\}_{\tau_1,\tau_2}$. Let the (h, Δ, χ)-motion $z^h_\Delta(\cdot)$ be defined in the interval $[t_0, \tau_i]$. At the moment $t = \tau_i$ we assume

$$u_i \in \mathcal{U}^e(\tau_i, \xi_i, \psi_i), \quad |\xi_i - x^h_\Delta(\tau_i)|_H \le h, \quad |\psi_i - p^h_\Delta(\tau_i)| \le h,$$

i.e., $u^h(t) = u_i$ for $t \in [\tau_i, \tau_{i+1})$. As the result of the action of this control and of an unknown disturbance $v_{\tau_i,\tau_{i+1}}(\cdot)$, the (h, Δ, χ)-motion of the system (1) $\{z^h_\Delta(\cdot)\}_{\tau_i,\tau_{i+1}} = \{x^h_\Delta(\cdot;\tau_i, x^h_\Delta(\tau_i), \mathcal{U}, v_{\tau_i,\tau_{i+1}}(\cdot)), p^h_\Delta(\cdot;\tau_i, p^h_\Delta(\tau_i), \mathcal{U}, v_{\tau_i,\tau_{i+1}}(\cdot))\}_{\tau_i,\tau_{i+1}}$ is realized in the interval $[\tau_i, \tau_{i+1}]$. The described above procedure of forming the (h, Δ, χ)-motion stops at the moment ϑ.

THEOREM 2 *The strategy $\mathcal{U}^e(t, x, p)$ of the form (18) solves the Problem 2.*

Scheme of the proof of this theorem is analogous to the proof of Theorem 1. At that we estimate the evolution of the function

$$\mu(t; x^h_\Delta(\cdot), p^h_\Delta(t), x_*(\cdot), z_*(t)) = \varepsilon(t, x^h_\Delta(\cdot), x_*(\cdot)) + 1/2|p^h_\Delta(t) - z_*(t)|^2.$$

Let a partition $\Delta = \{\tau_i\}_{i=0}^m$ of the interval T with a diameter $\delta(\Delta) = \delta$ and a value of the level of informational noise h be fixed. Analogous to (10) we deduce for a. a. $t \in [\tau_i, \tau_{i+1})$, $i \ge 1$, that the inequality

$$\frac{d}{dt}\mu(t; x^h_\Delta(\cdot), p^h_\Delta(t), x_*(\cdot), z_*(t)) \le \tag{19}$$

$$(f(t, u_i, v(t)) - u_1(t), x^h_\Delta(t) - x_*(t)) + \alpha_0|x^h_\Delta(t) - x_*(t)|^2_H +$$
$$(\chi(t, x^h_\Delta(t), u_i, v(t)) - u_2(t))(p^h_\Delta(t) - z_*(t))$$

holds. Here

$$u_i \in \mathcal{U}^e(\tau_i, \xi_i, \psi_i), \qquad |\xi_i - x^h_\Delta(\tau_i)|_H \le h, \quad |\psi_i - p^h_\Delta(\tau_i)| \le h, \tag{20}$$

$v_{\tau_i,\tau_{i+1}}(\cdot)$ is an unknown realization of disturbance, the strategy

$$\mathcal{U}^e(t, x, p)$$

is determined from (18). It follows from (19), (20) and the local Lipschitz property that

$$\frac{d}{dt}\mu(t, x^h_\Delta(\cdot), p^h_\Delta(t), x_*(\cdot), z_*(t)) \le (f(t, u_i, v(t)) - u_1(t), \xi_i - x_*(\tau_i)) +$$

$$(\chi(\tau_i, \xi_i, u_i, v(t)) - u_2(t))(p^h_\Delta(\tau_i) - z_*(\tau_i)) + \alpha_0|x^h_\Delta(t) - x_*(t)|^2_H +$$

$$k_1\Big(h + \omega_\chi(t - \tau_i) \int\limits_{\tau_i}^{t} \{|\dot{x}^h_\Delta(\tau)|_H + |\dot{x}_*(\tau)|_H + |\dot{p}^h_\Delta(\tau)| + |\dot{z}_*(\tau)|\}\, d\tau\Big),$$

$$t \in \delta_i = [\tau_i, \tau_{i+1}),$$

and constant k_1 can be explicitly written. Here

$$\omega_\chi(\delta) = \sup\{|\chi(t_1, x, u, v) - \chi(t_2, x, u, v)| : t_1,\ t_2 \in T,$$

$$|t_1 - t_2| \le \delta,\ u \in P,\ v \in Q,\ x \in X\}$$

is the modulo of continuity of a function $\chi(\cdot)$, $X \subset H$ is some bounded domain, which all solutions of the inequality (1) belong to. The conclusion of the theorem follows from Gronwall's lemma. Further argument repeats the one presented in the proof of Theorem 1. At that we take the function Φ instead of the function f. **The theorem is proved.**

References

[1] V. Barbu. *Optimal control of variational inequalities.* Pitman Advanced Publishing Program, London, 1984.

[2] N.N. Krasovskii. *Controlling of a dynamical system.* Nauka, Moscow, 1985. in Russian.

[3] N.N. Krasovskii and A.I. Subbotin. *Game-theoretical control problems.* Springer, Berlin, 1988.

[4] V. I. Maksimov. Feedback minimax control for parabolic variational inequality. *C.R.Acad.Sci., Paris*, Série II b:105–108, 2000.

[5] Ju. S. Osipov. Differential games for systems with hereditary. *Dokl. AN USSR*, 196(4):779–782, 1971. in Russian.

TRACKING CONTROL OF PARABOLIC SYSTEMS

Luciano Pandolfi*
Politecnico di Torino, Dip. di Matematica
Lucipan@polito.it

Enrico Priola
Università di Torino, Dip. di Matematica
priola@dm.unito.it

Abstract We consider the tracking problem for parabolic systems with boundary control. Assuming that the reference signal is bounded and measurable, we prove various regularity results as well representation formulas for the optimal control and the optimal trajectory.

Keywords: Regulator problem, distributed systems, boundary control

1. Introduction and Preliminaries

The quadratic regulator problem for distributed parameter systems was analyzed in the monographs [1, 5], in particular the (time and space) regularity properties of the Riccati equations, arising in boundary control of PDEs, is deeply studied in [5]. However variants of the quadratic regulator problem like the tracking and cheap control did not receive much attention in the boundary control case.

The aim of this paper is to partially fill this gap, by investigating the tracking problem. This consists in finding a control v to force the output z of a given system to follow a desired reference signal y; we refer to [3, 8]for an introduction to this problem in finite dimensions. We obtain regularity results for the optimal control v as well as useful representation formulas involving v and the optimal trajectory w, see in

*The research of both the authors was supported in part by the Italian Ministero dell'Università e della Ricerca Scientifica e Tecnologica. It fits the program of GNAMPA.

particular Theorems 8, 10 and 13. These are extensions of known finite dimensional formulas (see [3]) to the present boundary control case.

Since the presence of the reference signal y entails a lack of regularity in the solution and a different form of the optimal control, our theorems are not contained in the monograph [5]. Moreover the results which are presented here will be applied to the study of the cheap control problem in a forthcoming paper.

The system that we consider is described by

$$\dot{w}(t) = Aw(t) + Bv(t)\,, \quad t \in (0,T), \qquad w(0) = w_0, \tag{1}$$

where A generates an exponentially stable holomorphic semigroup on a Hilbert space X and $v \in L^2(0,T;X)$. Exponential stability is assumed only in order to simplify the notations. The operator B takes values in $(\text{dom}A^*)'$. Here A^* denotes the adjoint of A and $(\text{dom}A^*)'$ stands for the topological dual of $\text{dom}A^*$ (the space $\text{dom}A^*$ is endowed with the graph norm); see [5]for more details as well as for several applications of (1). We assume that for some $\gamma \in [0,1)$,

$$D = (-A)^{-\gamma}B \in \mathcal{L}(U,X), \tag{2}$$

i.e. D is a bounded linear operator from U to X, where U is a second Hilbert space (if A is not stable then the notation $(-A)^\gamma$ is to be replaced by $(-A-rI)^\gamma$ with r large enough). Note that (2) is equivalent to $B \in \mathcal{L}(U,[\text{dom}(-A^*)^\gamma]')$ and implies that, for some σ and $M>0$,

$$||B^*e^{A^*t}|| \le \frac{Me^{-\sigma t}}{t^\gamma}, \quad t>0\,. \tag{3}$$

The tracking problem is the following:

$$\min_v J_\alpha(w_0;v)\,, \qquad J_\alpha(w_0;v) = \int_0^T \{||z(t)-y(t)||^2 + \alpha||v(t)||^2\}\,dt\,,$$
$$z(t) = z(t;w_0,v) = Cw(t;w_0,v), \tag{4}$$

where $\alpha > 0$ is fixed, y is a prescribed reference signal and $w(t;w_0,v)$ denotes the solution to (1); further C is a linear and bounded operator from X to a third Hilbert space Y. The cheap control problem consisting in studying the limit for $\alpha \to 0^+$ will be studied in the sequel. The standing assumption on y is that it is measurable and bounded, i.e. $y \in L^\infty(0,T;Y)$.

Let us introduce the following operators

$$Lv(t) = \int_0^t e^{A(t-s)}Bv(s)ds\,, \qquad \Lambda = CL$$

$$L^*w(t) = B^*\int_t^T e^{A^*(s-t)}w(s)ds\,, \quad \Lambda^* = L^*C^*; \quad \Gamma w_0(t) = Ce^{At}w_0\,,$$

$w_0 \in X$. The properties of these operators have been precisely studied in [5]. Moreover we recall from [5, p. 13 and p. 23]:

THEOREM 1 *Let* $\gamma \in [0,1)$ *as in (2). We have:*

$$L \in \mathcal{L}(L^2(0,T;U), L^2(0,T;\mathrm{dom}(-A)^{1-\gamma}))$$

and $L \in \mathcal{L}(L^\infty(0,T;U), C([0,T];\mathrm{dom}(-A)^\theta))$ *for every* $0 \le \theta < 1-\gamma$

We observe that we are in the second smoothing case studied in [5]so that we can freely use all the regularity results in that book, which concern the solutions of the Riccati equation and the operators L and Λ. Using a result in [6], we can improve Lemma 1 as follows (as usual, C^σ denotes the space of Hölder continuous functions):

THEOREM 2 *For any* $\theta \in [0, 1-\gamma)$, *we have*

$$L \in \mathcal{L}(L^\infty(0,T;U), C^{1-\gamma-\theta}([0,T];\mathrm{dom}(-A)^\theta).$$

In particular $L \in \mathcal{L}(L^\infty(0,T;U), C^{1-\gamma}([0,T];X)$.

Proof We write:

$$Lv(t) = \int_0^t e^{A(t-s)} Bv(s)ds = (-A)^\gamma \int_0^t e^{A(t-s)} Dv(s)ds,$$

see (2). Now recall that in Proposition 4.2.2 in [6], see also Sec. 2.2.2 in [6], it is proved that the operator R,

$$Rf(t) = \int_0^t e^{A(t-s)} f(s)ds$$

belongs to $\mathcal{L}(L^\infty(0,T;X), C^{1-\alpha}([0,T];\mathrm{dom}(-A)^\alpha)$, for any $\alpha \in (0,1)$. Remark that this implies also that $R \in \mathcal{L}(L^\infty(0,T;X), C^{1-\alpha}([0,T];X))$, for any $\alpha \in (0,1)$. Applying this result to L we get the assertion.

2. The Tracking Problem

Let us consider (4). The existence of the optimal control v_α is clear,

$$v_\alpha = (\alpha + \Lambda^*\Lambda)^{-1}\Lambda^*(y - \Gamma w_0)\,. \tag{5}$$

Let w_α be the state produced by v_α, i.e. the solution of Eq. (1) when $v = v_\alpha$. Let moreover $z_\alpha = Cw_\alpha = \Lambda v_\alpha + \Gamma w_0$ be the corresponding output. We easily obtain from (5) a second representation formula for the optimal control:

$$v_\alpha = \frac{1}{\alpha}\Lambda^*[y - z_\alpha]\,. \tag{6}$$

Using Young inequalities, we see that $\Lambda^*(y - \Gamma w_0)$ is continuous on $[0, T]$ and $||\Lambda^*(y - \Gamma w_0)(t)|| \leq M(T-t)^{1-\gamma}$. Moreover from [5, Theorem 1.4.4.4]$(\alpha + \Lambda^*\Lambda)^{-1}$ is boundedly invertible on $C_\gamma(0, T; U)$, where

$$C_\gamma(0,T;U) = \left\{ f \in C[0,T); U \text{ uch that } \sup_{t\in(0,T)} (T-t)^\gamma |f(t)| < +\infty \right\}.$$

Hence we have that v_α is continuous on $[0, T)$. In addition the usual bootstrap argument, based on Young inequalities, shows:

THEOREM 3 *The optimal control v_α is continuous on $[0, T]$. Hence, also $w_\alpha(t)$ and $z_\alpha(t)$ are continuous too. Moreover, $||v_\alpha(t)|| = O(T-t)^{1-\gamma}$.*

Combining Lemmas 2 and 3 we obtain:

THEOREM 4 *The function w_α is Hölder continuous on every compact interval contained in $(0, T]$ with values in X. The Hölder exponent is $1 - \gamma$.*

Proof From

$$w_\alpha(t) = e^{At} w_0 + \int_0^t e^{A(t-s)} B v_\alpha(s) ds\,,$$

we need only to prove Hölder continuity of the integral (since the first addendum is continuously differentiable for $t > 0$, because e^{At} is a holomorphic semigroup). To this end it is enough to apply Lemma 2.

In the next result, we give two representations of the minimum value of the cost.

THEOREM 5 *We have:*

$$J_\alpha(w_0; v_\alpha) = \langle y - \Gamma w_0, y - z_\alpha\rangle = \langle y - \Gamma w_0, \alpha(\alpha + \Lambda\Lambda^*)^{-1}(y - \Gamma w_0)\rangle\,.$$

Proof In the following computation, $\tilde{y} = y - \Gamma w_0$ and norm and inner product are in L^2. We note that

$$\begin{aligned}
&||\Lambda v_\alpha - \tilde{y}||^2 + \alpha||v_\alpha||^2 = \langle \Lambda v_\alpha, \Lambda v_\alpha\rangle - 2\langle \Lambda v_\alpha, \tilde{y}\rangle + ||\tilde{y}||^2 + \alpha||v_\alpha||^2 \\
&= \langle v_\alpha, \Lambda^*\Lambda(\alpha + \Lambda^*\Lambda)^{-1}\Lambda^*\tilde{y}\rangle - 2\langle v_\alpha, \Lambda^*\tilde{y}\rangle + ||\tilde{y}||^2 + \alpha||v_\alpha||^2 \\
&= \langle v_\alpha, \Lambda^*\tilde{y} - \alpha(\alpha + \Lambda^*\Lambda)^{-1}\Lambda^*\tilde{y}\rangle - 2\langle v_\alpha, \Lambda^*\tilde{y}\rangle + ||\tilde{y}||^2 + \alpha||v_\alpha||^2 \\
&= -\langle v_\alpha, \Lambda^*\tilde{y}\rangle + ||\tilde{y}||^2 = \langle \tilde{y}, \tilde{y} - \Lambda v_\alpha\rangle
\end{aligned}$$

since

$$\tilde{y} - \Lambda v_\alpha = y - z_\alpha\,.$$

This is the first representation. The second representation is obtained from here, since

$$\Lambda(\alpha + \Lambda^*\Lambda)^{-1}\Lambda^* = (\alpha + \Lambda\Lambda^*)^{-1}\Lambda\Lambda^*\,.$$

The proof is complete.

We introduce now the explicit form of (6):

$$v_\alpha(t) = \frac{1}{\alpha}\int_t^T B^* e^{A^*(s-t)} C^*[y(s) - z_\alpha(s)]ds = -\frac{1}{\alpha}B^* p_\alpha(t) \qquad (7)$$

where

$$p_\alpha(t) = -\int_t^T e^{A^*(s-t)} C^*[y(s) - z_\alpha(s)]ds\,. \qquad (8)$$

We note that $p_\alpha \in C([0,T];X)$, since C is a bounded operator and

$$||p(t)|| \leq M(T-t)^{(1-\gamma)}\,.$$

For $t = 0$ we have the equality

$$p_\alpha(0) = -\int_0^T e^{A^* s} C^*[y(s) - z_\alpha(s)]ds = -\Gamma^*(y - z_\alpha)$$

so that we find a third representation for the optimal cost,

$$J_\alpha(w_0; v_\alpha) = \langle y, y - z_\alpha\rangle_{L^2(0,T;Y)} + \langle w_0, p_\alpha(0)\rangle_X\,.$$

The function p_α is the weak solution of

$$\dot{p} = -A^* p - C^*[Cw_\alpha - y]\,, \qquad p(T) = 0\,. \qquad (9)$$

We have the condition $p(T) = 0$ since the final value of w is not penalized. In this way we arrive at the usual hamiltonian system

$$\begin{cases} \dot{w} = Aw - \frac{1}{\alpha}BB^* p & w(0) = w_0, \\ \dot{p} = -C^*Cw - A^*p + C^*y & p(T) = 0\,. \end{cases} \qquad (10)$$

The functions p_α and w_α solve (10) in a weak sense. We improve the regularity of p_α in the next Lemma.

THEOREM 6 *We have* $p_\alpha \in C^{1-\theta}([0,T]; \mathrm{dom}(-A^*)^\theta)$, *for every* $0 < \theta < 1$, *and* $p_\alpha \in L^2(0,T; \mathrm{dom}\, A^*)$.

Proof The first assertion follows from (8), taking into account that the function $C^*[Cw_\alpha - y]$ is bounded and applying Proposition 4.2.2 in [6]. The second statement can be proved as in [4], see also [5, p. 4].

Let us consider now the special but important case when y is Hölder continuous. Using [6, Theorem 4.3.4]and [7, Theorem 3.5.], we get:

THEOREM 7 *If there exists* $\eta \in (0,1)$, $\eta \leq 1-\gamma$, *s.t.* $y \in C^\eta([\epsilon,T];Y)$ *for every* $\epsilon > 0$, *then* p_α *is continuously differentiable, and the derivative is Hölder continuous too, i.e.,* $p_\alpha \in C^{1+\eta}([\epsilon,T],X)$, *for every* $\epsilon > 0$.

In the next result we consider the regularity of the map $B^*p_\alpha = -\alpha v_\alpha$.

THEOREM 8 *Let $\eta \in (0,1)$ and $y \in C^\eta([\epsilon,T];Y)$, for any $\epsilon > 0$. The functions B^*p_α (and so also v_α) is Hölder continuous on compact intervals contained in $(0,T]$, of exponent $\min\{\eta, 1-\gamma\}$.*

Proof Let $f = -C^*[Cw_\alpha - y]$. The function f is bounded. Now fix $\epsilon > 0$ and take $t'' > t' \geq \epsilon$. We obtain

$$[p_\alpha(t'') - p_\alpha(t')] = \int_0^{T-t''} e^{A^*s} f(s+t'')ds - \int_0^{T-t'} e^{A^*s} f(s+t')ds\,.$$

Hence,

$$[B^*p(t'') - B^*p(t')] = B^* \int_{T-t''}^{T-t'} e^{A^*s} f(s+t')ds$$
$$+ B^* \int_0^{T-t''} e^{A^*s}[f(s+t'') - f(s+t')]ds\,.$$

The function f is bounded so that the first addendum is less then

$$\int_{T-t''}^{T-t'} \frac{M_0}{s^\gamma} ds \leq M_1 \left\{(T-t')^{1-\gamma} - (T-t'')^{1-\gamma}\right\} \leq M(t''-t')^{(1-\gamma)}\,.$$

The second integral is the sum of the following two terms:

$$B^* \int_0^{T-t''} e^{A^*s} C^* \left[y(s+t') - y(s+t'')\right] ds, \tag{11}$$

$$B^* \int_0^{T-t''} e^{A^*s} C^*C \left[w_\alpha(s+t'') - w_\alpha(s+t')\right] ds\,. \tag{12}$$

Hölder continuity of y and condition (3) imply that the norm of (11) is less then $M_\epsilon[t''-t']^\eta$. The second integral is treated analogously, and we get a similar estimate, with exponent $1-\gamma$ on every interval $[\epsilon, T]$, $\epsilon > 0$, see Lemma 4.

A further regularity result that is needed below is as follows:

THEOREM 9 *Let $x_0 \in \operatorname{dom} A^*$. The function:*

$$t \longrightarrow \langle x_0, \int_0^t e^{A(t-s)} Bv_\alpha(s)ds\rangle = \langle (-A^*)^\gamma x_0, \int_0^t e^{A(t-s)} Dv_\alpha(s)ds\rangle,$$

is differentiable on $[0,T]$ and moreover

$$\frac{d}{dt}\langle x_0, \int_0^t e^{A(t-s)} Bv_\alpha(s)ds\rangle$$
$$= \langle (-A^*)^\gamma x_0, Dv_\alpha(t)\rangle + \langle A^* x_0, \int_0^t e^{A(t-r)} Bv_\alpha(r)dr\rangle\,.$$

Proof First recall that the function $t \to Dv_\alpha(t)$ is continuous on $[0,T]$. Then let $\delta > 0$; because the semigroup is holomorphic, we have

$$\frac{d}{dt}\langle x_0, \int_0^{t-\delta} e^{A(t-s)} Bv_\alpha(s)ds\rangle = \frac{d}{dt}\langle (-A^*)^\gamma x_0, \int_0^{t-\delta} e^{A(t-s)} Dv_\alpha(s)ds\rangle$$
$$= \langle (-A^*)^\gamma x_0, e^{A\delta} Dv_\alpha(t-\delta)\rangle + \langle (-A^*)^\gamma x_0, A\int_0^{t-\delta} e^{A(t-s)} Dv_\alpha(s)ds\rangle$$
$$= \langle (-A^*)^\gamma x_0, e^{A\delta} Dv_\alpha(t-\delta)\rangle + \langle A^* x_0, (-A)^\gamma \int_0^{t-\delta} e^{A(t-s)} Dv_\alpha(s)ds\rangle\,.$$

We see from here that

$$\left\|\frac{d}{dt}\langle x_0, \int_0^{t-\delta} e^{A(t-s)} Bv_\alpha(s)ds\rangle\right\| \le M\|v_\alpha\|_\infty \left\{\|(-A^*)^\gamma x_0\| + \|A^* x_0\|\right\}\,.$$

Thanks to this estimate, we can use dominated convergence theorem and we can pass to the limit for $\delta \to 0^+$ in the following equality:

$$\langle (-A^*)^\gamma x_0, \int_0^{t-\delta} e^{A(t-s)} Dv_\alpha(s)ds\rangle = \int_\delta^{t-\delta} \left\{\langle (-A^*)^\gamma x_0, e^{A\delta} Dv_\alpha(s-\delta)\rangle\right.$$
$$\left. + \langle A^* x_0, \int_0^{s-\delta} e^{A(s-r)} Bv_\alpha(r)dr\rangle\right\} ds\,.$$

We differentiate both sides of the resulting equality and we get:

$$\frac{d}{dt}\langle x_0, \int_0^t e^{A(t-s)} Bv_\alpha(s)ds\rangle$$
$$= \langle (-A^*)^\gamma x_0, Dv_\alpha(t)\rangle + \langle A^* x_0, \int_0^t e^{A(t-r)} Bv_\alpha(r)dr\rangle\,.$$

The proof is complete.

If $y = 0$ it is well known that the optimal control can be put in feedback form. This is not possible if $y \neq 0$ since at a given time t the future values of y, which affect the optimal control, are unknown. Infact we have:

THEOREM 10 *We have $p_\alpha(t) = P_\alpha(t)w_\alpha(t) + d_\alpha(t)$ where P_α solves the Riccati differential equation*

$$\langle \frac{d}{dt} P_\alpha(t)x, y\rangle = -\langle P_\alpha(t)Ax, y\rangle - \langle P_\alpha(t)x, Ay\rangle \tag{13}$$
$$-\langle Cx, Cy\rangle + \frac{1}{\alpha}\langle B^* P_\alpha(t)x, B^* P_\alpha(t)y\rangle\,, \qquad P_\alpha(T) = 0\,.$$

Here x and y are arbitrary elements in $\operatorname{dom} A$. *The function d_α is continuous on $[0,T]$ and is zero for $t=T$. It depends on y but not on w_0 and it is given by (here $(\Lambda_t u)(r) = \int_t^r Ce^{A(r-s)}Bu(s)ds$)*

$$d_\alpha(t) = -\int_t^T e^{A^*(s-t)}C^*y(s)ds$$
$$+\int_t^T e^{A^*(s-t)}C^*C\int_t^s e^{A(s-r)}B[(\alpha+\Lambda_t\Lambda_t^*)^{-1}\Lambda_t^*y](r)dr\,ds\,.$$

Proof The operator valued function $P_\alpha(t)$ is the solution of the usual Riccati equation. For each $x \in X$, $P_\alpha(\cdot)x$ is continuous on $[0,T]$ and it is zero for $t=T$. Moreover, it is differentiable on $(0,T)$, with continuous and bounded derivative in closed subintervals, see Theorem [5, p. 19-20]. In order to prove the theorem, it is sufficient to show that the continuous function $d_\alpha(t) = p_\alpha(t) - P_\alpha(t)w_\alpha(t)$ only depends on the tracking signal y. We introduce the functions $v^+(s;t,x_0)$ and $w^+(s;t,x_0)$, $x_0 \in X$, the solutions of the optimization problem under study, in the case that $y=0$ and with initial condition x_0 at time t instead then 0. Hence,$v^+ = -(\alpha+\Lambda_t^*\Lambda_t)^{-1}\Lambda_t^*\Gamma_t x_0$, where $\Gamma_t x_0(s) = Ce^{(s-t)A}x_0$ (note that v^+ depends on α). Moreover, we use the following representation formula for $P_\alpha(t)$:

$$P_\alpha(t)x = \int_t^T e^{A^*(r-t)}C^*Cw^+(r;t,x)dr.$$

Hence,

$$p_\alpha(t) - P_\alpha(t)w_\alpha(t) = -\int_t^T e^{A^*(s-t)}C^*y(s)ds$$
$$+\int_t^T e^{A^*(s-t)}C^*C[w_\alpha(s;0,w_0) - w^+(s;t,w_\alpha(t;0,w_0))]ds\,.$$

For clarity, in this formula we indicated explicitly the initial time and initial value of w_α. Now we use dynamic programming: the optimal control on $[t,T]$, with initial condition $w_\alpha(t;0,w_0)$ is the restriction to $[t,T]$ of v_α. This holds for every given reference signal y. Hence,

$$\begin{aligned}
&w_\alpha(s;0,w_0) - w^+(s;t,w_\alpha(t;0,w_0))\\
&= w_\alpha(s;t,w_\alpha(t;0,w_0)) - w^+(s;t,w_\alpha(t;0,w_0))\\
&= \left(e^{A(s-t)}[w_\alpha(t;0,w_0) - w_\alpha(t;0,w_0)]\right)\\
&\quad + \int_t^s e^{A(s-r)}B[v_\alpha(r) - v^+(r)]dr
\end{aligned}$$

$$= \int_t^s e^{A(s-r)} B \left\{ (\alpha + \Lambda_t^* \Lambda_t)^{-1} [\Lambda_t^*(y - \Gamma_t w_0) + \Lambda_t^*(\Gamma_t w_0)] \right\} (r) dr$$

$$= \int_t^s e^{A(s-r)} B \left[(\alpha + \Lambda_t^* \Lambda_t)^{-1} \Lambda_t^* y \right] (r) dr ,$$

as wanted.

In fact, in finite dimensions d_α solves

$$\dot{d}_\alpha(t) = -\left(A^* - \frac{1}{\alpha} P_\alpha(t) B B^* \right) d_\alpha(t) + C^* y(t) , \qquad d_\alpha(T) = 0 . \quad (14)$$

We will show that an analogous result holds in general. We prove first:

THEOREM 11 *We have, for every* $x \in \text{dom} A$,

$$\frac{d}{dt} \langle x, d_\alpha(t) \rangle \quad (15)$$
$$= \langle -Ax, d_\alpha(t) \rangle + \frac{1}{\alpha} \langle B^* P_\alpha(t) x, B^* d_\alpha(t) \rangle + \langle x, C^* y(t) \rangle , \quad d_\alpha(T) = 0 .$$

Proof We already know the continuity of d_α and that $d_\alpha(T) = 0$. Moreover, from [5, p. 21, formula 1.2.2.19], $B^* P_\alpha \in \mathcal{L}(X; C([0,T]; U))$.

Recall that $d_\alpha(t) = p_\alpha(t) - P_\alpha(t) w_\alpha(t)$. First let us treat p_α. Using the Hamilton equation (10), we see that

$$\frac{d}{dt} \langle x, p_\alpha(t) \rangle = -\langle C^* C x, w_\alpha(t) \rangle - \langle Ax, p_\alpha(t) \rangle + \langle Cx, y(t) \rangle . \quad (16)$$

Now we consider differentiability of $P_\alpha(t)$. We use formula [5, (1.2.2.14)] and Lemma 9 in order to compute

$$\frac{d}{dt} \langle x, P_\alpha(t) w_\alpha(t) \rangle = \frac{d}{dr} \langle x, P_\alpha(r) w_\alpha(t) \rangle_{|r=t} + \frac{d}{dr} \langle x, P_\alpha(t) w_\alpha(r) \rangle_{|r=t} .$$

The first addendum is

$$\langle -A^* P_\alpha(t) x - P_\alpha(t) A x - C^* C x + \frac{1}{\alpha} P_\alpha(t) B B^* P_\alpha(t) x, w_\alpha(t) \rangle . \quad (17)$$

The second addendum is computed from Lemma 9 (recall that $P_\alpha(t) x \in \text{dom} A^*$ since $x \in \text{dom} A$, see [5, property vii), p. 20]). We get

$$\langle (-A^*)^\gamma P_\alpha(t) x, D v_\alpha(t) \rangle + \langle A^* P_\alpha(t) x, w_\alpha(t) \rangle \quad (18)$$
$$= -\frac{1}{\alpha} \langle B^* P_\alpha(t) x, B^* p_\alpha(t) \rangle + \langle A^* P_\alpha(t) x, w_\alpha(t) \rangle .$$

Now we subtract (17) and (18) from (16). The result follows.

Now we improve our information on the regularity of $d_\alpha(t)$:

THEOREM 12 *We have* $d_\alpha \in C^{1-\theta}([0,T];\mathrm{dom}(-A^*)^\theta)$, *for every* $0 < \theta < 1$, *and* $d_\alpha \in L^2(0,T;\mathrm{dom}\,A^*)$.

Proof We derive an integral representation formula for d_α, which displays the desired regularity properties. For any $s \in [0,T]$, set $[P_\alpha(s)B] = [B^*P_\alpha(s)]^*$. From [5, p. 21]it follows that $P_\alpha B$ is linear and continuous from U to $C([0,T];X)$. Moreover, $B^*d_\alpha(s) = B^*p_\alpha(s) - B^*P_\alpha(s)w_\alpha(s)$ is well defined and continuous, the first addendum from Lemma 3 and the second one from the continuity of $w_\alpha(s)$ and of $B^*P_\alpha(s)$. For the same reason, $s \to [P_\alpha(s)B]B^*d_\alpha(s)$ is continuous so that, from (15), the following representation formula holds:

$$\begin{aligned} d_\alpha(t) &= -\int_t^T e^{A^*(s-t)}\left\{\frac{1}{\alpha}[P_\alpha(s)B]B^*d_\alpha(s) + C^*y(s)\right\} ds \\ &= \int_t^T e^{A^*(s-t)} f(s)ds, \end{aligned}$$

where $f(s)$ is bounded on $[0,T]$ with values in X. Now we conclude as in Lemma 6.

We are going to prove a variation of constants formula for $d_\alpha(t)$. Namely, we want to prove

THEOREM 13 *The function* $d_\alpha(t)$ *is given by*

$$d_\alpha(t) = -\int_t^T U(T-t,T-s)C^*y(s)ds$$

where $U(t,s)$ *is an evolution operator which is exponentially bounded, strongly continuous and which transforms* X *into* $domB^* = dom(-A^*)^\gamma$, *for a.e.* $t > s$.

In order to reduce the notation to a more usual form, it is convenient to replace $\xi(t) = d_\alpha(T-t)$. A simple transformation shows that $\xi(t)$ solves

$$\xi(t) = \int_0^t e^{A^*(t-r)}\left\{[\tilde{P}(r)B]B^*\xi(r) - C^*\tilde{y}(r)\right\} dr$$

where $\tilde{P}(r) = -\frac{1}{\alpha}P_\alpha(T-r)$, $\tilde{y}(r) = y(T-r)$. To prove the previous theorem we need the next result.

THEOREM 14 *There exists a unique strongly continuous and exponentially bounded evolution family* $U(t,s)$ *which, for* $t > s$, *is defined by*

$$U(t,s)x = e^{A^*(t-s)}x + \int_s^t U(t,r)[\tilde{P}(r)B]B^*e^{A^*(r-s)}xdr\,,$$

for every $x \in X$. Moreover, for a.e. $t > s$, we have $U(t,s)X \subseteq \operatorname{dom} B^*$ and $[\tilde{P}(\cdot)B]B^*U(\cdot,s)x$ is locally integrable on $[s,+\infty)$, for every $x \in X$. The evolution family $U(t,s)$ verifies, for $t \geq s$ and $x \in X$,

$$U(t,s)x = e^{A^*(t-s)}x + \int_s^t e^{A^*(t-r)}[\tilde{P}(r)B]B^*U(r,s)x dr\,. \tag{19}$$

Proof We use Theorem 9.19, p. 487, in [2]. Let $U_0(t,s) = e^{A^*(t-s)}$ and let $\mathcal{B}(t) = [\tilde{P}(t)B]B^*$. Then, $\operatorname{dom}\mathcal{B}(t) = \operatorname{dom} B^*$ is constant and we have $\mathcal{B}(\cdot)U_0(\cdot,s)$ strongly continuous for $t > s$, with

$$||\mathcal{B}(t)U_0(t-s)|| = ||\,[\tilde{P}(t)B]B^*e^{A^*(t-s)}|| \leq \frac{M}{(t-s)^\gamma}\,,$$

locally integrable, from (3). The conclusion follows from this.
Proof of Theorem 13. We introduce

$$\tilde{\xi}(t) = -\int_0^t U(t,s)C^*\tilde{y}(s)ds\,.$$

We are going to prove that $\xi(t) = \tilde{\xi}(t)$. We see from (19) that $s \to B^*U(t,s)x$ is integrable on $[0,t]$, for any $x \in X$. Since B^* is closed, it is straightforward to check, by using suitable Riemann sums, that

$$B^*\tilde{\xi}(t) = -\int_0^t B^*U(t,s)C^*\tilde{y}(s)ds\,.$$

Moreover, we see from (19)

$$\tilde{\xi}(t) = -\int_0^t e^{A^*(t-s)}C^*\tilde{y}(s)ds + \int_0^t e^{A^*(t-s)}[\tilde{P}(s)B]B^*\tilde{\xi}(r)dr\,.$$

Hence, $\xi(t)$ and $\tilde{\xi}(t)$ solve the same Volterra integral equation, and

$$[\xi(t) - \tilde{\xi}(t)] = \int_0^t e^{A^*(t-r)}[\tilde{P}(r)B]B^*[\xi(r) - \tilde{\xi}(r)]dr \tag{20}$$

so that also

$$(B^*[\xi(t) - \tilde{\xi}(t)]) = \int_0^t B^*e^{A^*(t-r)}[\tilde{P}(r)B](B^*[\xi(r) - \tilde{\xi}(r)])dr\,.$$

Thanks to the inequality (3), Young inequalities and continuity of

$$[\tilde{P}(r)B]$$

the operator on $L^2(0,\tilde{T})$ defined by

$$\phi(\cdot) \longrightarrow \int_0^t B^* e^{A^*(t-r)}[\tilde{P}(r)B]\phi(r)dr$$

has norm less then MT^γ for a suitable number M; so it is a contraction on $L^2(0,T_1)$, $T_1 < (1/M)^{1/\gamma} = T_1$. Hence $B^*[\xi(t)-\tilde{\xi}(t)]$ is zero on $[0,T_1]$ and, for $t > T_1$,

$$(B^*[\xi(t) - \tilde{\xi}(t)]) = \int_{T_1}^t B^* e^{A^*(t-r)}\tilde{[}P(r)B](B^*[\xi(r) - \tilde{\xi}(r)])dr\,.$$

The same argument shows that $(B^*[\xi(t) - \tilde{\xi}(t)])$ is zero on $[T_1, 2T_1]$ too. In fact, it is easily seen that the norm of the operator

$$\phi(\cdot) \longrightarrow \int_{T_1}^t e^{A^*(t-r)}[\tilde{P}(r)B]\phi(r)dr$$

from $L^2(T_1, 2T_1)$ in itself, is less then MT_1^γ, with the same coefficient M as above. After a finite number of steps we see that $B^*[\xi(t) - \tilde{\xi}(t)]$ is zero on $[0,T]$ so that, from (20) we have $\xi(t) = \tilde{\xi}(t)$ on $[0,T]$. This finishes the proof.

References

[1] A. Bensoussan, G. Da Prato, M. C. Delfour, and S. K. Mitter. *Representation and control of infinite-dimensional systems.* Birkhäuser, Boston, 1992.

[2] K-J. Engel and R. Nagel. *One-parameter semigroups for linear evolution systems.* Springer-Verlag, Berlin, 2000.

[3] J. M. Grimble and M. A. Johnson. *Optimal control and stochastic estimation: theory and applications.* John Wiley & Sons, Chichester, 1988.

[4] I. Lasiecka, L. Pandolfi, and R. Triggiani. A singular control approach to highly damped second-order abstract equations and applications. *Applied Mathematics & Optimization*, 36:67–107, 1997.

[5] I. Lasiescka and R. Triggiani. *Control theory for partial differential equations: continuous and approximation theories: Abstract parabolic systems.* Cambridge University Press, cambridge, 2000.

[6] A. Lunardi. *Analytic Semigroups and Optimal Regularity in Parabolic Problems.* Birkhäuser, Basel, 1995.

[7] A. Pazy. *Semigroups of linear operators and applications to partial differential equations.* Springre-Verlag, Berlin, 1983.

[8] E. D. Sontag. *Mathematical Control Theory.* Springer-Verlag, Berlin, 1990.

MODELING OF TOPOLOGY VARIATIONS IN ELASTICITY

Serguei A. Nazarov*
Institute of Mechanical Engineering Problems, Laboratory of Mathematical Methods, Russian Academy of Sciences, V.O. Bol'shoi 61, 199178 St. Petersburg, Russia
serna@snark.ipme.ru

Jan Sokolowski†
Institut Elie Cartan, Laboratoire de Mathematiques, Université Henri Poincaré Nancy I, BP 239, 54506 Vandoeuvre-Les-Nancy Cedex, France
Jan.Sokolowski@iecn.u-nancy.fr

Abstract Two approaches are proposed for the modeling of deformation of elastic solids with small geometrical defects. The first approach is based on the theory of self adjoint extensions of differential operators. In the second approach function spaces with separated asymptotics and point asymptotic conditions are introduced, and the variational formulation is established. For both approaches the accuracy estimates are derived. Finally, the spectral problems are considered and the error estimates for eigenvalues are given.

Keywords: Shape optimization, topology optimization, asymptotic analysis, elliptic operators, singular perturbations

Introduction

It seems that in the literature on shape optimisation there is a lack of general numerical method or technique, beside the level set method, that can be applied in the process of optimisation of an arbitrary shape functional (SF) for simultaneous boundary and topology variations. In the paper [22] (see also [19]) the so–called topological derivative (TD) of an arbitrary SF is introduced. TD usually determines whether a

*Funding provided by grant from Institut franco-russe A.M.Liapunov d'informatique et de mathématiques appliquées
†Partially supported by the grant 4 T11A 01524 of the State Committee for the Scientific Research of the Republic of Poland

change of topology by nucleation of a small hole, or in similar setting of a small inclusion at a given point $x \in \Omega$, would result in improving the value $J(\Omega)$ of a given SF or not. In the paper the *boundary topology variations* are considered for boundary value problems for elastic solids. The singular perturbations of the geometrical domain Ω are defined by small arcs $\gamma_h^1, ..., \gamma_h^I$ of the length $O(h)$ on the boundary $\partial\Omega$.

We propose two efficient approaches to the modeling of topological variations. First approach is developped in the framework of the self-adjoint extensions of differential operators, the second uses the function spaces with the detached asymptotics. In both cases, the main idea consists in modeling of small defects or inhomogeneities by concentrated *actions*, the so-called potentials of zero-radii. In this way the solution $u(\varepsilon, h)$ with *singular* behaviour for $\varepsilon \to 0+$ is replaced by a function with the singularities at the centres $P^1, ..., P^I$ of the defects. The modern framework of analysis of elliptic boundary value problems in non smooth domains allows for the the relatively complete theory of singular solutions and provides the techniques of derivation of error estimates for asymptotic approximations. We can use the known results in this field for the solution of shape and topology optimization problems in an *inverse order*. First, the localization and integral atributes of openings are determined, followed by the appropriate changes of the topology of geometrical domains. The proposed two different approaches to topology optimization have some positive features. The first approach deals with selfadjoint operators, so can be readily extended to the evolution boundary value problems. The second approach, based on the *generalized Green's formulae*, results in the variational problem formulation with the solution given by a stationary point of an auxiliary functional close in its form to the energy functional.

1. Problem Formulation

Let us consider the deformations of plane heterogeneous anisotropic elastic body $\Omega \subset \mathbb{R}^2$ clamped on small parts of the boundary $\Gamma = \partial\Omega$ in the form of closed connected curves $\gamma_h^1, ..., \gamma_h^I$. Instead of tensor notation, we make use of the matrix notation which we describe briefly. The constitutive relations in the elasticity theory are written with the elastic fields in the form of columns. First, two matrices are introduced,

$$D(x)^\top = \begin{bmatrix} x_1 & 0 & \alpha x_2 \\ 0 & x_2 & \alpha x_1 \end{bmatrix}, \qquad d(x)^\top = \begin{bmatrix} 1 & 0 & -\alpha x_2 \\ 0 & 1 & \alpha x_1 \end{bmatrix}, \tag{1}$$

where $\alpha = 2^{-1/2}$ is the normalizing coefficients, and $\top$ stands for transposition. The first matrix is used to define the column of strains from the displacement column $u = (u_1, u_2)^\top$,

$$\varepsilon(u) := (\varepsilon_{11}(u), \varepsilon_{22}(u), \alpha^{-1}\varepsilon_{12}(u))^\top = D(\nabla)u \ . \tag{2}$$

Here $\nabla = (\frac{\partial}{\partial x_1}, \frac{\partial}{\partial x_2})$ is the gradient, and $\varepsilon_{jk}(u)$ are Cartesian components of the strain tensor (the multipliers α in (1) and (2) make the norms of vector and tensor of strains equal). The second matrix generates the rigid motions $d(x)b$ of the body Ω for any column $b \in \mathbb{R}^3$. The Hooke's law

$$\sigma(u; x) = A(x)\varepsilon(u; x) \ , \tag{3}$$

represents the column of stresses in function of the strains (2), and includes the symmetric and positively definite (3×3)-matrix function A of elastic moduli, which is supposed to be smooth function of the variable x. In view of (1)-(3), the equilibrium equations and the boundary conditions of traction free type are given as follows

$$L(x, \nabla)u(h, x) := D(-\nabla)^\top A(x)D(\nabla)u(h, x) = f(x) \ , \quad x \in \Omega \ , \tag{4}$$

$$B(x, \nabla)u(h, x) := D(n(x))^\top A(x)D(\nabla)u(h, x) = 0 \ , \quad x \in \Sigma_h \ . \tag{5}$$

Here, $n = (n_1, n_2)^\top$ is the unit column of external normal vector to the contour Γ, the contour is supposed to be sufficiently smooth for the sake of simplicity of the presentation. In (5)

$$\Sigma_h = \Gamma \setminus \{\gamma_h^1 \cup ... \cup \gamma_h^I\} \ ,$$

and γ_h^j are arcs of the length hl_j, with the centres $P^j \in \partial\Omega$, where $h \in (0, h_0]$ is a small parameter and $l_1, ..., l_I$ are fixed constants. The elastic body is clamped on the sets γ_h^j,

$$u(h, x) = 0 \ , \quad x \in \gamma_h^1 \cup ... \cup \gamma_h^I \ . \tag{6}$$

The Dirichlet condition (6) provides the Korn inequality

$$\|u \ ; \ H^1(\Omega)\| \le K(h)\|D(\nabla)u \ ; \ L_2(\Omega)\| \ , \tag{7}$$

however the dependence of the multiplier $K(h)$ on the parameter h is to be clarified.

Proposition 1 *Let $I \ge 2$. For any field $u \in H^1(\Omega)^2$ verifying the Dirichlet conditions* (6)*, the Korn inequality holds with the multiplier $K(h)$ such that*

$$K(h) \le c|\ln h| \ . \tag{8}$$

The estimate is asymptotically exact. In (8) *the constant c is independent of u and $h \in (0, h_0]$ with $h_0 < 1$.*

Note that in the case $I = 1$ the Korn inequality (7) is still valid, but the multiplier $K(h)$ becomes of order h^{-1} (cf. [21]) and thus, it does not satisfy estimate (8).

2. Modeling of Singularly Perturbed Boundary Value Problem

We consider the functional

$$\mathcal{F}(u;h) = \int_{\Omega} J(x;u(h,x))dx \ . \tag{9}$$

Asymptotic structures for specific problems with the logarithmic growth of fundamental solutions, turn out to be quite complex and therefore, of limited practical interest for analysis of functional (9). The main particularity of the asymptotic analysis, beside the presence of boundary layers near the arcs $\gamma_h^1, ..., \gamma_h^I$, is the form of asymptotic terms which are rational functions of the large parameter $|\ln h|$. Such phenomenon was discovered by Il'in for the scalar problem in [6] (see also [7] and [13]). A simplification caused regarding the asymptotics with respect to the parameter $|\ln h|^{-1}$ is not sufficient to provide the analysis, since the leading terms of asymptotics do not reflect the distribution of contact regions and do not exhibit the interactions between the regions.
We propose an appproach, based on the modeling of problem (4)-(6), by means of auxiliary boundary value problems with the boundary conditions of traction free type on the punctured contour $\partial\Omega \setminus \{P^1, ..., P^I\}$ and with the prescribed class of singularities at the points $P^1, ..., P^I$. Such singularities are obtained by an application of forces concentrated at the points, and therefore, imitate the reaction of the elastic body at the obstacles $\gamma_h^1, ..., \gamma_h^I$. Thus, in the setting, the models take into account the interaction between the elastic body with the rigid foundation. On the other hand, the proposed singularities are not included in the energy class $H^1(\Omega)^2$, however the resulting singular solutions are still in the space $L_q(\Omega)$, for $q \geq 1$. The main profit from our point of view for such modeling is the possibility, with the singular solutions, for asymptotically exact approximation of functional (9), under the condition that for some $q \in [1, \infty)$ and for any $u, v \in L_q(\Omega)^2$ the following inequality is valid

$$|\mathcal{J}(u;h) - \mathcal{J}(v;h)| \leq \tag{10}$$
$$\leq c_{\mathcal{F}} \|u - v \ ; \ L_q(\Omega)\| \Big(\|u \ ; \ L_q(\Omega)\|^{q-1} + \|v \ ; \ L_q(\Omega)\|^{q-1} \Big)$$

with the constant $c_{\mathcal{F}}$ independent of $h \in (0, h_0]$ and u, v.
Modeling defects in media by an application of extensions of differential operators which give rise to the singular solutions comes back to the work [3] and is developped in [20], [16], [18], [17], [9] and in other publications, for problems of mathematical physics and general elliptic systems. There are two possibilities for realization of such ideas. First

of all, the operator $L(x, \nabla)$ in (4), considered as an unbounded operator in the space $L_2(\Omega)^2$, subsists the restriction of the domain of definition, which becomes smaller compared to intrinsic domain $H^1(\Omega)^2$. In this way the domain of the adjoint becomes wider, and finally the selfadjoint operator is selected in the form of an intermediate operator. Under the proper choice of extension parameters, the selfadjoint operator asymptotically acquires the attributes of singularly perturbed problem such as the energy functional and the spectre (see [17], [9], [19]). Since the selected operator is selfadjoint, the classical semigroup theory can be used to construct solutions for the associated evolution problems. On the other hand, for our problem, the domain of selfadjoint extension depends on the large parameter $|\ln h|$, which could leads to ill posed problems for numerical methods when applied for solution of shape optimization or shape inverse problems (see [8], [22], [23], [5]). This difficulty can be avoided by application of slightly different technique, including the space with separated asymptotics (see [18] and others). Roughly speaking, the boundary value problem is defined in larger class, compared with the energy space $H^1(\Omega)^2$. In the class, the behaviour of functions at points $P^1, ..., P^I$ is prescribed a priori. The coefficients of asymptotic expansions satisfy some additional relations, in order to ensure the unique solvability of boundary value problems. Matching conditions for parameters of selfadjoint extensions, and the relations called asymptotic point conditions, result in the exactly same solutions obtained by the first and the second approach.

3. Modeling with Self Adjoint Extensions

We denote by $\chi_1, ..., \chi_J$ the cutoff functions with mutually disjoint supports, equal to one in neighbourhoods of the points $P^1, ..., P^I$, respectively, and by $T^j = (T^{j1}, T^{j2})$ the Poisson kernel, i.e., the (2×2)-matrix function, each column $T^{jk}, k = 1, 2$ is a solution of the elasticity boundary value problem in the half-plane $\{x \ : \ n(P^j)^\top x > 0\}$ under unit force concentrated at the point P^j and directed in the positive direction of $\mathcal{O}x_k$ (linear combinations of the Boussinesq-Cerruti solutions problems).

The unbounded operator $\mathcal{L}$ in $L_2(\Omega)^2$ defined by the differential expression $L(x, \nabla)$ with the domain

$$\mathcal{D}(\mathcal{L}) = \{v \in H^2(\Omega)^2 \ : \ B(x, \nabla)v = 0 \quad \text{on } \partial\Omega \ , \qquad v(P^1) = ... = v(P^I) = 0\} \tag{11}$$

is closed and symmetric, however the adjoint $\mathcal{L}^*$ has the larger domain compared to (11),

$$\mathcal{D}(\mathcal{L}^*) = \{v \in \mathfrak{D} \ : \ B(x,\nabla)v = 0 \quad \text{on } \partial\Omega \setminus \{P^1, ..., P^I\}\} \ , \tag{12}$$

$$\mathfrak{D} = \{v(x) = \widetilde{v}(x) + \sum_{i=1}^{I} \chi_i(x) \left[b^i + T^i(x - P^i)a^i\right]$$

$$\widetilde{v} \in H^2(\Omega)^3 \ , \ \widetilde{v}^1(P^1) = ... = \widetilde{v}^I(P^I) = 0 \ , a^i, b^i \in \mathbb{R}^3\} \ .$$

The following representation is well known

$$T^j(x) = -\mathcal{T}^{j0} \ln|x| + \mathcal{T}^{j1}(|x|^{-1}x) \ , \tag{13}$$

where $\mathcal{T}^{j1}$ is a smooth matrix function on the semisphere and $\mathcal{T}^{j0}$ is a constant (2×2)-matrix, symmetric and positive definite.
By comparison of formulae (11) and (12) we can see that the defect of the operator $\mathcal{L}$ is $(3I : 3I)$. The coefficients b_1^i, b_2^i, b_3^i and b_j from (12) are collected in the column $\mathbf{f} \in \mathbb{R}^{3I}$, the remaining coefficients are collected in the column $\mathbf{a}$.
Lemma 1 *Let* $\mathbf{S}$ *be a symmetric* $(3I \times 3I)$*-matrix. The restriction* $\mathbf{L}$ *of the operator* $\mathcal{L}^*$ *to the linear subset of* (12)

$$\mathcal{D}(\mathbf{L}) = \{v \in \mathcal{D}(\mathbf{L}^*) \ : \ \mathbf{f} = \mathbf{S}\mathbf{a}\} \tag{14}$$

is a selfadjoint operator in $L_2(\Omega)^2$. *If the matrix* $\mathbf{S}$ *is not singular, then under condition* $I > 1$ *the equation*

$$\mathbf{L}\mathbf{v} = f \tag{15}$$

admits the unique solution for each $f \in L_2(\Omega)^2$.
The proper choice of parameters of selfadjoint extension, i.e., the selection of the matrix $\mathbf{S}$ is performed in Section 9.5 in such a way that the solution $\mathbf{v}$ of equation (15) becomes an approximation of the solution to problem (4)-(6).

4. Modeling in Spaces with Separated Asymptotics

The linear set (12) with the norm

$$\|v \ ; \ \mathfrak{D}\| = (\|\widetilde{v} \ ; \ H^2(\Omega)\|^2 + \|\mathbf{a} \ ; \ \mathbb{R}^{3I}\|^2 + \|\mathbf{f} \ ; \ \mathbb{R}^{3I}\|^2)^{\frac{1}{2}} \ .$$

becomes the Hilbert space. Two projection operators are introduced $\pi^{\pm} \ : \ \mathfrak{D} \to \mathbb{R}^{3I}$, which take from the function v the columns of coefficients

$$\pi^- v = \mathbf{a} \ , \quad \pi^+ v = \mathbf{f} \ .$$

Let us consider the boundary value problem of linear elasticity with the asymptotic conditions at the points $P^1, ..., P^I$,

$$\begin{aligned} L(x,\nabla)\mathfrak{v}(x) = f(x)\,, \quad x \in \Omega\,, \\ B(x,\nabla)\mathfrak{v}(x) = 0\,, \quad x \in \partial\Omega \setminus \{P^1,...,P^I\}\,, \\ S\pi^-\mathfrak{v} - \pi^+\mathfrak{v} = 0 \in \mathbb{R}^{3I}\,. \end{aligned} \tag{16}$$

It is easy to see that for the same matrix S in (14) and in (16) the solutions $\mathbf{v} \in \mathcal{D}(\mathbf{L})$ of (15) and $\mathfrak{v} \in \mathfrak{D}$ coincide.

Proposition 2

1) *For functions* $\mathfrak{v}, \mathfrak{u} \in \mathfrak{D}$ *the generalized Green's formula is valid*

$$\begin{aligned} (L\mathfrak{v},\mathfrak{u})_\Omega + (B\mathfrak{v},\mathfrak{u})_{\partial\Omega} + \langle S\pi^-\mathfrak{v} - \pi^+\mathfrak{v}, \pi^-\mathfrak{u}\rangle = \\ = (\mathfrak{v}, L\mathfrak{u})_\Omega + (\mathfrak{v}, B\mathfrak{u})_{\partial\Omega} + \langle \pi^-\mathfrak{v}, S\pi^-\mathfrak{u} - \pi^+\mathfrak{u}\rangle\,, \end{aligned} \tag{17}$$

where $(\cdot,\cdot)_\Xi$ *and* $\langle\cdot,\cdot\rangle$ *are scalar products in the spaces* $L_2(\Xi)^2$ *and* $\mathbb{R}^{3I}$*, respectively.*

2) *The function* $\mathfrak{v} \in \mathfrak{D}$ *is a solution to problem* (16) *if and only if it is a stationary point of the functional*

$$\mathfrak{E}(\mathfrak{v}) = \frac{1}{2}(L\mathfrak{v},\mathfrak{v})_\Omega + \frac{1}{2}(B\mathfrak{v},\mathfrak{v})_{\partial\Omega} + \frac{1}{2}\langle S\pi^-\mathfrak{v} - \pi^+\mathfrak{v}, \pi^-\mathfrak{v}\rangle - (f,\mathfrak{v})_\Omega\,. \tag{18}$$

If $\det S \neq 0$ *and the condition* $I \geq 1$ *is satisfied, then the stationary point of functional* (18) *is uniquely determined.*

The symmetric generalized Green's formula shows that the boundary value problem is formally selfadjoint.

The second assertion in Proposition 1 furnishes the variational formulation of problem (16) over the Hilbert space $\mathfrak{D}$, and shows the uniqueness of solutions under the same conditions as in the case of equation (15).

5. How to Determine the Model Parameters

The solution $\mathbf{v} = \mathfrak{v}$ of equation (15) or of problem (16) satisfies system (4) and boundary conditions (5), however, in general, leaves a discrepancy in the boundary conditions (6). In order to construct an approximation for the solution $u(h,x)$ in the vicinity of the points $P^1, ..., P^I$, the method of matched asymptotic expansions is applied (see [7], [11], and cf. [13], [18]). Thus, selecting for the outer asymptotic expansion $\mathbf{v} = \mathfrak{v}$, we construct the inner expansions $w^j(\xi^j)$, employing the fast variables $\xi^j = h^{-1}(x - P^j)$. The dilatation of coordinates in the limit $h \to +0$ implies the rectifying of the boundary, freezing of coefficients at the point P^j, and the volume forces vanish from the equilibrium equations. In the other words, the boundary value problem for w^j consist of the homogeneous elasticity system

$$D(-\nabla_\xi)^\top A(P^j)D(\nabla_\xi)w^j(\xi) = 0\ , \quad \xi \in \mathbb{R}^2_j\ , \tag{19}$$

the boundary conditions of traction free type

$$D(n^j)^\top A(P^j)D(\nabla_\xi)w^j(\xi) = 0\ , \quad \xi \in \partial\mathbb{R}^2_j\ , \quad |\xi| > l_j/2\ , \tag{20}$$

the Dirichlet conditions for $j = 1,..,I$

$$w^j(\xi) = 0\ , \quad \xi \in \partial\mathbb{R}^2_j\ , \quad |\xi| < l_j/2\ . \tag{21}$$

Here $n^j = n^j(P^j)$ is normal vector on $\partial\Omega$ evaluated at points P^j; $\mathbb{R}^2_j$ is the half-plane $\{\xi \in \mathbb{R}^2 \ : \ \xi^\top n^j < 0\}$.
Since we are going to glue w^j with the singular solution $\mathbf{v} = \mathfrak{v}$ with the logarithmic singularity, it is necessary to allow for the logarithmic growth of $w^j(\xi)$ for $|\xi| \to +\infty$. Such solutions of homegeneous problem (19)-(21) are well known (see [2], [1] and others). The solutions ressemble capacitary potentials in the theory of harmonic functions (see e.g., [10]), belong to the space $H^2_{\text{loc}}(\overline{\mathbb{R}}^2_j)^2$ and admit the following asymptotic representation at the infinity

$$w^j(\xi) = T^j(\xi)a^j + c^j + O(|\xi^j|^{-1})\ , \quad |\xi^j| \to +\infty\ . \tag{22}$$

The column a^j in (22) can be arbitrary, however,

$$c_j = M^j a^j\ , \tag{23}$$

where the symetric (2×2)-matrix M^j is called *Wiener elastic capacity matrix* for the half-plane clamped along the interval $[-l_j/2, l_j/2]$. When we return to the coordinates x, by comparison of representations obtained from (22)-(23) and (13)

$$\begin{aligned} &w^i(h^{-1}(x - P^i)) = \\ &= T^i(x - P^i)a^i + (\mathcal{T}^{i0}\ln h + M^i)a^i + O(h|x - P^i|^{-1})\ , \end{aligned} \tag{24}$$

with the expansion of the field $v = \mathbf{v} = \mathfrak{v}$ given in (12), the following equalities arise

$$b^i = \{\mathcal{T}^{j0}\ln h + M^i\}a^i\ , \quad i = 1, ..., I\ . \tag{25}$$

$$\tag{26}$$

which in vector notation takes the form $\mathbf{b} = \mathbf{Sa}$, used already in (14) and indirectly in (16). Thus, the matrix $\mathbf{S}$ is diagonal by blocks and contains (3×3)-matrices separated in (25) by curly braces. In view of

the properties of $\mathcal{T}^{j0}$ listed after the formula (13), the matrix $\mathbf{S}$ is symetric and negative definite for sufficiently small $h \in (0, h_0]$.
The relations (25) are derived by matching the outer expansion $\mathbf{v} = \mathfrak{v}$ with the inner expansions $w^j(\xi^j), j = 1, ..., I$. Therefore, by the Korn inequality (7), (8), proximity to the true solution $u(h, x)$ of the global asymptotic approximation in the energy norm can be established. The global asymptotic approximation is obtained by glueing of the expansions in the standard way (cf. [7] and [13], [18]). However, in view of the assumption (10) for the modeling of functional (9) the estimate for the difference $u - \mathbf{v} = u - \mathfrak{v}$ in the norm $L_q(\Omega)^2$ is required. Such an estimate can be established, taking into account the embedding $H^1(\Omega) \subset L_q(\Omega)$, by direct evaluation of the $L_q(\Omega)$-norms of the remainders in the representations (24).

THEOREM 1 *If u and $\mathbf{v} = \mathfrak{v}$ are solutions to problems* (4) -(6) *and* (15)=(16), *respectively, with the same right-hand side $f \in L_2(\Omega)^2$, then*

$$\|u - \mathbf{v}; L_q(\Omega)\| \le c_\varkappa h |\ln h|^{\varkappa + 5/2} \|f; L_2(\Omega)\| \ . \tag{27}$$

Functional (9) *admits the estimate*

$$|\mathcal{F}(u; h) - \int_\Omega J(x; \mathbf{v}(\ln h)) dx| \le C_\varkappa \mu_q(h) \|f; L_2\|^q \ , \tag{28}$$

where $\varkappa$ is arbitrary positive, the constants $c_\varkappa$ and $C_\varkappa$ are independent of f and $h \in (0, h_0]$, and

$$\mu_q(h) = h |\ln h|^{q(\varkappa + 5/2)} \text{ for } q \in [1, 2]; \ \mu_q(h) = h^{2/q} \text{ for } q > 2 \ . \tag{29}$$

According to the Clapeyron's Theorem *the potential energy = the elastic energy − the work of external forces* takes the form

$$\mathcal{E}(u; f) = \frac{1}{2}(AD(\nabla)u, D(\nabla)u)_\Omega - (f, u)_\Omega = -\frac{1}{2}\int_\Omega u^\top f dx \ , \tag{30}$$

and Theorem 1 can be used to show that functional (30) evaluated on the solutions to problem (4)-(6), with the precision $O(\mu_q(h)\|f; L_2(\Omega)\|)$ is approximated by the energy functionals, for the problems (15), (16),

$$\begin{aligned} \mathbf{E}(\mathbf{v}; f) &= \frac{1}{2}(\mathbf{L}\mathbf{v}, \mathbf{v})_\Omega - (f, \mathbf{v})_\Omega \\ \mathfrak{E}(\mathfrak{v}; f) &= \frac{1}{2}(D(-\nabla)^\top AD(\nabla)\mathfrak{v}, \mathfrak{v})_\Omega + \frac{1}{2}\langle \mathbf{S}\pi^-\mathfrak{v} - \pi^+\mathfrak{v}, \pi^-\mathfrak{v}\rangle - (f, \mathfrak{v})_\Omega \end{aligned}$$

6. Spectral Problems

Let us assume that the elastic body is homogeneous, i.e., the Hooke's matrix A and the material density $\rho > 0$ are independent of the point $x \in \Omega$. The spectral boundary value problem includes the system of partial differential equations

$$L(x, \nabla)u(h, x) = \Lambda(h)u(h, x) \ , \quad x \in \Omega \ , \tag{31}$$

(compare with (4)) with the boundary conditions (5), (6), and admits the sequence of eigenvalues

$$0 < \Lambda_1(h) \leq \Lambda_2(h) \leq \ldots \leq \Lambda_n(h) \leq \ldots \to +\infty, \tag{32}$$

with an orthonormal in $L_2(\Omega)^2$ system of eigenfunctions $u^n(h, \cdot)$. Asymptotic expansion, for $h \to +0$, of eigenvalues in analogical scalar problems are characterized by holomorphic dependence upon the parameter $|\ln h|^{-1}$ (see [12] and [13]where such asymptotics were constructed and justified). For the boundary value problems in the elasticity the asymptotic constructions become more complicated (cf. [4], where, in particular, the mistake in [14]was corrected), and therefore the modeling of spectral problems are the most actual issue.

We are going to compare the spectral sequence (32) with the spectre

$$\sigma(\rho^{-1}\mathbf{L}) = \{\lambda_1, \lambda_2, ...\} \tag{33}$$

of the selfadjoint operator, indicated already in Lemma 1, or equivalently defined by the spectral problem with point conditions at $P^1, ..., P^I$:

$$\begin{aligned} L(\nabla)\mathfrak{v}(x) &= \lambda(h)\rho\mathfrak{v}(x) \ , \quad x \in \Omega \ , \\ B(x, \nabla)\mathfrak{v}(x) = 0 \ , &\quad x \in \partial\Omega \setminus \{P^1, ..., P^I\} \ , \\ \mathbf{S}\pi^-\mathfrak{v} - \pi^+\mathfrak{v} &= 0 \ . \end{aligned} \tag{34}$$

The space $\mathfrak{D}$ in (12) is compactly embedded in $L_2(\Omega)^3$, whence accordingly, the eigenvalues $\lambda_N(h)$ are of finite multiplicity, and the unique accumulation point at infinity. The operator $\mathbf{L}$ is not positive, since the matrix $\mathbf{S}$ determined in Section 9.5 (see (25)) is negative definite for small $h > 0$. Whence, the numbers from the collection (33), in contrast to (32), may be located as well in the negative part of the real axis. Nevertheless, by an application of the approach proposed in [9]the following result is obtained.

THEOREM 2 *For any $T > 0$ there exists $h_T > 0$ such that all eigenvalues $\lambda_1(h), ..., \lambda_{N(T)}(h) \in \sigma(\rho^{-1}\mathbf{L}) \cap (-T, T)$, for $h \in (0, h_T)$, become positive and satisfy the estimate*

$$|\lambda_n(h)\Lambda_n(h)| \leq c_{n,\varkappa}\mu_2(h) \ ,$$

where $\mu_2(h)$ *is defined in* (29) *and the constant* $c_{n,\varkappa}$ *depends on the eigenvalue number* $n = 1, ..., N(T)$ *and* $\varkappa > 0$ *but it is independent of* $h \in (0, h_T]$.

Remark 3

1) *The result remains valid for the spectres of problems* (31), (5), (6) *and* (34) *in the case of nonhomogeneous elastic material (*A *and* ρ *depend on* x*). For determination of the appropriate selfadjoint extension, the differential operator* $\rho(x)^{-1/2}L(x;\nabla)\rho(x)^{-1/2}$ *should be used or the weighted class* $L_2(\Omega)$ *(cf. [9], [15]).*

2) *The information on the asymptotic behaviour of the eigenfunctions is also available, but it is omitted here.*

References

[1] I.I. Argatov. Integral characteristics of rigid inclusions and cavities in the two-dimensional theory of elasticity. *Prikl. Mat. Mekh.*, 62:283–289, 1998 (English translation in : J. Appl. Maths Mechs, 62(1998) 263–268).

[2] V. M. Babich and M.I. Ivanov. Long-wave asymptotics in problems of the scattering of elastic waves. (Russian). Zap. Nauchn. Sem. Leningrad. Otdel. Mat. Inst. Steklov. (LOMI), 156 (1986), Mat. Voprosy Teor. Rasprostr. Voln. 16, 6–19, 184; (English translation in : J. Soviet Math. 50 (1990), no. 4, 1685–1693).

[3] F.A. Berezin and L.D.Faddeev. Remark on the Schrdinger equation with singular potential. *Dokl. Akad. Nauk SSSR*, 137:1011–1014, 1961 (English translation in : Soviet Math. Dokl 2(1961) 372–375).

[4] A. Campbell and S.A. Nazarov. Asymptotics of eigenvalues of a plate with small clamped zone. *Positivity*, 3:275–295, 2001.

[5] S. Garreau, P. Guillaume, and M. Masmoudi. The topological asymptotic for pde systems: the elasticity case. *SIAM Journal on Control and Optimization*, 39:1756–1778, 2001.

[6] A.M. Il'in. A boundary value problem for the elliptic equation of second order in a domain with a narrow slit. I. The two-dimensional case. *Mat. Sb.*, 99:514–537, 1976 (English translation in : Math. USSR Sbornik 28(1976)).

[7] A.M. Il'in. *Matching of Asymptotic Expansions of Solutions of Boundary Value Problems*, volume 102 of *Translations of Mathematical Monographs*. American Mathematical Society, Providence, RI, 1992.

[8] L. Jackowska-Strumiłło, J. Sokołowski, A. Żochowski, and A. Henrot. On numerical solution of shape inverse problems. *Computational Optimization and Applications*, 23:231–255, 2002.

[9] I.V. Kamotski and S.A. Nazarov. Spectral problems in singular perturbed domains and self adjoint extensions of differential operators. *Trudy St.-Petersburg Mat. Obshch.*, 6:151–212, 1998(English translation in : Proceedings of the St. Petersburg Mathematical Society, 6(2000) 127-181, Amer. Math. Soc. Transl. Ser. 2, 199, Amer. Math. Soc., Providence, RI).

[10] N.S. Landkoff. *Fundamentals of modern potential theory (Russian).* Izdat. Nauka, Moscow, 1966.

[11] D. Leguillon and E. Sánchez-Palencia. *Computation of singular solutions in elliptic problems and elasticity.* Masson, Paris, 1987.

[12] V.G. Maz'ya, S.A. Nazarov, and B.A. Plamenevskii. Asymptotic expansions of the eigenvalues of boundary value problems for the laplace operator in domains with small holes. *Izv. Akad. Nauk SSSR. Ser. Mat.*, 48:347–371, 1984 (English translation in : Math. USSR Izvestiya 24(1985) 321–345.

[13] V.G. Maz'ya, S.A. Nazarov, and B.A. Plamenevskii. *Asymptotic theory of elliptic boundary value problems in singularly perturbed domains.* Birkhuser Verlag, Basel, Vol. 1, 2, 2000.

[14] A.B. Movchan. Oscillations of elastic bodies with small holes. *Vestink Leningrad University*, 1:33–37, 1989 (English translation in : Vestnik Leningrad Univ. Math. 22 (1989), no. 1, 50–55.

[15] S.A. Nazarov. Selfadjoint extensions of the Dirichlet problem operator in weighted function spaces. *Mat. sbornik*, 137:224–241, 1988 (English translation in : Math. USSR Sbornik 65(1990) 229–247).

[16] S.A. Nazarov. Two-term asymptotics of solutions of spectral problems with singular perturbations. *Mat. sbornik.*, 69:291–320, 1991 (English translation in : Math. USSR. Sbornik 69(1991) 307–340).

[17] S.A. Nazarov. Asymptotic conditions at points, self adjoint extensions of operators and the method of matched asymptotic expansions. *Trudy St.-Petersburg Mat. Obshch.*, 5:112–183, 1996 (English translation in : Trans. Am. Math. Soc. Ser. 2. 193(1999) 77–126).

[18] S.A. Nazarov and B.A. Plamenevsky. *Elliptic Problems in Domains with Piecewise Smooth Boundaries.* Walter de Gruyter, De Gruyter Exposition in Mathematics 13, 1994.

[19] S.A. Nazarov and J. Sokolowski. Asymptotic analysis of shape functionals. *Journal de Mathématiques pures et appliquées*, 82:125–196, 2003.

[20] B.S. Pavlov. The theory of extension and explicitly soluble models. *Uspehi Mat. Nauk*, 42:99–131, 1987 (English translation in : Soviet Math. Surveys 42(1987) 127–168.

[21] E. Sánchez-Palencia. Forces appliquées à une petite region de la surface d'un corps élastique. application aux jonctions. *C. R. Acad. Sci. Paris Sér. II Méc. Phys. Chim. Sci. Univers Sci. Terre*, 30:689–694, 1988.

[22] J. Sokołowski and A. Żochowski. On topological derivative in shape optimization. *SIAM Journal on Control and Optimization*, 37:1251–1272, 1999.

[23] J. Sokołowski and A. Żochowski. Optimality conditions for simultaneous topology and shape optimization. *SIAM Journal on Control and Optimization*, 42:1198–1221, 2003.

FACTORIZATION BY INVARIANT EMBEDDING OF ELLIPTIC PROBLEMS IN A CIRCULAR DOMAIN

J. Henry
INRIA-Futurs
MAB Université Bordeaux 1
351, cours de la libération, 33405 Talence, France
jacques.henry@inria.fr

B. Louro
Departamento de Matemática, Faculdade de Ciências e Tecnologia
Universidade Nova de Lisboa, 2829-516 Caparica, Portugal
bjl@fct.unl.pt

M.C. Soares*
Departamento de Matemática, Faculdade de Ciências e Tecnologia
Universidade Nova de Lisboa, 2829-516 Caparica, Portugal
mcs@fct.unl.pt

Abstract We present a method to factorize a second order elliptic boundary value problem in a circular domain, in a system of uncoupled first order initial value problems. We use a space invariant embedding technique along the radius of the circle, in both an increasing and a decreasing way. This technique is inspired in the temporal invariant embedding used by J.-L. Lions for the control of parabolic systems. The singularity at the origin for the initial value problems is studied.

Keywords: Factorization, Riccati equation, invariant embedding.

Introduction

The technique of invariant embedding was first introduced by Bellman ([2]) and was formally used by Angel and Bellman ([1]) in the resolution

*Funding provided by FCT and FSE, Praxis XXI, BD/21443/99

of Poisson's problem defined over a rectangle. J.L. Lions ([5]) gave a justification for this invariant embedding in the computation of the optimal feedback in the framework of Optimal Control of evolution equations of parabolic type. Henry and Ramos ([3]) presented a justification for the invariant embedding of Poisson's problem in a cylindrical domain. The problem is embedded in a family of similar problems defined on sub-cylinders limited by a moving boundary. They obtained a factorization in two uncoupled problems of parabolic type, in opposite directions. In this paper, we want to generalize this method to other types of geometries and, in particular, to the case where the family of surfaces which limits the sub-domains, starts on the outside boundary of the domain and shrinks to a point. We present here the simple situation where Ω (resp Ω_s) is a disk of $\mathbb{R}^2$ with radius a (resp s) and centered on the origin and where the sub-domains defined by the invariant embedding are both the annuli $\Omega \setminus \Omega_s$, $s \in (0,a)$ ([4]) and the family of disks Ω_s, $s \in (0,a)$. This factorization can be viewed as an infinite dimensional extension of the block Gauss factorization for linear systems.

1. Motivation

Given $f \in L^2(0,1)$, $y_0, y_1 \in \mathbb{R}$, $q \in \mathbb{R}^+$, $p \in \mathbb{R}^+ \setminus \{0\}$, let y be the solution of the following boundary value problem:

$$\begin{cases} -p\frac{d^2y}{dx^2} + qy = f, \ x \in]0,1[\\ \frac{dy}{dx}(0) = y_0 \\ y(1) = y_1 \end{cases}$$

Considering the operator $A = -p\frac{d^2}{dx^2} + q$, the natural way to factorize it, is by searching α, β such that $A = -p\left(\frac{d}{dx} + \beta(x)\right)\left(\frac{d}{dx} - \alpha(x)\right)$.

Then, for each $\varphi \in C^2((0,1))$ we have

$$A(\varphi) = -p\frac{d^2\varphi}{dx^2} + p(\alpha - \beta)\frac{d\varphi}{dx} + p\left(\frac{d\alpha}{dx} + \alpha\beta\right)\varphi$$

Thus, we must have $\alpha = \beta$ and $\dfrac{d\beta}{dx} + \beta^2 = \dfrac{q}{p}$. If we set $\beta(0) = 0$ and $\xi = -\dfrac{dy}{dx} + \beta y$, we find $\xi(0) = -y_0$ and the following system of uncoupled equations:

$$\begin{cases} \dfrac{d\beta}{dx} + \beta^2 = \dfrac{q}{p}, & \beta(0) = 0 \\ \dfrac{d\xi}{dx} + \beta\xi = \dfrac{f}{p}, & \xi(0) = -y_0 \\ \dfrac{dy}{dx} - \beta y = -\xi, & y(1) = y_1 \end{cases}$$

We point out that the equation in β is a Riccati equation.

2. Formulation of the Problem and a Regularization Result

We consider the Dirichlet problem for the Poisson equation defined over Ω.

$$(\mathcal{P}) \qquad -\Delta u = f, \text{ in } \Omega;\ u|_{\Gamma_a} = u_0,$$

where Γ_s denotes the circle of radius s and center at the origin, $f \in L^2(\Omega)$ and $u_0 \in H^{1/2}(\Gamma_a)$. We assume the additional regularity around the origin $f \in C^{0,\alpha}(\mathcal{O})$, $\mathcal{O}$ being a neighborhood of the origin. Introducing polar coordinates, $\hat{u}(\rho, \theta) = u(x_1, x_2)$ satisfies

$$(\hat{\mathcal{P}}) \begin{cases} -\dfrac{1}{\rho}\dfrac{\partial}{\partial\rho}\left(\rho\dfrac{\partial \hat{u}}{\partial \rho}\right) - \dfrac{1}{\rho^2}\dfrac{\partial^2 \hat{u}}{\partial \theta^2} = \hat{f}, \text{ in }]0, a[\times\mathcal{I} \\ \hat{u}|_{\Gamma_a} = \hat{u}_0;\ \hat{u} \ \ 2\pi - \text{periodic with respect to } \theta, \end{cases}$$

where $\mathcal{I} =]0, 2\pi[$. However, by doing this, we introduce a singularity at the origin. Furthermore the analogous of the computation done in [3] would need to know $u(0)$ which is not a data of the problem.

In order to avoid this difficulty we start by defining the following intermediate problem:

$$(\mathcal{P}_\varepsilon) \begin{cases} -\Delta u_\varepsilon = f, \text{ in } \Omega \setminus \Omega_\varepsilon;\ u_\varepsilon|_{\Gamma_a} = u_0 \\ \displaystyle\int_{\Gamma_\varepsilon} \frac{\partial u_\varepsilon}{\partial n}\, d\Gamma = 0;\ u_\varepsilon|_{\Gamma_\varepsilon} \text{is constant} \end{cases}$$

where Ω_ε is a circular domain of radius $0 < \varepsilon < a$ and concentric with Ω. It's easy to see that this problem is well posed.

THEOREM 1 *When* $\varepsilon \to 0$, $\tilde{u}_\varepsilon$, *defined as* $\tilde{u}_\varepsilon = \begin{cases} u_\varepsilon, & in\ \Omega \setminus \Omega_\varepsilon \\ u_\varepsilon = u_\varepsilon|_{\Gamma_\varepsilon}, & in\ \Omega_\varepsilon \end{cases}$, *where* u_ε *is the solution of problem* $(\mathcal{P}_\varepsilon)$ *converges to* u, *solution of problem* $(\mathcal{P})$, *in* $H^1(\Omega)$.

We can write problem $(\mathcal{P}_\varepsilon)$ in polar coordinates restricting problem $(\hat{\mathcal{P}})$ over $]\varepsilon, a[\times\mathcal{I}$ and joining the boundary conditions $\hat{u}_\varepsilon|_{\Gamma_\varepsilon}$ constant, $\displaystyle\int_{\Gamma_\varepsilon} \frac{\partial \hat{u}_\varepsilon}{\partial \rho}\, d\theta = 0.$

3. Factorization by Invariant Embedding

We embed problem $(\mathcal{P}_\varepsilon)$ in a family of similar problems $(\mathcal{P}_{s,h})$ defined on the annulus $\Omega \setminus \Omega_s$, $s \in]\varepsilon, a[$ and satisfying an additional Neumann boundary condition in Γ_s; in terms of polar coordinates we find

$$(\hat{\mathcal{P}}_{s,h}) \begin{cases} -\dfrac{1}{\rho}\dfrac{\partial}{\partial\rho}\left(\rho\dfrac{\partial \hat{u}_s}{\partial\rho}\right) - \dfrac{1}{\rho^2}\dfrac{\partial^2 \hat{u}_s}{\partial\theta^2} = \hat{f}, \text{ in }]s,a[\times\mathcal{I} \\ \hat{u}_s|_{\Gamma_a} = \hat{u}_0, \hat{u}_s \quad 2\pi - \text{periodic with respect to } \theta \\ \dfrac{\partial \hat{u}_s}{\partial\rho}|_{\Gamma_s} = h \end{cases}$$

Since $\dfrac{\partial \hat{u}_\varepsilon}{\partial\rho}|_{\Gamma_\varepsilon}$ is well determined through the conditions "$\hat{u}_{\varepsilon_{|\Gamma_\varepsilon}}$ constant" and "$\displaystyle\int_{\Gamma_\varepsilon} \frac{\partial \hat{u}_\varepsilon}{\partial\rho}\, d\theta = 0$", it's clear that $(\hat{\mathcal{P}}_\varepsilon)$ belongs to the family $(\hat{\mathcal{P}}_{s,h})$ for $s = \varepsilon$.

Defining $H^1_{\rho,P}(\mathcal{I})$ as the space of periodic functions v of θ, verifying $v \in L^2(\mathcal{I})$ and $\frac{1}{\rho}\frac{\partial v}{\partial\theta} \in L^2(\mathcal{I})$, we take $h \in H^{1/2}_{\rho,P}(\mathcal{I})'$, where $H^{1/2}_{\rho,P}(\mathcal{I}) = [H^1_{\rho,P}(\mathcal{I}), L^2(\mathcal{I})]_{1/2}$.

For every $s \in [\varepsilon, a)$, $h \in H^{1/2}_{\rho,P}(\mathcal{I})'$ we define $P(s)h = \gamma_{s_{|\Gamma_s}}$, where γ_s is the solution of

$$\begin{cases} -\dfrac{1}{\rho}\dfrac{\partial}{\partial\rho}\left(\rho\dfrac{\partial \gamma_s}{\partial\rho}\right) - \dfrac{1}{\rho^2}\dfrac{\partial^2 \gamma_s}{\partial\theta^2} = 0, \text{ in }]s,a[\times\mathcal{I} \\ \gamma_{s_{|\Gamma_a}} = 0 \\ \dfrac{\partial \gamma_s}{\partial\rho}|_{\Gamma_s} = h \\ \gamma_s \quad 2\pi - \text{periodic with respect to } \theta \end{cases} \tag{1}$$

and $r(s) = \beta_{s_{|\Gamma_s}}$, where β_s is the solution of

$$\begin{cases} -\dfrac{1}{\rho}\dfrac{\partial}{\partial\rho}\left(\rho\dfrac{\partial \beta_s}{\partial\rho}\right) - \dfrac{1}{\rho^2}\dfrac{\partial^2 \beta_s}{\partial\theta^2} = \hat{f}, \text{ in }]s,a[\times\mathcal{I} \\ \beta_{s_{|\Gamma_a}} = \hat{u}_0 \\ \dfrac{\partial \beta_s}{\partial\rho}|_{\Gamma_s} = 0 \\ \beta_s \quad 2\pi - \text{periodic with respect to } \theta \end{cases} \tag{2}$$

By linearity of $(\hat{\mathcal{P}}_{s,h})$ we have

$$\hat{u}_s|_{\Gamma_s} = P(s)h + r(s), \tag{3}$$

where $P(s)$ is the Neumann to Dirichlet map for the annulus $\Omega \setminus \Omega_s$.

Let $X = \{\hat{v} | \hat{v} \in L^2_\rho(0,a; H^1_{\rho,P}(\mathcal{I})) \cap \frac{\partial \hat{u}}{\partial \rho} \in L^2_\rho(0,a; L^2(\mathcal{I}))\}$, where L^2_ρ stands for the space of functions of ρ square integrable with the weight ρ. After passing to the limit, when $\varepsilon \to 0$, the factorization of problem $(\hat{\mathcal{P}})$ is synthesized by the following theorem

THEOREM 2 *The solution $\hat{u}$ of $(\hat{\mathcal{P}})$ is the unique solution of the following system of uncoupled, first order in ρ, initial value problems*

1. *for every $h, \bar{h}$ in $L^2(\mathcal{I})$, the self-adjoint operator P, $P \leq 0$,*

$$P \in L^\infty\big(0,a;\ \mathcal{L}(L^2(\mathcal{I}), H^1_{\rho,P}(\mathcal{I})) \cap \mathcal{L}(H^{1/2}_{\rho,P}(\mathcal{I})', H^{1/2}_{\rho,P}(\mathcal{I})) \cap \mathcal{L}(H^1_{\rho,P}(\mathcal{I})', L^2(\mathcal{I}))\big),$$

   satisfies the Riccati equation

$$\left(\frac{\partial P}{\partial \rho} h, \bar{h}\right) + \left(\frac{1}{\rho^2}\frac{\partial}{\partial \theta} Ph, \frac{\partial}{\partial \theta} P\bar{h}\right) - \left(\frac{1}{\rho} h, P\bar{h}\right) = (h, \bar{h}) \qquad (4)$$

   in $\mathcal{D}'(0,a)$, with the initial condition $P(a) = 0$;

2. *for every h in $L^2(\mathcal{I})$, $r \in X$ satisfies the equation*

$$\left(\frac{\partial r}{\partial \rho}, h\right) + \left(\frac{1}{\rho^2}\frac{\partial r}{\partial \theta}, \frac{\partial}{\partial \theta} Ph\right) = \left(\hat{f}, Ph\right) \qquad (5)$$

   in $\mathcal{D}'(0,a)$, with the initial condition $r(a) = \hat{u}_0$;

3. *for every h in $H^1_{\rho,P}(\mathcal{I})'$, $\hat{u} \in X$ satisfies the equation*

$$-\left(\frac{\partial \hat{u}}{\partial \rho}, Ph\right) + \langle \hat{u}, h \rangle_{H^1_{\rho,P}(\mathcal{I}), H^1_{\rho,P}(\mathcal{I})'} = \langle r, h \rangle_{H^1_{\rho,P}(\mathcal{I}), H^1_{\rho,P}(\mathcal{I})'} \qquad (6)$$

   in $\mathcal{D}'(0,a)$, with the initial condition $\hat{u}(0) = \lim_{\rho \to 0} r(\rho)$ in $L^2(\mathcal{I})$ which is constant.

P, r and $\hat{u}$ thus defined are unique. Equations (4) and (5) are well posed for ρ decreasing from a to 0 and (6) is well posed for ρ increasing from 0 to a.

The formal analogy between this result and the *LU* Gauss factorization of a matrix should be emphasized. The Riccati equation (4) for P is the analogous of the block *LU* factorization of a block tridiagonal matrix, and the initial value problems (5) and (6) are the analogous of the lower and upper block triangular systems. This factorization inherits the well known property of the Gauss factorization for multiple right hand sides: if $(\mathcal{P})$ has to be solved for different f and u_0, (4) is to be solved only once, then the initial value problems (5) and (6) are solved for each value of the data. Furthermore, if one considers a finite difference discretization of $(\mathcal{P})$ in polar coordinates, for example, one can show that the Gauss factorization of the obtained linear system can be obtained by *one particular discretization of* (4), (5), (6). But other possible discretizations exist with their own interest. Also this equivalent formulation of problem $(\mathcal{P})$ furnishes the Neumann to Dirichlet operator $P(s)$ for the annulus $\Omega \setminus \Omega_s$ which is of interest for various kinds of problems as domain decomposition or the definition of transparent boundary conditions.

4. Sketch of the Proof of Theorem 2

From (3), the solution $\hat{u}_\varepsilon$ of $(\hat{\mathcal{P}}_\varepsilon)$ in polar coordinates, satisfies the relation $\hat{u}_\varepsilon(\rho) = P(\rho)\frac{\partial \hat{u}_\varepsilon}{\partial \rho}|_{\Gamma_\rho} + r(\rho), \forall \rho \in [\varepsilon, a]$. From this last equality, taking the derivative, in a formal way, with respect to ρ and considering $\frac{\partial \hat{u}_\varepsilon}{\partial \rho}$ arbitrary, we obtain $\frac{\partial P}{\partial \rho} - P\frac{1}{\rho^2}\frac{\partial^2}{\partial \theta^2}P - P\frac{1}{\rho} = I$ and $-P\hat{f} - P\frac{1}{\rho^2}\frac{\partial^2 r}{\partial \theta^2} + \frac{\partial r}{\partial \rho} = 0$, and considering the boundary condition on Γ_a in $(\mathcal{P}_\varepsilon)$, we obtain $P(a) = 0$ and $r(a) = \hat{u}_0$.

From the two equations above, and respective initial conditions, we can obtain P and r. Let M and N be defined by $M = \{v \in \left(H^{1/2}_{\rho,P}(\mathcal{I})\right)' \mid \int_0^{2\pi} v \, d\theta = 0\}$ and $N = M^\perp = \{v \in H^{1/2}_{\rho,P}(\mathcal{I}) | v \text{ is constant}\}$. They are invariant by P. One has $L^2(\mathcal{I}) = (M \cap L^2(\mathcal{I})) \oplus N$ and let Π_M and Π_N be the projection, for the $L^2(\mathcal{I})$ metrics, on each subspace respectively. The following theorem provides the initial condition for $\hat{u}_\varepsilon$ on Γ_ε.

THEOREM 3 *Given $r(\varepsilon) \in L^2(\mathcal{I})$, there exists a unique solution $\hat{u}_\varepsilon(\varepsilon) \in N$ and $\frac{\partial \hat{u}_\varepsilon}{\partial \rho}(\varepsilon) \in M$ for the equation $\hat{u}_\varepsilon(\varepsilon) = P(\varepsilon)\frac{\partial \hat{u}_\varepsilon}{\partial \rho}(\varepsilon) + r(\varepsilon)$. In particular, $\hat{u}_\varepsilon(\varepsilon) = \Pi_N r(\varepsilon)$.*

We use the Galerkin method as in [5], [3], and adequate properties on the operator P and function r. In finite dimension, we can prove the existence of a global solution of the decoupled system. Then we can justify the preceding formal calculation and we obtain, after passing to the limit when the dimension tends to infinity, the following result:

THEOREM 4 *For every $h, \bar{h}$ in $L^2(\mathcal{I})$, the operator P belongs to $L^\infty\left(\varepsilon, a; \mathcal{L}(H^{1/2}_{\rho,P}(\mathcal{I})', H^{1/2}_{\rho,P}(\mathcal{I}))\right)$ and satisfies the following equation*

$$\left(\frac{\partial P}{\partial \rho} h, \bar{h}\right) + \left(\frac{1}{\rho^2}\frac{\partial}{\partial \theta} Ph, \frac{\partial}{\partial \theta} P\bar{h}\right) - \left(\frac{1}{\rho} h, P\bar{h}\right) = \left(h, \bar{h}\right), \tag{7}$$

in $\mathcal{D}'(\varepsilon, a)$, with $P(a) = 0$. The function r belongs to $X|_{\Omega\setminus\Omega_\varepsilon}$, satisfies $r(a) = \hat{u}_0$ and for every h in $L^2(\mathcal{I})$, satisfies in $\mathcal{D}'(\varepsilon, a)$ the following equation

$$\left(\frac{\partial r}{\partial \rho}, h\right) + \left(\frac{1}{\rho^2}\frac{\partial r}{\partial \theta}, \frac{\partial}{\partial \theta} Ph\right) = \left(\hat{f}, Ph\right). \tag{8}$$

Since P and r do not depend on ε and thanks to the estimates on $P(\rho)$ and $r(\rho)$, we can take ε arbitrarily small and consequently consider the previous equalities defined on $\mathcal{D}'(0, a)$. Let $\|.\|_\rho$ denote the norm in $\mathcal{L}\left(H^{1/2}_{\rho,P}(\mathcal{I})', H^{1/2}_{\rho,P}(\mathcal{I})\right)$. The following theorem gives the behavior of P and r around the origin which provides the regularity claimed in Theorem 2:

THEOREM 5 *For P satisfying (7) and $P(a) = 0$, we have $\lim_{\rho \to 0} \|P(\rho)\|_\rho = 1$. Furthermore $\lim_{\rho \to 0} \|P(\rho) - \rho(P_\infty \, o \, \Pi_M)\|_\rho = 0$, where P_∞ is the negative self-adjoint operator satisfying $-P_\infty \dfrac{\partial^2}{\partial \theta^2} P_\infty = I$.*

The solution r of (8) and $r(a) = \hat{u}_0$, has a limit $r(0)$ constant with respect to θ : $\lim_{\rho \to 0} \|r(\rho) - r(0)\|_{L^2(\mathcal{I})} = 0$.

It should be also remarked that, measured with the fixed norm $\|.\|_{\mathcal{L}(L^2(\mathcal{I}), L^2(\mathcal{I}))}$, $P(\rho)$ goes to 0 as ρ goes to 0. Concerning the equation on $\hat{u}_\varepsilon$ we have

THEOREM 6 *For every h in $H^1_{\rho,P}(\mathcal{I})'$, $\hat{u}_\varepsilon$ satisfies in $\mathcal{D}'(\varepsilon, a)$ the following equation*

$$-\left(\frac{\partial \hat{u}_\varepsilon}{\partial \rho}, Ph\right) + \langle \hat{u}_\varepsilon, h \rangle_{H^1_{\rho,P}(\mathcal{I}), H^1_{\rho,P}(\mathcal{I})'} = \langle r, h \rangle_{H^1_{\rho,P}(\mathcal{I}), H^1_{\rho,P}(\mathcal{I})'}, \tag{9}$$

with the initial condition $\hat{u}_\varepsilon(\varepsilon) = \Pi_N r(\varepsilon)$.

Using Theorem 1 we obtain the convergence in X of $\hat{u}_\varepsilon$ to $\hat{u}$ satisfying (6). Furthermore, thanks to the local regularity assumption on f near the origin which implies that $u \in \mathcal{C}^{2,\alpha}(\mathcal{O})$, one can prove that $\lim_{\varepsilon \to 0} \hat{u}_\varepsilon(\varepsilon) = \hat{u}(0)$ (the proof for the uniqueness of the solution of (6) uses the determination of $\hat{u}(0)$).

5. Factorization by Invariant Embedding: Dual Case

Another factorization could be obtained by using an invariant embedding defined by the family of disks Ω_s. Here the main difficulty is to define the initial conditions for P and r at the origin.

We embed problem $(\mathcal{P}_\varepsilon)$ in a family of similar problems $(\tilde{\mathcal{P}}_{\varepsilon,h})$ defined on the annulus $\Omega_s \setminus \Omega_\varepsilon$, $s \in]\varepsilon, a[$ and satisfying an additional Robin boundary condition in Γ_s. We find the following problem in polar coordinates:

$$(\hat{\tilde{\mathcal{P}}}_{\varepsilon,h}) \begin{cases} -\dfrac{1}{\rho}\dfrac{\partial}{\partial\rho}\left(\rho\dfrac{\partial\hat{u}_\varepsilon}{\partial\rho}\right) - \dfrac{1}{\rho^2}\dfrac{\partial^2\hat{u}_\varepsilon}{\partial\theta^2} = \hat{f}, \text{ in }]\varepsilon, s[\times\mathcal{I} \\ \hat{u}_\varepsilon|_{\Gamma_\varepsilon} \text{ constant}, \hat{u}_\varepsilon \;\; 2\pi-\text{periodic with respect to } \theta \\ \displaystyle\int_0^{2\pi} \dfrac{\partial\hat{u}_\varepsilon}{\partial\rho}|_{\Gamma_\varepsilon}\, d\theta = 0 \\ \dfrac{\partial\hat{u}_\varepsilon}{\partial\rho}|_{\Gamma_s} + \alpha\hat{u}_{\varepsilon|_{\Gamma_s}} = h, \quad \alpha > 0 \end{cases}$$

It is clear that $(\hat{\mathcal{P}}_\varepsilon)$ is exactly $(\hat{\tilde{\mathcal{P}}}_{\varepsilon,h})$ for $s = a$ and $h = \dfrac{\partial\hat{u}_\varepsilon}{\partial\rho}|_{\Gamma_s} + \alpha\hat{u}_0$, and so we use the same notation $\hat{u}_\varepsilon$ for the solution of $(\hat{\mathcal{P}}_\varepsilon)$ and the family of solutions of $(\hat{\tilde{\mathcal{P}}}_{\varepsilon,h})$. This should not make confusion with the solutions of $(\hat{\mathcal{P}}_{s,h})$ in the previous section. For every $s \in (\varepsilon, a]$ and $h \in H^{1/2}_{\rho,P}(\mathcal{I})'$ we define $\tilde{P}_\varepsilon(s)h = \tilde{\gamma}_{\varepsilon|_{\Gamma_s}}$, where $\tilde{\gamma}_\varepsilon$ is the solution of

$$\begin{cases} -\dfrac{1}{\rho}\dfrac{\partial}{\partial\rho}\left(\rho\dfrac{\partial\tilde{\gamma}_\varepsilon}{\partial\rho}\right) - \dfrac{1}{\rho^2}\dfrac{\partial^2\tilde{\gamma}_\varepsilon}{\partial\theta^2} = 0, \text{ in }]\varepsilon, s[\times\mathcal{I} \\ \tilde{\gamma}_{\varepsilon|_{\Gamma_\varepsilon}} \text{ constant}, \tilde{\gamma}_\varepsilon \;\; 2\pi-\text{periodic with respect to } \theta \\ \displaystyle\int_0^{2\pi} \dfrac{\partial\tilde{\gamma}_\varepsilon}{\partial\rho}|_{\Gamma_\varepsilon}\, d\theta = 0 \\ \dfrac{\partial\tilde{\gamma}_\varepsilon}{\partial\rho}|_{\Gamma_s} + \alpha\,\tilde{\gamma}_\varepsilon|_{\Gamma_s} = h \end{cases}$$

and $\tilde{r}_\varepsilon(s) = \tilde{\beta}_{\varepsilon|_{\Gamma_s}}$, where $\tilde{\beta}_\varepsilon$ is the solution of

$$\begin{cases} -\dfrac{1}{\rho}\dfrac{\partial}{\partial\rho}\left(\rho\dfrac{\partial\tilde{\beta}_\varepsilon}{\partial\rho}\right) - \dfrac{1}{\rho^2}\dfrac{\partial^2\tilde{\beta}_\varepsilon}{\partial\theta^2} = \hat{f}, \text{ in }]\varepsilon, s[\times\mathcal{I} \\ \tilde{\beta}_{\varepsilon|_{\Gamma_\varepsilon}} \text{ constant}, \tilde{\beta}_\varepsilon \;\; 2\pi-\text{periodic with respect to } \theta \\ \displaystyle\int_0^{2\pi} \dfrac{\partial\tilde{\beta}_\varepsilon}{\partial\rho}|_{\Gamma_\varepsilon}\, d\theta = 0 \\ \dfrac{\partial\tilde{\beta}_\varepsilon}{\partial\rho}|_{\Gamma_s} + \alpha\,\tilde{\beta}_\varepsilon|_{\Gamma_s} = 0 \end{cases}$$

By linearity of $(\tilde{\mathcal{P}}_{\varepsilon,h})$ we have

$$\hat{u}_{\varepsilon_{|\Gamma_s}} = \tilde{P}_\varepsilon(s)\left(\frac{\partial \hat{u}_\varepsilon}{\partial \rho}|_{\Gamma_s} + \alpha\, \hat{u}_\varepsilon|_{\Gamma_s}\right) + \tilde{r}_\varepsilon(s), \forall s \in [\varepsilon, a]. \tag{10}$$

The following theorem gives another factorization of problem $(\hat{\mathcal{P}})$:

THEOREM 7 *The solution $\hat{u}$ of $(\hat{\mathcal{P}})$ is the unique solution of the following system of uncoupled, first order in ρ, initial value problem:*

1. *for every $h, \bar{h}$ in $L^2(\mathcal{I})$, the self-adjoint operator $\tilde{P}$, $\tilde{P} \geq 0$,*

$$\tilde{P} \in L^\infty(0,a; \quad \mathcal{L}(L^2(\mathcal{I}), H^1_{\rho,P}(\mathcal{I})) \cap \mathcal{L}(H^{1/2}_{\rho,P}(\mathcal{I})', H^{1/2}_{\rho,P}(\mathcal{I})) \\ \cap \mathcal{L}(H^1_{\rho,P}(\mathcal{I})', L^2(\mathcal{I}))),$$

satisfies the Riccati equation

$$\left(\frac{\partial \tilde{P}}{\partial \rho}h, \bar{h}\right) - \left(\frac{1}{\rho}h, \tilde{P}\bar{h}\right) + \frac{1}{\rho}\alpha\left(\tilde{P}h, \tilde{P}\bar{h}\right) + 2\alpha\left(\tilde{P}h, \bar{h}\right) \\ + \left(\frac{1}{\rho^2}\frac{\partial}{\partial \theta}\tilde{P}h, \frac{\partial}{\partial \theta}\tilde{P}\bar{h}\right) - \alpha^2\left(\tilde{P}h, \tilde{P}\bar{h}\right) = (h, \bar{h}) \tag{11}$$

in $\mathcal{D}'(0,a)$, with the initial condition $\tilde{P}(0) = \frac{1}{\alpha}\Pi_N$;

2. *for every h in $L^2(\mathcal{I})$, $\tilde{r} \in X$ satisfies the equation*

$$\left(\frac{1}{\rho}\alpha\tilde{r}, \tilde{P}h\right) + \left(\frac{1}{\rho^2}\frac{\partial \tilde{r}}{\partial \theta}, \frac{1}{\partial \theta}\tilde{P}h\right) - \alpha^2\left(\tilde{r}, \tilde{P}h\right) + \left(\frac{\partial \tilde{r}}{\partial \rho}, h\right) \\ +\alpha(\tilde{r}, h) = \left(f, \tilde{P}h\right) \tag{12}$$

in $\mathcal{D}'(0,a)$, with the initial condition $\tilde{r}(0) = 0$;

3. *for every h in $H^1_{\rho,P}(\mathcal{I})'$, $\hat{u} \in X$ satisfies the equation*

$$-\left(\frac{\partial \hat{u}}{\partial \rho} + \alpha\hat{u}, \tilde{P}h\right) + \langle \hat{u}, h\rangle_{H^1_{\rho,P}(\mathcal{I}), H^1_{\rho,P}(\mathcal{I})'} \\ = \langle \tilde{r}, h\rangle_{H^1_{\rho,P}(\mathcal{I}), H^1_{\rho,P}(\mathcal{I})'} \tag{13}$$

in $\mathcal{D}'(0,a)$, with the initial condition $\hat{u}(a) = \hat{u}_0$.

Equations (11) and (12) are well posed for ρ increasing from 0 to a and (13) is well posed for ρ decreasing from a to 0. $\tilde{P}$, $\tilde{r}$ and $\hat{u}$ thus defined are unique.

6. Sketch of the Proof of Theorem 7

From (10), the solution $\hat{u}_\varepsilon$ of $(\tilde{\mathcal{P}}_\varepsilon)$ satisfies

$$\hat{u}_\varepsilon(\rho) = \tilde{P}_\varepsilon(\rho)\left(\frac{\partial \hat{u}_\varepsilon}{\partial \rho}|_{\Gamma_\rho} + \alpha\, \hat{u}_{\varepsilon|_{\Gamma_\rho}}\right) + \tilde{r}_\varepsilon(\rho),\ \forall \rho \in [\varepsilon, a].$$

Taking the derivative in a formal way with respect to ρ and considering $\frac{\partial \hat{u}_\varepsilon}{\partial \rho} + \alpha\, \hat{u}_\varepsilon$ arbitrary, we obtain the following system

$$\begin{cases} \dfrac{\partial \tilde{P}_\varepsilon}{\partial \rho} - \dfrac{\tilde{P}_\varepsilon}{\rho} + \dfrac{1}{\rho}\alpha \tilde{P}_\varepsilon^2 - \dfrac{1}{\rho^2}\tilde{P}_\varepsilon \dfrac{\partial^2}{\partial \theta^2}\tilde{P}_\varepsilon + 2\alpha \tilde{P}_\varepsilon - (\alpha \tilde{P}_\varepsilon)^2 = I \\ -\tilde{P}_\varepsilon f + \dfrac{1}{\rho}\tilde{P}_\varepsilon \alpha \tilde{r}_\varepsilon - \dfrac{1}{\rho^2}\tilde{P}_\varepsilon \dfrac{\partial^2 \tilde{r}_\varepsilon}{\partial \theta^2} - \alpha^2 \tilde{P}_\varepsilon \tilde{r}_\varepsilon + \dfrac{\partial \tilde{r}_\varepsilon}{\partial \rho} + \alpha \tilde{r}_\varepsilon = 0 \\ \hat{u}_\varepsilon = \tilde{P}_\varepsilon\left(\dfrac{\partial \hat{u}_\varepsilon}{\partial \rho} + \alpha \hat{u}_\varepsilon\right) + \tilde{r}_\varepsilon \end{cases}$$

We have $\hat{u}_\varepsilon(a) = \hat{u}_0$. Further, considering the sets $M = \{v \in \left(H^{1/2}_{\tau,P}(\mathcal{I})\right)' \mid \int_0^{2\pi} v \, d\theta = 0\}$, $N = M^\perp = \{v \in H^{1/2}_{\tau,P}(\mathcal{I}) | v \text{ is constant}\}$ and the operators Π_M, Π_N as previously, one has $\tilde{P}_\varepsilon : M \to M$, $\tilde{P}_\varepsilon : N \to N$ and from (10) we obtain

$$\begin{aligned} &\Pi_M \hat{u}_\varepsilon(\varepsilon) = \tilde{P}_\varepsilon(\varepsilon)\left(\Pi_M \frac{\partial \hat{u}_\varepsilon}{\partial \rho}(\varepsilon) + \alpha \Pi_M \hat{u}_\varepsilon(\varepsilon)\right) + \Pi_M \tilde{r}_\varepsilon(\varepsilon) \\ \Rightarrow\ & 0 = \tilde{P}_\varepsilon(\varepsilon)\Pi_M \frac{\partial \hat{u}_\varepsilon}{\partial \rho}(\varepsilon) + \Pi_M \tilde{r}_\varepsilon(\varepsilon) \Rightarrow \Pi_M \tilde{P}_\varepsilon(\varepsilon) = 0,\ \Pi_M \tilde{r}_\varepsilon(\varepsilon) = 0 \end{aligned} \tag{14}$$

$$\begin{aligned} &\Pi_N \hat{u}_\varepsilon(\varepsilon) = \tilde{P}_\varepsilon(\varepsilon)\left(\Pi_N \frac{\partial \hat{u}_\varepsilon}{\partial \rho}(\varepsilon) + \alpha \Pi_N \hat{u}_\varepsilon(\varepsilon)\right) + \Pi_N \tilde{r}_\varepsilon(\varepsilon) \\ \Rightarrow\ & \hat{u}_\varepsilon(\varepsilon) = \alpha \tilde{P}_\varepsilon(\varepsilon)\hat{u}_\varepsilon(\varepsilon) + \Pi_N \tilde{r}_\varepsilon(\varepsilon) \Rightarrow \Pi_N \tilde{P}_\varepsilon(\varepsilon) = \frac{I}{\alpha},\ \Pi_N \tilde{r}_\varepsilon(\varepsilon) = 0 \end{aligned} \tag{15}$$

From (14) and (15) we obtain $\tilde{r}_\varepsilon(\varepsilon) = \Pi_M \tilde{r}_\varepsilon(\varepsilon) + \Pi_N \tilde{r}_\varepsilon(\varepsilon) = 0$. In the same way, since $\tilde{P}_\varepsilon(\varepsilon)h = \tilde{P}_\varepsilon(\varepsilon)\Pi_M h + \tilde{P}_\varepsilon(\varepsilon)\Pi_N h = \frac{1}{\alpha}\Pi_N h$ we obtain $\tilde{P}_\varepsilon(\varepsilon) = \frac{1}{\alpha}\Pi_N$.

Using again the Galerkin method, we can justify these formal calculations through the adequate proprieties on $\tilde{P}_\varepsilon$ and $\tilde{r}_\varepsilon$. After passing to the limit when the dimension tends to infinity, we find the following result, by the same reasoning as in section 4:

THEOREM 8 *1. For every $h, \bar{h}$ in $L^2(\mathcal{I})$, the operator $\tilde{P}_\varepsilon$ belongs to $L^\infty\left(\varepsilon, a; \mathcal{L}(H^{1/2}_{\rho,P}(\mathcal{I})', H^{1/2}_{\rho,P}(\mathcal{I}))\right)$ and satisfies the following equation*

$$\left(\frac{\partial \tilde{P}_\varepsilon}{\partial \rho}h, \bar{h}\right) - \left(\frac{1}{\rho}h, \tilde{P}_\varepsilon \bar{h}\right) + \left(\frac{1}{\rho^2}\frac{\partial}{\partial \theta}\tilde{P}_\varepsilon h, \frac{\partial}{\partial \theta}\tilde{P}_\varepsilon \bar{h}\right)$$
$$+\frac{1}{\rho}\alpha\left(\tilde{P}_\varepsilon h, \tilde{P}_\varepsilon \bar{h}\right) + 2\alpha\left(\tilde{P}_\varepsilon h, \bar{h}\right) - \alpha^2\left(\tilde{P}_\varepsilon h, \tilde{P}_\varepsilon \bar{h}\right) = \left(h, \bar{h}\right),$$

in $\mathcal{D}'(\varepsilon, a)$, considering $\tilde{P}_\varepsilon(\varepsilon) = \dfrac{1}{\alpha}\Pi_N$.

2. The function $\tilde{r}_\varepsilon$ belongs to $X_{|\Omega_s \setminus \Omega_\varepsilon}$, satisfies $\tilde{r}_\varepsilon(\varepsilon) = 0$, and for every h in $L^2(\mathcal{I})$ verifies, in $\mathcal{D}'(\varepsilon, a)$, the following equation

$$\left(\frac{1}{\rho}\alpha\tilde{r}_\varepsilon, \tilde{P}_\varepsilon h\right) + \left(\frac{1}{\rho^2}\frac{\partial \tilde{r}_\varepsilon}{\partial \theta}, \frac{1}{\partial \theta}\tilde{P}_\varepsilon h\right) - \alpha^2\left(\tilde{r}_\varepsilon, \tilde{P}_\varepsilon h\right) + \left(\frac{\partial \tilde{r}_\varepsilon}{\partial \rho}, h\right)$$
$$+\alpha\left(\tilde{r}_\varepsilon, h\right) = \left(f, \tilde{P}_\varepsilon h\right).$$

3. For every h in $H^1_{\rho,P}(\mathcal{I})'$, $\hat{u}_\varepsilon$ satisfies the following equation

$$-\left(\frac{\partial \hat{u}_\varepsilon}{\partial \rho} + \alpha\hat{u}_\varepsilon, \tilde{P}_\varepsilon h\right) + \langle \hat{u}_\varepsilon, h\rangle_{H^1_{\rho,P}(\mathcal{I}), H^1_{\rho,P}(\mathcal{I})'}$$
$$= \langle \tilde{r}_\varepsilon, h\rangle_{H^1_{\rho,P}(\mathcal{I}), H^1_{\rho,P}(\mathcal{I})'}$$

in $\mathcal{D}'(\varepsilon, a)$, with $\hat{u}_\varepsilon(a) = \hat{u}_0$.

Using Theorem 1 we obtain the convergence in X of $\hat{u}_\varepsilon$ to $\hat{u}$ satisfying (13). Now, since $\tilde{P}_\varepsilon$ and $\tilde{r}_\varepsilon$ depend on ε and considering $\hat{u}(\rho) = \tilde{P}(\rho)h + \tilde{r}(\rho)$ we use the following consequence of Theorem 1:

COROLLARY 9 *For all $\rho \in [0, a]$, $\tilde{r}_\varepsilon(\rho) \to \tilde{r}(\rho)$ strongly in $H^{1/2}_{\rho,P}(\mathcal{I})$, when $\varepsilon \to 0$. Also, for all $\rho \in [0, a]$ and for a fixed h, $\tilde{P}_\varepsilon(\rho)h \to \tilde{P}(\rho)h$, strongly in $H^{1/2}_{\rho,P}(\mathcal{I})$ and weakly in $H^{3/2}_{\rho,P}(\mathcal{I})$, when $\varepsilon \to 0$.*

Therefore, passing to the limit when $\varepsilon \to 0$ we find $\tilde{P}$ and $\tilde{r}$ satisfying (11) and (12), respectively.

Using again the appropriate conditions of regularity around the origin (that is, $f \in \mathcal{C}^{0,\alpha}(\Omega)$), we can define the value of $\hat{u}(0)$ (as a constant), and consequently we have $\hat{u}(0) \in N$. Also, since $\dfrac{\partial \hat{u}}{\partial \rho} = \dfrac{\partial u}{\partial x}\cos(\theta) + \dfrac{\partial u}{\partial y}\sin(\theta)$

and we have assumed enough regularity around the origin, we have $\int_0^{2\pi} \frac{\partial \hat{u}}{\partial \rho}(0)\, d\theta = \int_0^{2\pi} c_1 \cos(\theta) + c_2 \sin(\theta)\, d\theta = 0$, from which we conclude that $\frac{\partial \hat{u}}{\partial \rho}(0) \in M$. Therefore, from $\hat{u}_{|\Gamma_s} = \tilde{P}(s) \left(\frac{\partial \hat{u}}{\partial \rho}_{|\Gamma_s} + \alpha\, \hat{u}_{|\Gamma_s} \right) + \tilde{r}(s), \forall s \in [0, a]$ we obtain

$$\begin{aligned} &\Pi_M \hat{u}(0) = \tilde{P}(0) \left(\Pi_M \frac{\partial \hat{u}}{\partial \rho}(0) + \alpha \Pi_M \hat{u}(0) \right) + \Pi_M \tilde{r}(0) \\ \Rightarrow\quad & 0 = \tilde{P}(0) \Pi_M \frac{\partial \hat{u}}{\partial \rho}(0) + \Pi_M \tilde{r}(0) \Rightarrow \Pi_M \tilde{P}(0) = 0,\ \Pi_M \tilde{r}(0) = 0, \end{aligned} \tag{16}$$

$$\begin{aligned} &\Pi_N \hat{u}(0) = \tilde{P}(0) \left(\Pi_N \frac{\partial \hat{u}}{\partial \rho}(0) + \alpha \Pi_N \hat{u}(0) \right) + \Pi_N \tilde{r}(0) \\ \Rightarrow\quad & \hat{u}(0) = \alpha \tilde{P}(0) \hat{u}(0) + \Pi_N \tilde{r}(0) \Rightarrow \Pi_N \tilde{P}(0) = \frac{1}{\alpha} I,\ \Pi_N \tilde{r}(0) = 0. \end{aligned} \tag{17}$$

From (16) and (17) we obtain $\tilde{r}(0) = \Pi_M \tilde{r}(0) + \Pi_N \tilde{r}(0) = 0$. In the same way, since $\tilde{P}(0)h = \tilde{P}(0)\Pi_M h + \tilde{P}(0)\Pi_N h = \frac{1}{\alpha}\Pi_N h$ we obtain $\tilde{P}(0) = \frac{1}{\alpha}\Pi_N$.

7. Final Remarks

We believe that this method is much more general than presented here and that it can be extended to higher dimensions, to other operators than the Laplacian and more general domains. For example for star shaped domains the embedding can be done by homothety, taking the angle θ and the homothety factor as independent variables. Then the singularity at the origin is treated in the same way.

References

[1] E. Angel and R. Bellman. *Dynamic Programming and Partial Differential Equations.* Academic Press, London, 1972.

[2] R. Bellman. *Dynamic Programming.* Princeton University Press, Princeton, 1957.

[3] J. Henry and A.M. Ramos. Factorization of second order elliptic boundary value problems by dynamic programming. *to appear in Nonlinear Analysis, T.M.A.*.

[4] J. Henry, B. Louro and M.C. Soares. A factorization method for elliptic problems in a circular domain. *to appear in CRAS.*

[5] J.-L. Lions. *Contrôle optimal de systèmes gouvernés par des équations aux dérivées partielles.* Dunod, Paris, 1968.

ON IDENTIFIABILITY OF LINEAR INFINITE-DIMENSIONAL SYSTEMS

Yury Orlov
Electr. Dept., CICESE Research Center
Km. 107, Carretera Tijuana-Ensenada, Ensenada, B.C., Mexico 22860
yorlov@cicese.mx

Abstract Identifiability analysis is developed for linear dynamic systems evolving in a Hilbert space. Finite-dimensional sensing and actuation are assumed to be only available. Identifiability conditions for the transfer function of such a system is constructively addressed in terms of sufficiently nonsmooth controlled inputs. The introduced notion of a sufficiently nonsmooth input does not relate to a system and it can therefore be verified independently of any particular underlying system.

Keywords: Identifiability, Hilbert space, Markov parameters, sufficiently rich input.

Introduction

A standard approach to identifying a linear system implies that the structure of the system is deduced by using physical laws and the problem is in finding the values of parameters in the state equation. The ability to ensure this objective is typically referred to as parameter identifiability. For complex systems, however, it may not be possible to model all of the system and a black box approach should be brought into play. Here knowledge of the input-output map comes from controlled experiments. The questions then arise as to if the input-output map and the transfer function of the system have a one-to-one relation and in which sense a state space model, if any, is unique.

Many aspects of the identifiability of linear finite-dimensional systems are now well-understood. In this regard, we refer to [1], [5], and [7] to name a few monographs.

Recently [8], [9], the identifiability analysis was proposed in an infinite-dimensional setting for linear time-delay systems with delayed states, control inputs and measured outputs, all with a finite number of lumped delays. It was demonstrated that the transfer function of such a system can be re-constructed on-line whenever a sufficiently nonsmooth input signal is applied to the system. The present work extends this result to linear dynamic systems evolving in a Hilbert space. The transfer function identifiability conditions for these systems are also addressed in terms of sufficiently nonsmooth input signals which are constructively introduced in the Hilbert space. The notion of a sufficiently nonsmooth input does not relate to a system and it can therefore be verified independently of any particular underlying system.

Similar to linear finite-dimensional systems, the parameter identifiability in a Hilbert space requires the system with unknown parameters to be specified in a form such that if confined to this form all the unknown parameters are uniquely determined by the transfer function. In contrast to [2], [3], and [10] where the parameter identifiability is established for linear distributed parameter systems with sensing and actuation distributed over the entire state space, the present development deals with a practical situation where finite-dimensional sensing and actuation are only available.

1. Basic Definitions

We shall study linear infinite-dimensional systems

$$\dot{x} = Ax + Bu, \quad x(0) = x^0 \tag{1}$$

$$y = Cx, \tag{2}$$

defined in a Hilbert space H, where $x \in H$ is the state, x^0 is the initial condition, $u \in R^m$ is the control input, $y \in R^p$ is the measured output, A is an infinitesimal operator with a dense domain $\mathcal{D}(A)$, B is the input operator, C is the measurement operator. All relevant background materials on infinite-dimensional dynamic systems in a Hilbert space can be found, e.g., in [4].

The following assumptions are made throughout:

1 A generates an analytical semigroup $S_A(t)$ and has compact resolvent;

2 $B \in \mathcal{L}(R^m, H)$ and $C \in \mathcal{L}(H, R^p)$;

3 $x(0) = x^0 \in \bigcap_{n=1}^{\infty} \mathcal{D}(A^n)$;

4 $\mathcal{R}(B) \in \bigcap_{n=1}^{\infty} \mathcal{D}(A^n)$.

Hereafter, the notation is fairly standard. The symbol $\mathcal{L}(U, H)$ stands for the set of linear bounded operators from a Hilbert space U to H; $\mathcal{D}(A)$ is for the domain of the operator A; $\mathcal{R}(B)$ denotes the range of the operator B.

The above assumptions are made for technical reasons. It is well-known that under these assumptions, the Hilbert space-valued dynamic system (1), driven by a locally integrable input $u(t)$, has a unique strong solution $x(t)$, globally defined for all $t \geq 0$. The spectrum $\sigma(A) = \{\lambda_n\}_{n=1}^{\infty}$ of the operator A is discrete, $\lambda_n \to -\infty$ as $n \to \infty$, and the solution of (1) can be represented in the form of the Fourier series

$$x(t) = \Sigma_{n=1}^{\infty}\{\prec x^0, r_n \succ e^{\lambda_n t} + \int_0^t \prec Bu(\tau), r_n \succ e^{\lambda_n (t-\tau)} d\tau\} r_n \qquad (3)$$

written in terms of the eigenvectors r_n, $n = 1, \dots$ of the operator A and the inner product $\prec \cdot, \cdot \succ$ in the Hilbert space H. Furthermore, this solution is regular enough in the sense that $x(t) \in \bigcap_{n=1}^{\infty} \mathcal{D}(A^n)$ for all $t \geq 0$.

The identifiability concept is based on the comparison of system (1), (2) and its reference model

$$\dot{\hat{x}}(t) = \hat{A}\hat{x} + \hat{B}u, \ \hat{x}(0) = \hat{x}^0 \qquad (4)$$

$$\hat{y}(t) = \hat{C}\hat{x}, \qquad (5)$$

defined on a Hilbert space $\hat{H}$ with the initial condition $\hat{x}^0$ and operators $\hat{A}, \hat{B}, \hat{C}$, substituted in (1), (2) for x^0 and A, B, C, respectively. Certainly, Assumptions 1-4 remain in force for the reference model (4), (5).

The transfer function $T(\lambda) = C(\lambda I - A)^{-1}B$ of system (1), (2) is completely determined by means of the Markov parameters $CA^{n-1}B$, $n = 1, 2, \dots$ through expanding into the Laurent series

$$T(\lambda) = \Sigma_{n=1}^{\infty} \lambda^{-n} C A^{n-1} B,$$

and the identifiability of the transfer function is introduced as follows.

DEFINITION 1 *The transfer function of (1), (2), or equivalently, the Markov parameters of system (1), (2) are said to be identifiable on* $\bigcap_{n=1}^{\infty} \mathcal{D}(A^n)$ *iff there exists a locally integrable input function* $u(t)$ *to be sufficiently rich for the system in the sense that the identity* $y(t) \equiv \hat{y}(t)$ *implies that*

$$CA^{n-1}B = \hat{C}\hat{A}^{n-1}\hat{B}, \ n = 1, 2, \dots \qquad (6)$$

(and consequently $T(\lambda) = \hat{T}(\lambda)$*), regardless of a choice of the initial conditions* $x^0 \in \bigcap_{n=1}^{\infty} \mathcal{D}(A^n)$, $\hat{x}^0 \in \bigcap_{n=1}^{\infty} \mathcal{D}(\hat{A}^n)$. *In that case the identifiability is said to be enforced by the input* $u(t)$.

The identifiability of the system itself (rather than its transfer function) is addressed in a similar manner.

DEFINITION 2 *System (1), (2) is said to be identifiable on* $\bigcap_{n=1}^{\infty} \mathcal{D}(A^n)$ *iff there exists a locally integrable input function* $u(t)$ *such that the identity* $y(t) \equiv \hat{y}(t)$ *implies that*

$$C = \hat{C},\ A = \hat{A},\ B = \hat{B}, \tag{7}$$

regardless of a choice of the initial conditions $x^0 \in \bigcap_{n=1}^{\infty} \mathcal{D}(A^n)$, $\hat{x}^0 \in \bigcap_{n=1}^{\infty} \mathcal{D}(\hat{A}^n)$.

For later use, we also define sufficiently nonsmooth inputs, which form an appropriate subset of sufficiently rich inputs. Indeed, while being applied to a finite-dimensional system, a nonsmooth input has enough frequencies in the corresponding Fourier series representation and hence it turns out to be sufficiently rich in the conventional sense [6].

Let $u(t) = \Sigma_{i=1}^{m} u_i(t) e_i$ be a piece-wise smooth input, expanded in the basis vectors $e_i \in R^m$, $i = 1, \ldots m$, and let D_i be the set of discontinuity points $t \geq 0$ of $u_i(t)$.

DEFINITION 3 *The input* $u(t) = \Sigma_{i=1}^{m} u_i(t) e_i$ *is said to be sufficiently discontinuous iff for any* $i = 1, \ldots, m$ *there exists* $t_i \in D_i$ *such that* $t_i \notin D_j$ *for all* $j \neq i$.

DEFINITION 4 *The input* $u(t) = \Sigma_{i=1}^{m} u_i(t) e_i$ *is said to be sufficiently nonsmooth of class* C^l *iff there exists* l*-th order derivative of the input* $u(t)$ *and* $u^{(l)}(t)$ *is sufficiently discontinuous.*

It is worth noticing that the notion of a sufficiently nonsmooth (discontinuous) input does not relate to a system and it can therefore be verified independently of any particular underlying system.

2. Identifiability Analysis

Once the infinite-dimensional system (1), (2) is enforced by a sufficiently nonsmooth input, its transfer function is unambiguously determined by the input-output map. The transfer function identifiability in that case is guaranteed by the following result.

THEOREM 5 *Consider a Hilbert space-valued system (1), (2) with the assumptions above. Then the Markov parameters* $CA^{n-1}B$, $n = 1, 2, \ldots$ *of (1), (2) are identifiable and their identifiability can be enforced by an arbitrary sufficiently nonsmooth (particularly, sufficiently discontinuous) input* $u(t) = \Sigma_{i=1}^{m} u_i(t) e_i$.

Proof: For certainty, we assume that the input $u(t)$ is sufficiently discontinuous. The general proof in the case where $u(t)$ is sufficiently nonsmooth is nearly the same and it is therefore omitted.

According to Definition 1, we need to prove that the output identity

$$Cx(t) \equiv \hat{C}\hat{x}(t) \tag{8}$$

implies the equivalence (6) of the Markov parameters.

By differentiating (8) along the solutions of (1) and (4), we obtain that

$$CAx(t) + CB\Sigma_{i=1}^{m} u_i(t)e_i \equiv \hat{C}\hat{A}\hat{x}(t) + \hat{C}\hat{B}\Sigma_{i=1}^{m} u_i(t)e_i. \tag{9}$$

It follows that

$$CB = \hat{C}\hat{B} \tag{10}$$

because otherwise identity (9) could not remain true at the discontinuity instants t_i, $i = 1, \ldots, m$. Indeed, in spite of the discontinuous behavior of the input $u(t)$, the solution (3) of the differential equation (1) is continuous for all $t \geq 0$. Thus, by taking into account that $u(t)$ is sufficiently discontinuous (see Definition 3), $[CB - \hat{C}\hat{B}]u_i(t)e_i$ is the only term in (9), discontinuous at $t = t_i$. Hence, $[CB - \hat{C}\hat{B}]e_i = 0$ for each $i = 1, \ldots, m$, thereby yielding (10).

Now (9) subject to (10) is simplified to

$$CAx(t) \equiv \hat{C}\hat{A}\hat{x}(t) \tag{11}$$

and differentiating (11) along the solutions of (1) and (4), we arrive at

$$CA^2x(t) + CAB\Sigma_{i=1}^{m} u_i(t)e_i \equiv \hat{C}\hat{A}^2\hat{x}(t) + \hat{C}\hat{A}\hat{B}\Sigma_{i=1}^{m} u_i(t)e_i. \tag{12}$$

Following the same line of reasoning as before, we conclude from (12) that

$$CAB = \hat{C}\hat{A}\hat{B}. \tag{13}$$

Finally, the required equivalence (6) of the Markov parameters for $n \geq 3$ is obtained by iterating on the differentiation. Theorem 5 is thus proven.

In a fundamental term, Theorem 5 says that the input-output map (1), (2) and Markov parameters of the Hilbert space-valued system have a one-to-one relation. In general, the identifiability of the Markov parameters does not imply the identifiability of the system. Indeed, let $Q \in \mathcal{L}(H)$ be such that $\mathcal{D}(A)$ is invariant under Q and $Q^{-1} \in \mathcal{L}(H)$. Then the system

$$\begin{aligned} \dot{x} &= QAQ^{-1}x + QBu, \ x(0) = x^0 \\ y &= CQ^{-1}x, \end{aligned}$$

has the same Markov parameters CA^nB, $n = 1, 2, \ldots$.

Thus, in analogy to the finite-dimensional case, the identifiability of the infinite-dimensional system is guaranteed if it is in a canonical form such that if confined to this form the operators A, B, C are uniquely determined by the Markov parameters. The major challenge will be an explicit definition of the canonical form of a linear Hilbert space-valued system (1), (2).

References

[1] B. Astrom, K.J.and Wittenmark. *Adaptive Control.* Addison-Wesley, Reading, MA, 1989.

[2] J. Bentsman and Y. Orlov. Reference adaptive control of spatially varying distributed parameter systems of parabolic and hyperbolic types. *Internat. J. Adaptive Control and Signal Processing*, 15:679–696, 2001.

[3] M. Böhm, M.A. Demetriou, S. Reich, and I.G. Rosen. Model reference adaptive control of distributed parameter systems. *SIAM Journal on Control and Optimization*, 35:678–713, 1997.

[4] R. F. Curtain and H. J. Zwart. *An Introduction to Infinite-Dimensional Linear Systems Theory.* Springer-Verlag, New York, 1995.

[5] Y. D. Landau. *Adaptive Control - The Model Reference Approach.* Marcel Dekker, New York, 1979.

[6] R. K. Miller and A. N. Michel. An invariance theorem with applications to adaptive control. *IEEE Trans. Automat. Contr.*, 35:744–748, 1990.

[7] K. S. Narendra and A. Annaswamy. *Stable Adaptive Systems.* Prentice Hall, Englewood Cliffs, NJ, 1989.

[8] Y. Orlov, L. Belkoura, J.P. Richard, and M. Dambrine. On identifiability of linear time-delay systems. *IEEE Trans. Automat. Contr.*, 47:1319–1323, 2002.

[9] Y. Orlov, L. Belkoura, J.P. Richard, and M. Dambrine. Adaptive identification of linear time-delay systems. *Internat. J. Robust and Nonlinear Control*, 13:857–872, 20023.

[10] Y. Orlov and J. Bentsman. Adaptive distributed parameter systems identification with enforceable identifiability conditions and reduced spatial differentiation. *IEEE Trans. Automat. Contr.*, 45:203–216, 2000.

AN INVERSE PROBLEM FOR THE TELEGRAPH EQUATION

A.B. Kurzhanski*
Moscow State University, Russia
kurzhans@mail.ru

M.M. Sorokina
Moscow State University, Russia
masha_sorokina2@mail.ru

Abstract This paper deals with the problem of state estimation for a hyperbolic equation in the presence of unknown, but bounded disturbances, on the basis of information from sensors with finite-dimensional outputs. The object of investigation is the hyperbolic telegraph equation with energy dissipation. Observability properties similar to those introduced earlier for parabolic systems ([8]) are checked for various types of measurement sensors. Further on recurrent guaranteed minmax filtering procedures are introduced which give dynamic estimates of the current state of the system and dual control problems are indicated as well.

Keywords: Minmax filtering, telegraph equation, information set, sensors, observability.

Introduction

In this paper we consider the problem of state estimation for a system described by a hyperbolic equation of the "telegraph" type, with energy dissipation. This is to be done through available sensor measurements in the presence of unknown, but bounded disturbances.

We start with the the problem of observability which is the inverse problem of finding the final state for this system through available observations in the absence of any information on initial and boundary

*Funding provided by A.M. Liapunov French-Russian Institute, Project 02-01 and Russian Foundation for Basic Research (RFBR) grant N 03-01-00663.

conditions and in the absence of disturbances. This problem may turn to be solvable depending on the paricular type of sensor applied. Conditions for observability are therefore investigated.

We further derive a filtering equation which gives a set-membership estimate of the state of the system under unknown disturbances subjected to a given quadratic bound. These equations also produce a vector-valued estimate with respective bounds on the estimation error.

Finally some dual controllability problems are indicated.

1. The Telegraph Equation and the Estimation Problem

The telegraph equation is a PDE which describes, for example, an electric current transmission in the presence of wave aberration and depletion, namely ([2])

$$v_{xx} = lcv_{tt} + (lg + rc)v_t + rgv \tag{1}$$

$$i_{xx} = lci_{tt} + (lg + rc)i_t + rgi \tag{2}$$

where i is the current intensity, v is the voltage, r is the resistance, l is the induction, c is the capacity and g the conductivity. For $r, g = 0$ it turns into a wave equation.

We further consider the following system:

$$\begin{array}{ll} u_{xx} = \frac{1}{v_0^2}u_{tt} - \left(\frac{\sigma_1+\sigma_2}{v_0^2}\right)\frac{\partial u}{\partial t} + \frac{\sigma_1\sigma_2}{v_0^2}u + f & \text{in } Q_T \\ u|_{t=0} = \Phi(x) & \text{in } \Omega \\ u_t|_{t=0} = \Psi(x) & \text{in } \Omega \\ u|_{x=0} = \mu_1(t) \quad u|_{x=l} = \mu_2(t) & t \in [0,T], \quad x \in [0,l] = \Omega. \end{array} \tag{3}$$

Here $Q_T = \Omega \times (0,T)$, $S_T = \partial\Omega \times (0,T)$ and f is either a control or a disturbance (given or unknown).

An observation of the system performance is available through *measurement sensors* taken to be of the following types.

Examples of sensors

1 Spatially averaged $y(t) = \int_D g(\mathbf{x})u(t,\mathbf{x})d\mathbf{x} + \xi(t)$.

2 Pointwise $y(t) = u(t,\mathbf{x}^0) + \xi(t) \quad t \in [t_1,t_2] \; t_1 > 0$.

3 Dynamic pointwise $y(t) = u(t,\mathbf{x}(t)) + \xi(t) \quad t \in [t_1,t_2] \; t_1 > 0$.

4 Distributed observation $y(\mathbf{x}) = u(\theta,\mathbf{x}) + \xi(\mathbf{x})$.

5 Dynamic spatially averaged $y(t) = \int\limits_{O_\varepsilon(\mathbf{x}(t))} g(t,\mathbf{x})u(t,\mathbf{x})d\mathbf{x} + \xi(t)$.

Then the measurement equation may be presented as

$$y(t) = G_1(t)u(\cdot,t) + \xi(t), \quad t_0 \le t \le T \tag{4}$$

or

$$y(x) = G_2 u(x,\cdot) + \xi(x), \quad 0 \le x \le l, \tag{5}$$

where $G_1(t)(u(\cdot,t) \to \mathbf{R}^m), G_2(u(\cdot,\cdot) \to \{L_2^m(0,l), H_1(0,l)\})$ are the measurement maps given by one of the sensors of the above or a combination of these. The disturbance in equations (4, 5) is bounded in the space of observations $\mathcal{Y}$:

$$||\xi||_{\mathcal{Y}}^2 \le \mu^2 \tag{6}$$

To formulate the observation problem we need the notion of information or consistency set ([7], [10]).

DEFINITION 1 *The information set $U(\tilde{t}; y(\cdot))$ is the union of all states $\{u(\cdot,\tilde{t}),\ u_t(\cdot,\tilde{t})\}$ of system (3) at time $\tilde{t}$, for which there exists a tuple $\mathcal{D} = \{\Phi(\cdot), \Psi(\cdot), \mu_1(\cdot), \mu_2(\cdot), f(\cdot,\cdot), \xi(\cdot)\}$ (initial conditions, boundary conditions and input and measurement disturbance), satisfying (4, 6) (or (5, 6)) and consistent with observation $y(\cdot)$ due to equations (3, 4) (or (3, 5)).*

Problem A Find $U(0, y(\cdot))$ – *the set-valued estimate*, or produce a pointwise estimate for the state $u(0,\cdot)$.

With no bounds given for initial and boundary conditions and no disturbance f the information set $U(0, y(\cdot))$ may turn to be unbounded even with bounded measurement noise. This may happen, for example when the observation is pointwise, at a rational point x^0. Nevertheless, with bound (6) modified to include *all elements* of the tuple $\mathcal{D}$, the estimation Problem A makes sense even in the latter case.

A preferable type of solution to Problem A is a recurrent "guaranteed filtering" equation which describes the evolution of the estimate in time, on one hand, and also ensures numerical stability of the corresponding algorithm. A problem closely connected with Problem A is the one of *observability* of system (3, 4) (or (3, 5)).

Problem B Assume disturbances $f(\cdot,\cdot) = 0,\ \xi(\cdot) = 0$. In the absence of information on *input* – initial conditions $\{\Phi(\cdot), \Psi(\cdot)\}$ and boundary values $\mu_1(\cdot), \mu_2(\cdot)$ – determine conditions for solvability of the problem: given measurement $y(t),\ t \in [0,T]$, find *output* – the solution $u(\cdot,T)$ at time T.

The solution to this Problem B gives the so-called "observability condition" for system (3, 4) (or (3, 5)) with given type of measurement sensor G. It is important to understand which types of sensors ensure observability.

2. Some Properties of the Telegraph Equation

To produce our solutions we need some properties of the telegraph equation

The solution formula

THEOREM 2 *(Ladyzhenskaya) Let* $\Phi \in H_0^1$, $\Psi \in L_2(\Omega)$, $f \in L_{2,1}(Q_T)$, $\mu_1(t) = \mu_2(t) \equiv 0 \Rightarrow$ *there exists a unique solution of (3) from* $H_1(Q_T)$.

Here $H_1(\Omega) = \{\varphi | \varphi, \frac{\partial \varphi}{\partial x} \in L_2(\Omega)\}$, and H_0^1 is a subspace of $H_1(\Omega)$ where smooth functions with compact support form a dense set.

The solution of equation (3) can be written out through Green function, the latter being equal to

$$G(x, \xi, t) = -\frac{2}{l}\sum_{n=1}^{\infty} \sin\frac{\pi n x}{l} \sin\frac{\pi n \xi}{l} \frac{v_0^2 e^{\frac{\sigma_1+\sigma_2}{2}t}}{\sqrt{\nu_n}} \mathrm{sh}\sqrt{\nu_n}t$$

Here the frequencies ν_n are equal to: $\nu_n = \left(\frac{\sigma_1-\sigma_2}{2}\right)^2 - \left(v_0\frac{\pi n}{l}\right)^2$ due to wave dispersion. Using [1], the following theorem can be proved

THEOREM 3 *For* $\sigma < \sigma_{\max}(1)$ *the system* $\{e^{i\lambda_n t}\}$, $\lambda_{\pm n} = \pm\sqrt{-\nu_n}$ *is a Riesz basis in* $L_2(0, T_0)$, $T_0 = \frac{2v_0}{l}$;
For other values of σ $\exists n_0 : \sigma_{\max}(n_0) \leq \sigma < \sigma_{\max}(n_0+1)$, *in this case the system* $\{e^{i\lambda_n t}\}$, *where* $\lambda_{\pm n} = \pm\sqrt{-\nu_n}$, $n \geq n_0 + 1$; $\lambda_{\pm n} = \pm\frac{\pi n v_0}{l}$, $n = 0..n_0$ *will be a Riesz basis in* $L_2(0, T_0)$.

A biorthogonal system

Let $\sigma < \sigma_{\max}(1)$. We denote by

$$\varphi_n = e^{i\lambda_n t} = \cos\lambda_n t + i\sin\lambda_n t; \ \varphi_{-n} = e^{-i\lambda_n t} = \cos\lambda_n t - i\sin\lambda_n t \quad (7)$$

the system that forms Riesz basis. According to Bari theorem ([1]) there exists a biorthogonal system $\{\varphi_n'\}_{n\in Z}$ with uniformly bounded norms, and if $\{\varphi_n^0\}_{n\in Z}$ is the orthonormal system from Riesz basis definition (i.e. $\varphi_n = V\varphi_n^0$), then: $\varphi_n' = (V^{-1})^*\varphi_n^0$ $n \in Z$. It is possible to show that orthonormal system can be taken as

$$\{\varphi_n^0\}_{-\infty}^{\infty} = \{\cos\frac{v_0\pi n}{l}t + i\sin\frac{v_0\pi n}{l}t\}_{-\infty}^{\infty}$$

and V^{-1} will be bounded according to the same Bari theorem. In this case biorthogonal system elements $\{\varphi_n'\}_{n\in Z}$ can be constructed, and so it is also possible to construct system biorthogonal to $\{\sin\lambda_n t, \cos\lambda_n t\}$:

$$\begin{array}{ll} \psi_n = \cos\lambda_n t = \frac{\varphi_n+\varphi_{-n}}{2},\ n \in N & \phi_n = \sin\lambda_n t = \frac{\varphi_n-\varphi_{-n}}{2i},\ n \in N \\ \Rightarrow \psi_n' = \varphi_n' + \varphi_{-n}' & \Rightarrow \phi_n' = (\varphi_n' - \varphi_{-n}')i \Rightarrow \\ < \psi_n, \psi_k' >= T_0\delta_n^k,\ n,k \in N & < \phi_n, \phi_k' >= T_0\delta_n^k,\ n,k \in N \end{array} \quad (8)$$

3. Observability

In this section the system (3) is taken, with $f = \mu_1 = \mu_2 = 0$, coupled with observation equation (4) or (5) and bound (6). Let us first introduce several definitions (see [8]).

DEFINITION 4 *The system (3),(4) (or (3),(5)) with $f = \mu_1 = \mu_2 = 0$ is said to be* **weakly observable** *if for any signal $y(\cdot)$ observed due to equation (4) (or (5)) with zero disturbance ($\xi = 0$) there exists only one possible couple of initial conditions $\{\Phi, \Psi\}$ that generate the solution $u(\cdot,\cdot)$ which provides the signal $y(\cdot)$.*

Distributed observation of state and velocity at time t_1

This is the simplest situation. Let $f = 0$, $\mu_i = 0$ and the observation equations be as follows:

$$\begin{array}{ll} y_1(\cdot) = u(\cdot, t_1) + \xi_1(\cdot) & y_1,\ \xi_1 \in \mathcal{Y}_1 = L_2(0,l) \\ y_2(\cdot) = u_t(\cdot, t_1) + \xi_2(\cdot) & y_2,\ \xi_2 \in \mathcal{Y}_2 = L_2(0,l) \end{array} \tag{9}$$

Expanding Φ, Ψ, y_i and ξ_i into Fourier series over functions $\{\sin(\pi n x/l)\}$, and denoting corresponding Fourier coefficients by Φ^n, Ψ^n, y_n^1, y_n^2, ξ_n^1, ξ_n^2, we have:

$$\begin{aligned} y_n^1 &= \tfrac{2}{l}\tfrac{1}{\sqrt{\nu_n}} e^{\frac{\sigma_1+\sigma_2}{2}t_1}\{\mathrm{sh}\sqrt{\nu_n}t_1\,\Psi^n + (\sqrt{\nu_n}\mathrm{ch}\sqrt{\nu_n}t_1 - \\ &\quad - \tfrac{\sigma_1+\sigma_2}{2}\mathrm{sh}\sqrt{\nu_n}t_1)\Phi^n\} + \xi_n^0 \\ y_n^2 &= \tfrac{2}{l}\tfrac{1}{\sqrt{\nu_n}} e^{\frac{\sigma_1+\sigma_2}{2}t_1}\Big\{\tfrac{\sigma_1+\sigma_2}{2}\mathrm{sh}\sqrt{\nu_n}t_1\,\Psi^n + \sqrt{\nu_n}\mathrm{ch}\sqrt{\nu_n}t_1\,\Psi^n - \\ &\quad - \Big(\Big(\tfrac{v_0\pi n}{l}\Big)^2 + \sigma_1\sigma_2\Big)\mathrm{sh}\sqrt{\nu_n}t_1)\Phi^n\Big\} + \xi_n^1 \end{aligned}$$

The last system is always solvable and the Fourier components of initial state Φ, Ψ will be

$$\begin{aligned} \Phi^n &= -\frac{l}{2\sqrt{\nu_n}} e^{-\frac{\sigma_1+\sigma_2}{2}t_1}\Big((y_n^0 - \xi_n^0)(-\frac{\sigma_1+\sigma_2}{2}\mathrm{sh}\sqrt{\nu_n}t_1 - \\ &\quad -\sqrt{\nu_n}\mathrm{ch}\sqrt{\nu_n}t_1) + (y_n^1 - \xi_n^1)\mathrm{sh}\sqrt{\nu_n}t_1\Big) \\ \Psi^n &= -\frac{l}{2\sqrt{\nu_n}} e^{-\frac{\sigma_1+\sigma_2}{2}t_1}\Big((y_n^1 - \xi_n^1)(\frac{\sigma_1+\sigma_2}{2}\mathrm{sh}\sqrt{\nu_n}t_1 - \\ &\quad -\sqrt{\nu_n}\mathrm{ch}\sqrt{\nu_n}t_1) - (y_n^0 - \xi_n^0)(\sigma_1\sigma_2 + \Big(\frac{v_0\pi n}{l}\Big)^2)\mathrm{sh}\sqrt{\nu_n}t_1\Big) \end{aligned}$$

Hence, it follows that

THEOREM 5 *For sensor (9) the system is weakly observable.*

It is also interesting to investigate the property of strong observability.

DEFINITION 6 *The system (3), (4), (6) (or (3), (5), (6)) with* $f = \mu_1 = \mu_2 = 0$ *is said to be* **strongly observable** *if the information set of system initial states* $U(0, y(\cdot))$ *(p. 179) is a bounded set in* $L_2(\Omega)$, *whatever be the measurement* $y(\cdot)$.

To prove strong observability it is necessary that the series with components (9) converge. For $y_1,\ y_2 \in L_2(0,l)$ the series $\sum_{n=1}^{\infty} (\Psi^n)^2$ may not converge because in the expression for Ψ^n there is a component $\frac{n^2}{\sqrt{\nu_n}} \mathrm{sh}\sqrt{\nu_n} t_1 \sim n \sin\sqrt{-\nu_n} t_1$. In general this series does not converge. But if t_1 is such that:

$$t_1 : \frac{v_0 t_1}{l} \in \mathbf{Z} \tag{10}$$

where $\mathbf{Z}$ is the set of integers, then due to the properties of the eigenvalues ν_n we have:

$$\sin\sqrt{-\nu_n} t_1 = -\left(\frac{\sigma_1 - \sigma_2}{2}\right)^2 \frac{l t_1}{2 v_0 \pi n} + O\left(\frac{1}{n^3}\right)$$

THEOREM 7 *For sensor (9) the information domain of initial states* Φ *is bounded in* $L_2(0,l)$. *If instant* t_1 *satisfies (10), then the information domain for initial values of derivatives* u_t (Ψ) *will also be bounded in* $L_2(0,l)$. *If (10) does not hold, then one can only claim that the last set is bounded in* $V^* = (H^1(0,l))^*$

Distributed observation of state at two instants of time t_1, t_2 Take $f = \mu_1 = \mu_2 = 0$ with observation equations as

$$\begin{aligned} y_1(\cdot) &= u(\cdot, t_1) + \xi_1(\cdot) \quad y_1,\ \xi_1 \in \mathcal{Y}_1 = L_2(0,l) \\ y_2(\cdot) &= u(\cdot, t_2) + \xi_2(\cdot) \quad y_2,\ \xi_2 \in \mathcal{Y}_2 = L_2(0,l) \end{aligned} \tag{11}$$

Here a system similar to the one in previous section can be written down, but now its discriminant $D = \frac{4}{l^2\sqrt{\nu_n}} e^{(\sigma_1+\sigma_2)t_1} \mathrm{sh}\sqrt{\nu_n}(t_2 - t_1)$. For weak observability it is necessary and sufficient that discriminant $D \neq 0$.

$$t_2 - t_1 \notin \left\{\frac{\pi m}{-\sqrt{\nu_n}}\right\}_{n,m=1}^{\infty} \tag{12}$$

THEOREM 8 *For sensor (11) system (3) is weakly observable* $\Leftrightarrow$ *(12).*

Spatially averaged observations Let $f = \mu_1 = \mu_2 = 0$ with observation equation as

$$y(t) = \int_0^l u(x,t) w(x) dx + \xi(t) \quad y,\ \xi \in \mathcal{Y}_1 = L_2(0, T_0) \tag{13}$$

If now w_n are the Fourier coordinates of the weighting function $w(x)$ and $T \geq T_0$, multiplying observation equation by function $e^{-\frac{\sigma_1+\sigma_2}{2}t}$, and calculating scalar products with biorthogonal system functions (8) on $[0,\ T_0]$, we result in

$$\begin{aligned} y_{n,1} &= \tfrac{2}{l} w_n \cdot (\nu_n)^{-1/2} \cdot T_0 \{\Psi^n - \tfrac{\sigma_1+\sigma_2}{2}\Phi^n\} + \xi_{n,1} \\ y_{n,2} &= \tfrac{2}{l} w_n \cdot (\nu_n)^{-1/2} \cdot T_0 \{\sqrt{\nu_n}\Phi^n\} + \xi_{n,2} \end{aligned}$$

Here $y_{n,i}, \xi_{n,i}$ are scalar products of observations and disturbances with biorthogonal system functions with weight $e^{-\frac{\sigma_1+\sigma_2}{2}t}$. The initial state can be calculated as:

$$\begin{aligned} \Phi^n &= \frac{y_{n,1} - \xi_{n,1}}{T_0 w_n} \cdot \frac{l}{2} \\ \Psi^n &= \frac{\sqrt{\nu_n}(y_{n,0} - \xi_{n,0}) - \frac{\sigma_1+\sigma_2}{2}(y_{n,1} - \xi_{n,1})}{T_0 w_n} \cdot \frac{l}{2} \end{aligned}$$

THEOREM 9 *System (3) with sensor (13) having its coefficients $w_n \neq 0$ for all n is weakly observable for $T \geq T_0$.*

DEFINITION 10 *The system (3), (4), (6) (or (3), (5), (6)) with $f = \mu_1 = \mu_2 = 0$ is said to be* **ε-observable** *if the projection of the information set $U(0, y(\cdot))$ (p. 179) on any finite-dimensional subspace $X_r(0,l) = \mathrm{Span}\{w_{n_j}(\cdot)\}_{j=1}^r$ is bounded, whatever be the measurement $y(\cdot)$.*

Here system $\{w_{n_j}\}_{j=1}^r$ is the set of r arbitrary different functions from the system of eigenfunctions $\{\sqrt{2/l}\sin(\pi n x/l)\}_{n=1}^\infty$.

THEOREM 11 *The system with sensor (13) which satisfies $w_n \neq 0$ $n \leq N$ is observable in its first N Fourier components (its first N "harmonics") for $T \geq T_0$.*

COROLLARY 12 *If the coefficients w_n are non-zero $\forall n$, and $T \geq T_0$, then system (3) is observable the in first N Fourier components for any N and therefore ε-observable.*

A pointwise sensor at point x_0
Here the sensor equation is

$$y(\cdot) = u(x_0, \cdot) + \xi(\cdot) \quad y,\ \xi \in \mathcal{Y}_1 = L_2(0,T) \tag{14}$$

Along the lines of previous procedures, we also come to the next propositions.

THEOREM 13 *System (3) with sensor (14) where x_0/l is irrational is weakly observable for $T \geq T_0$.*

THEOREM 14 *System (3) with sensor (14) which satisfies $\cos\frac{\pi n x_0}{l} \neq 0$, $n \leq N$, is observable in its first N Fourier components for $T \geq T_0$. Moreover, if x_0/l is an irrational point, then the system is observable in its first N Fourier components for any N, and therefore is ε-observable.*

4. The Filtering Equations

State Estimation

Consider the problem of dynamic state estimation for the telegraph equation

$$\begin{aligned} &u_{xx} = \tfrac{1}{v_0^2}u_{tt} - \left(\tfrac{\sigma_1+\sigma_2}{v_0^2}\right)\tfrac{\partial u}{\partial t} + \tfrac{\sigma_1\sigma_2}{v_0^2}u \\ &u|_{t=0} = \Phi(x) - ? \\ &u_t|_{t=0} = \Psi(x) - ? \\ &u|_{x=0} = 0 \quad u|_{x=l} = 0 \end{aligned} \tag{15}$$

$$\begin{aligned} &y(t) = G(t)u(\cdot,t) + \xi(t) \\ &||\xi|| \leq \mu^2 \end{aligned} \tag{16}$$

The functional describing *the measure of uncertainty* in the system is taken as:

$$\begin{aligned} F(T) =&< \Phi - \Phi^0, N_1(\Phi - \Phi^0) > + < \Psi - \Psi^0, N_2(\Psi - \Psi^0) > + \\ &+ < y(\cdot) - G(\cdot)u(\cdot,\cdot), M(\cdot)(y(\cdot) - G(\cdot)u(\cdot,\cdot)) >_{L_2((0,l)\times(0,T))} \end{aligned} \tag{17}$$

Here operators N_1, N_2 may be interpreted as regularizers. Introduce operators $S_1(\cdot)$ (Green function), $S_3(\cdot)$:

$$\begin{aligned} &S_1(t) = G(x,\xi,t)\ ;\ S_3(t) = \frac{\partial G(x,\xi,t)}{\partial t} - (\sigma_1 + \sigma_2)G(x,\xi,t) \\ &\Rightarrow u(\cdot,t) = S_3(t)\Phi(\cdot) + S_1(t)\Psi(\cdot) \end{aligned} \tag{18}$$

then functional (17) can be rewritten as (in the last scalar product dots are omitted):

$$\begin{aligned} F(T) =&< \Phi - \Phi^0, N_1(\Phi - \Phi^0) > + < \Psi - \Psi^0, N_2(\Psi - \Psi^0) > + \\ &+ < y - GS_3\Phi - GS_1\Psi, M(y - GS_3\Phi - GS_1\Psi) >_{L_2((0,l)\times(0,T))} \end{aligned} \tag{19}$$

If $u_0(T)$ and $u_1(T)$ are the minimizers Φ, Ψ of (19), and if we denote $u(t,T)$ to be the backward solution generated by these minimizers, *i.e.*,

$u(t,T) = S_3(t)u_0(T) + S_1(t)u_1(T)$, then the minimizers should satisfy

$$
\begin{aligned}
u_0(T) &= \Phi^0 + N_1^{-1}\int_0^T S_3^*(t)G^*(t)M(t)[y(t) - G(t)u(t,T)]dt \\
u_1(T) &= \Psi^0 + N_2^{-1}\int_0^T S_1^*(t)G^*(t)M(t)[y(t) - G(t)u(t,T)]dt
\end{aligned}
\tag{20}
$$

Now denoting $u(T,T)$ by $\tilde{u}(T)$, differentiating (20) with respect to T, we get for $u(t,T)$:

$$
\begin{aligned}
\frac{\partial u(t,T)}{\partial T} &= [S_3(t)N_1^{-1}S_3^*(T) + S_1(t)N_2^{-1}S_1^*(T)]\times \\
&\times G^*(T)M(T)[y(T) - G(T)\tilde{u}(T)] - \\
&-\int_0^T [S_3(t)N_1^{-1}S_3^*(\tau) + S_1(t)N_2^{-1}S_1^*(\tau)]G^*(\tau)M(\tau)G(\tau)\frac{\partial u(\tau,T)}{\partial T}d\tau
\end{aligned}
\tag{21}
$$

If $K(t,T)$ is an operator solving the previous Fredholm equation, then

$$
\frac{\partial u(t,T)}{\partial T} = K(t,T)G^*(T)M(T)[y(T) - G(T)\tilde{u}(T)] \tag{22}
$$

and

$$
\begin{aligned}
K(t,T) = \quad & [S_3(t)N_1^{-1}S_3^*(T) + S_1(t)N_2^{-1}S_1^*(T)] - \int_0^T [S_3(t)N_1^{-1}S_3^*(\tau) + \\
& + S_1(t)N_2^{-1}S_1^*(\tau)]G^*(\tau)M(\tau)G(\tau)K(\tau,T)d\tau
\end{aligned}
\tag{23}
$$

Using notation $P(T) = K(T,T)$, and denoting

$$
\Xi(T) = G^*(T)M(T)(y(T) - G(T)\tilde{u}(T)) \qquad \mathcal{D}\cdot = v_0^2\frac{\partial^2\cdot}{\partial x^2} - \sigma_1\sigma_2\cdot \tag{24}
$$

we finally come to the following system describing the dynamics of the state estimate and the estimates of initial conditions:

$$
\begin{aligned}
&\frac{\partial^2\tilde{u}}{\partial T^2} - (\sigma_1+\sigma_2)\frac{\partial\tilde{u}}{\partial T} = \mathcal{D}\tilde{u} + P(T)\frac{\partial\Xi(T)}{\partial T} + \\
&\qquad + \left[2\frac{\partial P(T)}{\partial T} - \left.\frac{\partial K(t,T)}{\partial T}\right|_{t=T} - (\sigma_1+\sigma_2)P(T)\right]\Xi(T) \\
&\frac{\partial u_0}{\partial T} = N_1^{-1}\left\{S_3^*(T) - \int_0^T S_3^*(t)G^*(t)M(t)G(t)K(t,T)dt\right\}\cdot\Xi(T) \\
&\frac{\partial u_1}{\partial T} = N_2^{-1}\left\{S_1^*(T) - \int_0^T S_1^*(t)G^*(t)M(t)G(t)K(t,T)dt\right\}\cdot\Xi(T) \\
&\tilde{u}(0) = 0 \qquad u_0(0) = \Phi^0 \qquad u_1(0) = \Psi^0
\end{aligned}
\tag{25}
$$

The equation for $P(T)$ can also be written as follows

$$\begin{aligned}
&\frac{\partial^2 P}{\partial T^2} - (\sigma_1+\sigma_2)\frac{\partial P}{\partial T} = \mathcal{D}P + P[\mathcal{D}^* + (\sigma_1+\sigma_2)G^*MGP- \\
&\quad - (G^*MGP)'_T - G^*MGPG^*MGP] - \left.\frac{\partial K(t,T)}{\partial T}\right|_{t=T} G^*MGP \\
&P(0) = N_1^{-1} \qquad \left.\frac{\partial P}{\partial T}\right|_{T=0} = N_1^{-1}G^*(0)M(0)G(0)N_1^{-1}
\end{aligned} \tag{26}$$

And for $K(t,T)$:

$$\begin{aligned}
&\frac{\partial^2 K(t,T)}{\partial T^2} - (\sigma_1+\sigma_2)\frac{\partial K(t,T)}{\partial T} + \frac{\partial^2 K(t,T)}{\partial t^2} - (\sigma_1+\sigma_2)\frac{\partial K(t,T)}{\partial t} = \\
&\quad = \mathcal{D}K(t,T) + K(t,T)\mathcal{D}^* + K(t,T)[(\sigma_1+\sigma_2)G^*MGP- \\
&\quad - (G^*MGP)'_T - G^*MGPG^*MGP] - \frac{\partial K(t,T)}{\partial T}G^*MGP
\end{aligned} \tag{27}$$

The initial conditions for $K(t,T)$ can be obtained from equation (23).

The Dynamic Estimate of Initial Conditions. Consider the problem of dynamic initial state estimation for the telegraph equation. This problem is directly related to the observability property, since for the formulation of the filtering equations the existence of bounded inverse for certain operators is necessary. The related operators are invertible if the property of strong observability is true. These conditions are precisely the ones discussed in the observability sections. Such types of conditions were earlier introduced by J.L.Lions ([4]) and further studied in [3].

Consider problem (15,16). Let the measure of uncertainty in the system be the same as in (6):

$$\begin{aligned}
F(T) &= ||\xi||^2 = \\
&=< y(\cdot) - G(\cdot)u(\cdot,\cdot), M(\cdot)(y(\cdot) - G(\cdot)u(\cdot,\cdot)) >_{L_2((0,l)\times(0,T))}
\end{aligned} \tag{28}$$

If we again use the notations $S_1(\cdot)$, $S_3(\cdot)$ (see (18)), the functional (28) can be rewritten as:

$$\begin{aligned}
F(T) =< y(\cdot) - G(\cdot)S_3(\cdot)\Phi - G(\cdot)S_1(\cdot)\Psi, M(\cdot)(y(\cdot)- \\
- G(\cdot)S_3(\cdot)\Phi - G(\cdot)S_1(\cdot)\Psi) >
\end{aligned} \tag{29}$$

If $u_0(T)$ and $u_1(T)$ are the functions Φ, Ψ which minimize (29) for each given T, they satisfy relations

$$\begin{array}{l} \int_0^T S_3^*(t)G^*(t)M(t)[y - G(t)S_3(t)u_0(T) - G(t)S_1(t)u_1(T)]dt = 0 \\ \int_0^T S_1^*(t)G^*(t)M(t)[y - G(t)S_3(t)u_0(T) - G(t)S_1(t)u_1(T)]dt = 0 \end{array} \quad (30)$$

If the system (15-16) is strongly observable, then expressing the minimizers from (30) and differentiating them with respect to T it is possible to write down the following system of evolution equations:

$$\begin{array}{l} \frac{\partial u_0}{\partial T} = K_3(T)u_0(T) + \mathcal{K}_3(T)\{y(T) - G(T)S_1(T)u_1(T)\} - \mathcal{M}_3(T)\frac{\partial u_1(T)}{\partial T} \\ \frac{\partial u_1}{\partial T} = K_1(T)u_1(T) + \mathcal{K}_1(T)\{y(T) - G(T)S_3(T)u_0(T)\} - \mathcal{M}_1(T)\frac{\partial u_0(T)}{\partial T} \\ u_0(0) = 0 \qquad u_1(0) = 0 \end{array}$$

Here $K_i(t) = S_i^*(t)G^*(t)M(t)G(t)S_i(t)$, $\mathcal{K}_i(T) = (\int_0^T K_i(t)dt)^{-1}S_i^*(T)\times$ $\times G(T)M(T)$, $\mathcal{M}_i(T) = (\int_0^T K_i(t)dt)^{-1}\int_0^T S_i^*(t)G^*(t)M(t)G(t)S_j(t)dt$ where $i, j = 1, 3$ and $i \neq j$.

If system (15,16) is only ε-observable, then the projections of the informational domain on any finite-dimensional subspace are bounded. It is therefore possible to indicate the same types of filtering equations in finite-dimensional space. These would be written in terms of respective Fourier coefficients.

5. The Duality of Optimal Control and Observation problems

Consider the observability problem for the telegraph equation:

$$\begin{array}{l} \frac{\partial^2 u}{\partial t^2} - (\sigma_1 + \sigma_2)\frac{\partial u}{\partial t} = \mathcal{D}u \\ u|_{\partial\Omega} = 0 \quad u|_{t=0} = \Phi \quad u_t|_{t=0} = \Psi \\ y = Gu + \xi \\ ||\xi|| \leq \mu \end{array} \quad (31)$$

Here operator $\mathcal{D}$ is defined by (24). If the solution is presented as (18), the support function of the information set U is as follows:

$$\rho(l, U(0,T)) = \sup\Big(< \Phi, S_3^*(T)l > + < \Psi, S_1^*(T)l > + < \psi(\cdot), y(\cdot) -$$

$$- G(\cdot)S_1(\cdot)\Psi - G(\cdot)S_3(\cdot)\Phi - \xi(\cdot) >_{L_2(Y\times[0,T])} \Big| \; ||Gu - y|| \leq \mu \Big) =$$

$$= \inf_{\psi\in\mathcal{Y}^*} \left(< \psi, y > + \mu||\psi|| \;\middle|\; \begin{array}{ll} S_3^*(T)l = \int_0^T S_3^*(t)G^*(t)\psi(t)dt & (*) \\ S_1^*(T)l = \int_0^T S_1^*(t)G^*(t)\psi(t)dt & (**) \end{array} \right)$$

Since for the telegraph equation the Green function is $S_1(t,\cdot)$, equation (*) is the solution to the adjoint equation in inverse time with right-hand side $G^*\psi$:

$$\begin{aligned} &\frac{\partial^2 v}{\partial t^2} + (\sigma_1+\sigma_2)\frac{\partial v}{\partial t} = \mathcal{D}v + G^*(t)\psi(t,\cdot) \\ &v|_{\partial\Omega} = 0 \quad v|_{t=T} = 0 \quad v_t|_{t=T} = 0 \\ &v|_{t=0} = S_1^*(T,\cdot)l(\cdot). \end{aligned} \tag{32}$$

This leads to a control problem for system (32), where $\psi(t,\cdot) \in \mathcal{Y}^*$ is the control. Due to $S_1(0) = 0$ the controllability of the following system will be equivalent to the solvability of equation (**):

$$\begin{aligned} &\frac{\partial^2 v}{\partial t^2}F + (\sigma_1+\sigma_2)\frac{\partial v}{\partial t} = \mathcal{D}v - \frac{\partial}{\partial t}\{G^*(t)\psi(t,\cdot)\} - (\sigma_1+\sigma_2)G^*(t)\psi(t,\cdot) \\ &v|_{\partial\Omega} = 0 \quad v|_{t=T} = 0 \quad v_t|_{t=T} = 0 \\ &v|_{t=0} = S_3^*(T,\cdot)l(\cdot) - S_1^*(T)G^*(T)\psi(T), \\ &\text{where } \psi(t,\cdot) \in \mathcal{Y}^* \text{ is the control.} \end{aligned} \tag{33}$$

This is a control problem for system (33) in backward time.

Thus, the solution of the observation problem for system (31) under disturbances (noise) is equivalent to finding the control $\psi(t,\cdot) \in \mathcal{Y}^*$, which simultaneously steers the adjoint systems (32), (33) in backward time to the prescribed end-points, as given in (32), (33), under minimum of the norm conjugate to the one that bounds the observation noise ([4]). The prerequisite of such properties for finite-dimensional systems was given in [5], [6] and for infinite-dimensional time lag systems in [9].

References

[1] S.A. Avdonin and S.A. Ivanov. *Families of exponentials: the method of moments in controllability problems for distributed parameter systems.* Cambridge University Press, 1995.

[2] R. Feynmann, R.B. Leighton, and M. Sands. *The Feynmann Lectures in Physics.* Addison-Wesley, Reading, 1963.

[3] L.H. Ho. Observabilité frontière des equations des ondes. *Comptes Rendus de lÁcademie des Sciences de Paris*, 302:443–446, 1986.

[4] Lions J-L. *Contrôlabilité exacte, perturbations et stabilisation de systèmes distribués*, volume Tome 1 Controlabilité Exacte. Masson, 1988.

[5] R.E. Kalman. On the general theory of control systems, 1960. Proc. of the 1-st IFAC CONGRESS, v.1, Butterworths, London.

[6] N.N. Krasovski. *The theory of control of motion (in Russian).* M.: Nauka, 1968.

[7] A.B. Kurzhanski and A.Yu. Khapalov. An observation theory for distributed-parameter systems. *Journal of Math.Systems, Estimation and Control*, 1, 1990.

[8] A.B. Kurzhanski and I.F. Sivergina. ϵ-observability of distributed parameter systems, 1992. Trudy IMM UrO RAN, v. 1, pp. 122–137.

[9] A.B. Kurzhansky. On duality of problems of optimal control and observation. *J. Appl. Math Mechs*, 34, No. 3:..., 1970.

[10] A.B. Kurzhansky and I.F. Sivergina. On noninvertible evolutionary systems: guaranteed estimates and the regularization problem. *Sov.Math.Doklady*, 42, No. 2:451–455, 1991.

SOLVABILITY AND NUMERICAL SOLUTION OF VARIATIONAL DATA ASSIMILATION PROBLEMS

Victor Shutyaev*
Institute of Numerical Mathematics
Russian Academy of Sciences, Russia
shutyaev@inm.ras.ru

Abstract A class of quasilinear variational data assimilation problems on the identification of the the initial-value functions is considered for the models governed by evolution equations. The optimality system is reduced to the equation for the control function. The properties of the control equation are studied and the solvability theorems are proved for linear and quasilinear data assimilation optimality systems. The iterative algorithms for solving the problem are formulated and justified.

Keywords: Optimal control, data assimilation, iterative algorithms

1. Statement of Data Assimilation Problem

Let H and X be real separable Hilbert spaces such that X is imbedded into H continuously and densely, H^*, X^* are the spaces adjoint to H, X, respectively. We assume that $H \equiv H^*$, $(\cdot,\cdot)_{L_2(0,T;H)} = (\cdot,\cdot)$, $\|\cdot\| = (\cdot,\cdot)^{1/2}$. Let us consider also the spaces $Y^0 = L_2(0,T;\ H)$, $Y = L_2(0,T;\ X)$, $Y^* = L_2(0,T;\ X^*)$ of abstract functions $\varphi(t)$ with the values in H, X, X^*, respectively, and the space

$$W = \{\varphi \in L_2(0,T;\ X) : \frac{d\varphi}{dt} \in L_2(0,T;\ X^*)\}$$

with the norm

$$\|\varphi\|_W = (\|\frac{d\varphi}{dt}\|^2_{L_2(0,T;\ X^*)} + \|\varphi\|^2_{L_2(0,T;\ X)})^{1/2}\ .$$

*Funding provided by grant 03-01-00779 of the Russian Foundation for Basic Research

Let $a(t;\varphi,\psi)$ be a bilinear form defined for any $t \in [0,T]$, $\varphi,\psi \in X$ and satisfied the inequalities:

$$|a(t;\varphi,\psi)| \leq c_1 \|\varphi\|_X \, \|\psi\|_X \,, \quad c_1 = const > 0, \tag{1.1}$$

$$c_2 \|\varphi\|_X^2 \leq a(t;\varphi,\varphi) \,, \quad c_2 = const > 0 \,, \quad \forall\, t \in [0,T] \,, \quad \forall\, \varphi,\psi \in X \,. \tag{1.2}$$

By $A(t) \in \mathcal{L}(Y,Y^*)$ we denote the operator generated by this form:

$$(A(t)\varphi,\psi)_H = a(t;\varphi,\psi) \qquad \forall\, \varphi,\psi \in X \,. \tag{1.3}$$

Consider the following quasilinear evolution problem:

$$\begin{cases} \frac{d\varphi}{dt} + A(t)\varphi + \tau F(\varphi) = f(t) \,, \quad t \in (0,T) \\ \varphi(0) = u, \end{cases} \tag{1.4}$$

where $f \in Y^*$, $u \in H$, $\tau \in [-\tau_0, \tau_0]$ is a parameter, $\tau_0 \in \mathbf{R}^+$, $F(\varphi)$ is a nonlinear Frechet differentiable operator, $F : Y \to Y^*$. Introduce a functional of $u \in H$ of the form:

$$S(u) = \frac{\alpha}{2} \|u - \widehat{\varphi}^0\|_H^2 + \frac{1}{2} \|B\varphi - \widehat{\varphi}\|_Z^2, \tag{1.5}$$

where $\alpha = const \geq 0$, Z is a Hilbert space (observational space) with the scalar product $(\cdot,\cdot)_Z$ and the norm $\|\cdot\|_Z = (\cdot,\cdot)_Z^{1/2}$, $B : \; Y \to Z$ is a linear bounded operator, $\widehat{\varphi}^0 \in H$, $\widehat{\varphi} \in Z$. The functions $\widehat{\varphi}^0$, $\widehat{\varphi}$ are generally determined by *a priory* observational data. The coefficient α is a regularization parameter [1].

Consider the following data assimilation problem: for given $f \in Y^*$, $\widehat{\varphi} \in Z$, find $u \in H$, $\varphi \in W$ such that

$$\begin{cases} \frac{d\varphi}{dt} + A(t)\varphi + \tau F(\varphi) &= f, \;\; t \in (0,T) \\ \varphi(0) &= u \\ S(u) &= \min\limits_{\tilde{u} \in H} S(\tilde{u}). \end{cases} \tag{1.6}$$

The problems of the form (1.6) were studied by L.S.Pontryagin [7], J.-L.Lions [5], [2]and many others (see Refs.)

The necessary optimality condition [5]reduces the problem (1.6) to the system for finding the functions $\varphi, \varphi^* \in W$, $u \in H$, of the form:

$$\frac{\varphi}{dt} + A(t)\varphi + \tau F(\varphi) = f, \; t \in (0,T); \quad \varphi(0) = u \tag{1.7}$$

$$-\frac{d\varphi^*}{dt} + A^*(t)\varphi^* + \tau (F'(\varphi))^* \varphi^* = C\widehat{\varphi} - K\varphi, \; t \in (0,T); \quad \varphi^*(T) = 0, \tag{1.8}$$

$$\alpha(u - \widehat{\varphi}^0) - \varphi^*(0) = 0, \tag{1.9}$$

where $(F'(\varphi))^* : Y \to Y^*$ is the operator adjoint to the Frechet derivative of F at the point $\varphi \in W$, $A^*(t) : Y \to Y^*$ is adjoint to $A(t)$, $K : Y \to Y^*$, $C : Z \to Y^*$ are linear bounded operators, $K = CB$, C is defined by the equality $(C\theta, \psi) = (\theta, B\psi)_Z \ \forall\, \theta \in Z, \psi \in Y$, and equations (1.7), (1.8) are considered in the space Y^*.

2. Linear Data Assimilation Problem

Consider the problem (1.7)–(1.9) for $\tau = 0$. The solutions of problems (1.7), (1.8) for $\tau = 0$ may by represented [6]as

$$\varphi = G_0 u + G_1 f, \ \ \varphi^* = G_1^{(T)}(C\widehat{\varphi} - K\varphi), \tag{2.1}$$

where $G_0 : H \to W$, $G_1 : Y^* \to W$, $G_1^{(T)} : Y^* \to W$ are linear bounded operators. Eliminating φ, φ^* from (1.7)–(1.9) for $\tau = 0$, we come to the equation for the control u:

$$Lu = P, \tag{2.2}$$

where the operator $L : H \to H$ and the right-hand side P are defined by

$$L = \alpha E + T_0 G_1^{(T)} K G_0, \ \ P = \alpha\widehat{\varphi}^0 + T_0 G_1^{(T)} C\widehat{\varphi} - T_0 G_1^{(T)} K G_1 f, \tag{2.3}$$

E is the identity operator, $T_0 : \ W \to H$ is the trace operator: $T_0\varphi = \varphi|_{t=0}$.

Consider the operator L for $\alpha = 0$ and denote it by $\bar{L}$. Let $G_0 : H \to W$ be the operator from (2.1), where the element $G_0 u$ is defined as the solution of (1.7) for $\tau = 0, f = 0$. The following statement holds.

LEMMA 1 *The operator $\bar{L} : \ H \to H$ is continuous, self-adjoint, and positive semi-definite:*

$$(\bar{L}v, v)_H \geq 0 \ \ \forall v \in H.$$

If the operator $BG_0 : H \to Z$ is invertible, the operator $\bar{L}$ is positive: $(\bar{L}v, v)_H > 0 \ \ \forall v \in H, v \neq 0$.

Proof. Let $\rho \in H$ and $\varphi = G_0\, \rho$. Then

$$\bar{L}\rho = T_0 G_1^{(T)} K\varphi.$$

The first assertion of Lemma 1 was proved in [12]. The positive definiteness or semi-definiteness of $\bar{L}$ follow from the equalities:

$$(\bar{L}\rho, \rho)_H = (T_0 G_1^{(T)} K G_0\, \rho, \rho)_H = (K\varphi, \varphi) = (B\varphi, B\varphi)_Z = \|BG_0\rho\|_Z^2.$$

The lemma is proved.

From Lemma 1, we get

LEMMA 2 *If the operator $BG_0 : H \to Z$ is invertible, then the range $R(\bar{L})$ of the operator $\bar{L}$ is dense in H, and the equation $\bar{L}u = P$ is solvable uniquely and densely in H.*

Remark 1. In case of "complete observation", when $Z = Y^0, B = E$ (the identity operator), we have $C = E, K = E$, and the operator $\bar{L}$ is positive.

Introduce the following additional restriction on the operator $A(t)$:

Hypothesis (A): *For any $p \in Y^0$ the solution φ^* of the adjoint problem*

$$-\frac{d\varphi^*}{dt} + A^*(t)\varphi^* = p,\ t \in (0,T); \quad \varphi^*(T) = 0$$

satisfies the inequality $\|\varphi^*(0)\|_X \le c\|p\|_{Y^0}$, $c = const > 0$.

Remark 2. The hypothesis (A) is satisfied for a wide class of operators $A(t)$, among them – the second-order elliptic operators in uniformly parabolic problems [4], [12].

LEMMA 3 *Let X be compactly imbedded into H, the hypothesis* (A) *be satisfied, and the operator $K : Y^0 \to Y^0$ be bounded. Then the operator $\bar{L} : H \to H$ is compact.*

Proof. Let us prove that $\bar{L}$ maps a bounded set of H into a compact set. Consider $u \in H$ such that $\|u\|_H \le c_0$, $c_0 = const > 0$. Let $\varphi = G_0 u$, $\varphi^* = G_1^{(T)} K\varphi$, then $\bar{L}u = \varphi^*(0)$. Since

$$\|\varphi\|_W \le c_1 \|u\|_H \,, \ \|\varphi^*\|_W \le c_2 \|K\varphi\|_{Y^*} \,, \quad c_1, c_2 = const > 0 \,,$$

and by the hypothesis (A),

$$\|\varphi^*(0)\|_X \le c\|K\varphi\|_{Y^0},\ c = const > 0,$$

then, due to the boundedness of $K : Y^0 \to Y^0$, we get

$$\|\bar{L}u\|_X \le c_3\|u\|_H \le c_3 c_0,$$

where $c_3 = const > 0$. However, X is compactely imbedded into H, hence the set $M = \{\bar{L}u : \|u\|_H \le c_0\}$ is compact in H, i.e. the operator $\bar{L} : H \to H$ is compact.

LEMMA 4 *The spectrum $\sigma(\bar{L})$ of the operator $\bar{L}$ satisfies*

$$0 \le \sigma(\bar{L}) \le \nu^2 \|B\|^2 \tag{2.5}$$

with the constant ν *from the inequality* $\|\varphi\|_Y \leq \nu \; \|u\|_H$, *where* $u \in H$, *and* $\varphi = G_0 u$ *is the solution of the problem* $\frac{d\varphi}{dt} + A(t)\varphi = 0, \; t \in (0,T); \;\; \varphi(0) = u$.

Proof. To estimate the spectrum of the self-adjoint operator $\bar{L}$ consider $(\bar{L}u, u)$ for $u \in H$. Let $\varphi = G_0 u, \; \varphi^* = G_1^{(T)} \varphi$, then

$$(\bar{L}u, u)_H = (\varphi^*(0), u)_H = (K\varphi, \varphi) = \|B\varphi\|_Z^2$$
$$\leq \|B\|^2 \; \|\varphi\|_Y^2 \leq \nu^2 \|B\|^2 \; \|u\|_H^2.$$

Hence,

$$\sigma(\bar{L}) \leq \sup_{u \in H, \, u \neq 0} \frac{(\bar{L}u, u)}{(u, u)} \leq \nu^2 \; \|B\|^2.$$

This ends the proof.

For case of complete observation, from Lemmas 1–3 we have the following

LEMMA 5 *Let* $Z = Y^0$, $B = E$ *(the identity operator),* X *be compactly imbedded into* H *and the hypothesis* (A) *be satisfied. Then the operator* $\bar{L}^{-1} : H \to H$ *exists, being unbounded; zero is the point of the continuous spectrum of the operator* $\bar{L}$; *the equation* $\bar{L}u = P$ *is solvable in* H *if and only if*

$$\sum_{k=1}^{\infty} \mu_k^{-2} (P, u_k)_H^2 < \infty,$$

where u_k *is the orthonormal system of the eigenfunctions of the compact operator* $\bar{L}$, *corresponding to the eigenvalues* μ_k.

The spectrum bounds of the operator L are very important for justification and optimization of iterative algorithms for solving the original data assimilation problem. Some estimates for the spectrum bounds may be derived using Lemma 4. If $K = E$, for the spectrum $\sigma(L)$ of the operator L defined by (2.2) the following estimates hold [11]:

$$m \leq \sigma(L) \leq M, \tag{2.7}$$

where

$$m = \alpha + \int_0^T e^{-\int_0^t \lambda_{\max}(\tau) d\tau} dt, \quad M = \alpha + \int_0^T e^{-\int_0^t \lambda_{\min}(\tau) d\tau} dt,$$

and $\lambda_{\min}, \lambda_{\max}$ are the lower and upper bounds, respectively, of the spectrum of the operator $A + A^*$.

If $K = E$, and $A(t) = A : H \to H$ is a linear closed operator independent of time, being unbounded self-adjoint positive definite operator in H with the compact inverse, then the eigenvalues μ_k of the operator $\bar{L}$ are defined by the formula [11]:

$$\mu_k = \frac{1 - e^{-2\lambda_k T}}{2\lambda_k},$$

where λ_k are the eigenvalues of the operator A. In this case the estimates (2.7) are exact, because in (2.7) $\lambda_{\min} = 2\lambda_1$, $\lambda_{\max} = \infty$, and m, M are given in the explicit form:

$$m = \alpha, \quad M = \alpha + \frac{1 - e^{-2\lambda_1 T}}{2\lambda_1}. \tag{2.8}$$

where λ_1 is the least eigenvalue of the operator A.

From Lemma 1 it follows that for $\alpha > 0$ the operator $L : H \to H$ is positive definite (i.e. coercive). Then, using the well-known results on solvability of linear optimal control problems [5]we come to the solvability theorem for the linear problem (1.7)–(1.9):

THEOREM 6 *Let $f \in Y^*$, $\widehat{\varphi}^0 \in H, \widehat{\varphi} \in Z$. Then for $\alpha > 0$ the problem (1.7)–(1.9) for $\tau = 0$ has a unique solution $\varphi_0 \in W$, $\varphi_0^* \in W$, $u_0 \in H$, and the following estimate holds:*

$$\|\varphi_0\|_W + \|\varphi_0^*\|_W + \|u_0\|_H \le c_0(\|\widehat{\varphi}^0\|_H + \|C\widehat{\varphi}\|_{Y^*} + \|f\|_{Y^*}), \; c_0 = const > 0 \tag{2.9}$$

3. Solvability of Nonlinear Problem

Let c_0 be the constant from (2.9). The following theorem holds:

THEOREM 7 *Let $f \in Y^*$, $\widehat{\varphi}^0 \in H$; $\widehat{\varphi} \in Z$ and for some $R > 0$ the inequalities*

$$\|F'(\xi)\|_{Y \to Y^*} \le k_1, \quad \|F'(\xi) - F'(\eta)\|_{Y \to Y^*} \le k_2 \|\xi - \eta\|_W \tag{3.2}$$

are satisfied for any $\xi, \eta \in B(\varphi_0, R) = \{\varphi \in Y : \|\varphi - \varphi_0\|_W \le R\}$, where $k_i = k_i(\varphi_0, R) = const > 0$. Then for $|\tau| \le \tau_0$, with

$$\tau_0 = 1/c_0[k_1 + k_2(R + \|\varphi_0^*\|_W) + \frac{1}{R}(\|F(\varphi_0)\|_{Y^*} + k_1\|\varphi_0^*\|_W)]^{-1}, \tag{3.3}$$

the problem (1.2)–(1.4) has a unique solution $(\varphi, \varphi^*, u) \in W \times W \times H$.

Proof. Consider the problem for the remainders $\tilde{\varphi} = \varphi - \varphi_0$, $\tilde{\varphi}^* = \varphi^* - \varphi_0^*$, $\tilde{u} = u - u_0$, where $(\varphi_0, \varphi_0^*, u_0)$ is the solution to the problem

(1.7)–(1.9) for $\tau = 0$. The problem for $\tilde{\varphi}, \tilde{\varphi}^*, \tilde{u}$ reads:

$$\frac{d\tilde{\varphi}}{dt} + A(t)\tilde{\varphi} + \tau F(\varphi_0 + \tilde{\varphi}) = 0,\ t \in (0,T); \quad \tilde{\varphi}(0) = \tilde{u}, \tag{3.4}$$

$$-\frac{d\tilde{\varphi}^*}{dt} + A^*(t)\tilde{\varphi}^* + \tau(F'(\varphi_0 + \tilde{\varphi}))^*(\varphi_0^* + \tilde{\varphi}^*) = -K\tilde{\varphi},\ t \in (0,T);$$

$$\tilde{\varphi}^*(T) = 0, \tag{3.5}$$

$$\alpha\tilde{u} - \tilde{\varphi}^*(0) = 0. \tag{3.6}$$

Consider the following iterative process:

$$\frac{d\tilde{\varphi}^{(n+1)}}{dt} + A(t)\tilde{\varphi}^{(n+1)} + \tau F(\tilde{\varphi}^{(n)} + \varphi_0) = 0,\ t \in (0,T);$$

$$\tilde{\varphi}^{(n+1)}(0) = \tilde{u}^{(n+1)}, \tag{3.7}$$

$$-\frac{d\tilde{\varphi}^{*(n+1)}}{dt} + A^*(t)\tilde{\varphi}^{*(n+1)} + \tau(F'(\tilde{\varphi}^{(n)} + \varphi_0))^*(\tilde{\varphi}^{*(n)} + \varphi_0^*) = -K\tilde{\varphi}^{(n+1)},$$

$$\tilde{\varphi}^{*(n+1)}(T) = 0, \tag{3.8}$$

$$\alpha\tilde{u}^{(n+1)} - \tilde{\varphi}^{*(n+1)}(0) = 0 \tag{3.9}$$

for $\|\tilde{\varphi}^{(0)}\|_W + \|\tilde{\varphi}^{*(0)}\|_W \leq R$. Since (for a fixed n) $\tilde{\varphi}^{(n+1)}$, $\tilde{\varphi}^{*(n+1)}$, $\tilde{u}^{(n+1)}$ is the solution of the linear problem, then, in view of (3.1), it is easily seen that

$$\|\tilde{\varphi}^{(n+1)}\|_W + \|\tilde{\varphi}^{*(n+1)}\|_W + \|\tilde{u}^{(n+1)}\|_H \leq k|\tau|(\|\tilde{\varphi}^{(n)}\|_W + \|\tilde{\varphi}^{*(n)}\|_W) + f_0,$$

where

$$k = c_0(k_1 + k_2(R + \|\varphi_0^*\|_W)),\quad f_0 = c_0|\tau|(\|F(\varphi_0)\|_{Y^*} + k_1\|\varphi_0^*\|_W).$$

By successive use of the last inequality, we get

$$\|\tilde{\varphi}^{(n)}\|_W + \|\tilde{\varphi}^{*(n)}\|_W + \|\tilde{u}^{(n)}\|_H \leq (k|\tau|)^n(\|\tilde{\varphi}^{(0)}\|_W + \|\tilde{\varphi}^{*(0)}\|_W) +$$

$$+ \frac{1-(k|\tau|)^n}{1-k|\tau|} f_0 \leq (k|\tau|)^n R + \frac{1-(k|\tau|)^n}{1-k|\tau|} f_0 \leq R \tag{3.10}$$

if $|\tau| \leq \tau_0$. Then, consider the problem for $\tilde{\varphi}^{(n+1)} - \tilde{\varphi}^{(n)}, \tilde{\varphi}^{*(n+1)} - \tilde{\varphi}^{*(n)}, \tilde{u}^{(n+1)} - \tilde{u}^{(n)}$. This leads to the estimate:

$$\|\tilde{\varphi}^{(n+1)} - \tilde{\varphi}^{(n)}\|_W + \|\tilde{\varphi}^{*(n+1)} - \tilde{\varphi}^{*(n)}\|_W + \|\tilde{u}^{(n+1)} - \tilde{u}^{(n)}\|_H \leq$$

$$\leq k|\tau|(\|\tilde{\varphi}^{(n)} - \tilde{\varphi}^{(n-1)}\|_W + \|\tilde{\varphi}^{*(n)} - \tilde{\varphi}^{*(n-1)}\|_W),$$

which implies

$$\tilde{\varphi}^{(n)} \to \tilde{\varphi}, \quad \tilde{\varphi}^{*(n)} \to \tilde{\varphi}^*, \quad \tilde{u}^{(n)} \to \tilde{u} \text{ as } n \to \infty, \quad \text{for } |\tau| \le \tau_0,$$

where $\tilde{\varphi}, \tilde{\varphi}^*, \tilde{u}$ is the solution to the problem (3.4)–(3.6), and the convergence rate estimate holds:

$$\|\tilde{\varphi}^{(n)} - \tilde{\varphi}\|_W + \|\tilde{\varphi}^{*(n)} - \tilde{\varphi}^*\|_W + \|\tilde{u}^{(n)} - \tilde{u}\|_H \le c\frac{(k|\tau|)^n}{1 - k|\tau|} \tag{3.11}$$

with $c = const > 0$. It is easily seen that for $|\tau| \le \tau_0$ this solution is unique and satisfies the condition $\|\tilde{\varphi}\|_W + \|\tilde{\varphi}^*\|_W + \|\tilde{u}\|_H \le R$. Thus, under the hypotheses of Theorem, there exists a unique solution of the problem (1.7)–(1.9). Theorem is proved.

If the operator $F(\varphi)$ is analytic, then the functions (φ, φ^*, u) are represented as the series in the powers of τ:

$$\varphi = \varphi_0 + \sum_{i=1}^{\infty} \tau^i \varphi_i\,, \quad \varphi^* = \varphi_0^* + \sum_{i=1}^{\infty} \tau^i \varphi_i^*\,, \quad u = u_0 + \sum_{i=1}^{\infty} \tau^i u_i\,,$$

convergent for $|\tau| < \tau_0$ in W, W, H, respectively, where $\varphi_i, \varphi_i^*, u_i$ may be found by the small parameter method [8].

4. Iterative Algorithms

To solve (1.7)–(1.9) one may use the successive approximation method (3.7)–(3.9). Each step of this method involves a linear data assimilation problem of the form (1.7)–(1.9) for $\tau = 0$. To solve it we consider a class of iterative algorithms:

$$\frac{d\varphi^k}{dt} + A(t)\varphi^k = f, \; t \in (0,T); \quad \varphi^k(0) = u^k, \tag{4.1}$$

$$-\frac{d\varphi^{*k}}{dt} + A^*(t)\varphi^{*k} = C\widehat{\varphi} - K\varphi^k, \; t \in (0,T); \quad \varphi^{*k}(T) = 0, \tag{4.2}$$

$$u^{k+1} = u^k - \alpha_{k+1} B_k(\alpha(u^k - \widehat{\varphi}^0) - \varphi^{*k}|_{t=0}) + \beta_{k+1} C_k(u^k - u^{k-1}), \tag{4.3}$$

where $B_k, C_k : H \to H$ are some operators, and $\alpha_{k+1}, \beta_{k+1}$ the iterative parameters.

Let $\gamma = \nu^2 \|B\|^2$ with ν defined in (2.5). We introduce the following notations:

$$\tau_{opt} = 2(2\alpha + \gamma)^{-1}, \quad \theta = (2\alpha + \gamma)\gamma^{-1}, \tag{4.4}$$

$$\tau_k = 2(2\alpha + \gamma - \gamma \cos \omega_k \pi)^{-1}, \quad k = 1, 2, \ldots, s, \tag{4.5}$$

$$\alpha_{k+1} = \begin{cases} 2(2\alpha+\gamma)^{-1}, & k=0 \\ 4\gamma^{-1}\dfrac{T_k(\theta)}{T_{k+1}(\theta)}, & k>0 \end{cases};$$

$$\beta_{k+1} = \begin{cases} 0, & k=0 \\ \dfrac{T_{k-1}(\theta)}{T_{k+1}(\theta)}, & k>0, \end{cases} \tag{4.6}$$

$$e_k = \begin{cases} 0, & k=0 \\ p_k\|\xi^k\|_H^2/\|\xi^{k-1}\|_H^2, & k>0, \end{cases} \tag{4.7}$$

$$p_{k+1} = \alpha + (K\eta^k,\eta^k)/\|\xi^k\|_H^2 - e_k, \quad k=0,1,\dots, \tag{4.8}$$

where $\omega_k = (2i-1)/2s$, T_k is the k-th degree Chebyshev polynomial of the first kind, $\xi^k = \alpha(u^k - \widehat{\varphi}^0) - \varphi^{*k}(0)$, and η^k is the solution of the problem $\dfrac{d\eta^k}{dt} + A\eta^k = 0,\ t \in (0,T);\quad \eta^k(0) = \xi^k$.

THEOREM 8 (I) *If* $\alpha_{k+1} = \tau$, $B_k = E$, $\beta_{k+1} = 0$, $0 < \tau < 2/(\alpha+\gamma)$, *then the iterative process* (4.1)–(4.3) *is convergent. For* $\tau = \tau_{opt}$ *defined by* (4.4) *the following convergence rate estimates are valid:*

$$\|\varphi - \varphi^k\|_W \le c_1 q_k, \quad \|\varphi^* - \varphi^{*k}\|_W \le c_2 q_k, \quad \|u - u^k\|_H \le c_3 q_k, \tag{4.9}$$

where $q_k = 1/\theta^k, \theta$ *is given by* (4.4), *and the constants* c_1, c_2, c_3, c_4 *do not depend on the number of iterations and on the functions* $\varphi, \varphi^k, \varphi^*, \varphi^{*k}$, $u, u^k, k > 0$.

(II) *If* $B_k = E$, $\beta_{k+1} = 0$, *and* $\alpha_{k+1} = \tau_k$, *where the parameters* τ_k *are defined by* (4.5) *and repeated cyclically with the period* s, *then the error in the iterative process* (4.1)–(4.3) *is suppressed after each cycle of the length* s. *After* $k = ls$ *iterations the error estimates* (4.9) *are valid with* $q_k = (T_s(\theta))^{-l}$.

(III) *If* $B_k = C_k = E$ *and* $\alpha_{k+1}, \beta_{k+1}$ *are defined by* (4.6), *then the error in the algorithm* (4.1)–(4.3) *is suppressed for each* $k \ge 1$, *and the estimates* (4.9) *hold for* $q_k = (T_k(\theta))^{-1}$.

(IV) *If* $B_k = C_k = E$ *and* $\alpha_{k+1} = 1/p_{k+1}$, $\beta_{k+1} = e_k/p_{k+1}$, *where* e_k, p_{k+1} *are defined by* (4.7), (4.8), *then the iterative process* (4.1)–(4.3) *is convergent, and the convergence rate estimates* (4.9) *are valid with* $q_k = (T_k(\theta))^{-1}$.

Proof.

The iterative process (4.1)–(4.3) is equivalent to the following iterative algorithm [12]:

$$u^{k+1} = u^k - \alpha_{k+1} B_k (Lu^k - P) + \beta_{k+1} C_k (u^k - u^{k-1}) \tag{4.10}$$

for solving the control equation $Lu = P$, where L and P are defined in (2.2).

According to Lemma 4, the bounds of the spectrum of the control operator L are given by

$$m \stackrel{\text{def}}{=} \inf_{u \in H,\, u \neq 0} \frac{(Lu, u)}{(u, u)} \geq \alpha, \quad M \stackrel{\text{def}}{=} \sup_{u \in H,\, u \neq 0} \frac{(Lu, u)}{(u, u)} \leq \alpha + \nu^2 \|B\|^2. \tag{4.11}$$

Thus, for $\alpha > 0$ for solving the equation $Lu = P$ we may use the well-known iterative algorithms with optimal choice of parameters. The theory of these methods is well developed [9]. Taking into account the explicit form of the bounds for m and M from (4.11) and applying for the equation $Lu = P$ the simple iterative method, the Chebyshev acceleration methods (s-cyclic and two-step ones), and the conjugate gradient method in the form (4.10), we arrive at the conclusions of Theorem, using the well-known convergence results [9]for these methods. Theorem is proved.

In case $\alpha_k = 1/\alpha$, $B_k = E$, $\beta_k = 0$, the iterative algorithm (4.1)–(4.3) coincides with the Krylov-Chernousko method [3].

The numerical analysis of the above-formulated iterative algorithms has been done in [10]for the data assimilation problem with a linear parabolic state equation.

References

[1] A.N.Tikhonov. On the solution of ill-posed problems and the regularization method. *Dokl. Akad. Nauk SSSR*, 151:501–504, 1963.

[2] J.-L.Lions. On controllability of distributed system. *Proc. Natl. Acad. Sci. USA*, 94:4828–4835, 1997.

[3] I.A. Krylov and F.L. Chernousko. On a successive approximation method for solving optimal control problems. *Zh. Vychisl. Mat. Mat. Fiz.*, 2:1132–1139, 1962.

[4] O.A. Ladyzhenskaya, V.A. Solonnikov, and N.N. Uraltseva. *Linear and Quasilinear Equations of Parabolic Type.* Nauka, Moscow, 1967.

[5] J.L. Lions. *Contrôle Optimal des Systèmes Gouvernés par des Équations aux Dérivées Partielles.* Dunod, Paris, 1968.

[6] J.L. Lions and E.Magenes. *Problémes aux Limites non Homogenes et Applications.* Dunod, Paris, 1968.

[7] L.S.Pontryagin, V.G.Boltyanskii, R.V.Gamkre lid ze, and E.F.Mi schen ko. *The Mathematical Theory of Optimal Processes.* John Wiley, New York, 1962.

[8] G.I. Marchuk, V.I. Agoshkov, and V.P. Shutyaev. *Adjoint Equations and Perturbation Algorithms in Nonlinear Problems.* CRC Press Inc., New York, 1996.

[9] G.I. Marchuk and V.I. Lebedev. *Numerical Methods in the Theory of Neutron Transport.* Harwood Academic Publishers, New York, 1986.

[10] E.I. Parmuzin and V.P. Shutyaev. Numerical analysis of iterative methods for solving evolution data assimilation problems. *Russ. J. Numer. Anal. Math. Modelling*, 14:265–274, 1999.

[11] V.P. Shutyaev. Some properties of the control operator in a data assimilation problem and algorithms for its solution. *Differential Equations*, 12:2035–2041, 1995.

[12] V.P.Shutyaev. *Control operators and iterative algorithms for variational data assimilation problems.* Nauka, Moscow, 2001.

EXISTENCE OF SOLUTIONS TO EVOLUTION SECOND ORDER HEMIVARIATIONAL INEQUALITIES WITH MULTIVALUED DAMPING

Zdzisław Denkowski *
Jagiellonian University
Faculty of Mathematics and Computer Science
Institute of Computer Science
ul. Nawojki 11, PL-30072 Krakow, Poland
denkowski@softlab.ii.uj.edu.pl

Stanisław Migórski *
Jagiellonian University
Faculty of Mathematics and Computer Science
Institute of Computer Science
ul. Nawojki 11, PL-30072 Krakow, Poland
migorski@softlab.ii.uj.edu.pl

Abstract In this paper we examine an evolution problem which describes the dynamic contact of a viscoelastic body and a foundation. The contact is modeled by a general normal compliance condition and a friction law which are nonmonotone, possibly multivalued and of the subdifferential form while the damping operator is assumed to be coercive and pseudomonotone. We derive a formulation of the model in the form of a multidimensional hemivariational inequality. Then we establish the a priori estimates and the existence of weak solutions by using a surjectivity result.

Keywords: Contact problem, hemivariational inequality, subdifferential, damping, nonconvex, friction, hyperbolic, viscoelasticity.

*Supported in part by the State Committee for Scientific Research of the Republic of Poland (KBN) under Research Grants no. 2 P03A 003 25 and 4 T07A 027 26.

Introduction

In this paper we investigate the class of evolution second order hemivariational inequalities. By a hemivariational inequality we mean an evolution variational inequality involving a nonmonotone multivalued map of the Clarke subdifferential type. The problem under consideration is as follows

$$\begin{cases} u''(t) + A(t, u'(t)) + Bu(t) + \partial J(t, u(t)) \ni f(t) \quad \text{a.e. } t \in (0, T) \\ u(0) = u_0, \ u'(0) = u_1, \end{cases} \tag{1}$$

where $A\colon (0,T) \times V \to 2^{V^*}$ is a nonlinear multivalued damping operator, V being a reflexive Banach space with its dual V^*, $B\colon V \to V^*$ is a bounded linear operator, not necessary coercive, ∂J denotes the Clarke subdifferential of a locally Lipschitz function J and f, u_0 and u_1 are prescribed data. The motivation for the study of the problem (1) comes from mechanics and engineering where hemivariational inequalities express the principle of virtual work or power, e.g. unilateral contact problems in nonlinear elasticity and viscoelasticity, problems describing frictional and adhesive effects, problem of delamination of plates, loading and unloading problems in engineering structures (cf. Panagiotopoulos [26–27] and Naniewicz and Panagiotopoulos [24]).

The notion of hemivariational inequality was introduced by P.D. Panagiotopoulos in the early eighties as variational expressions for several classes of mechanical problems with nonsmooth and nonconvex energy superpotentials. In the case of convex superpotentials the hemivariational inequalities reduce to variational inequalities considered earlier by many authors (see e.g. Duvaut and Lions [7] and the references therein). The recent mathematical results on the stationary hemivariational inequalities can be found in Naniewicz and Panagiotopoulos [24], Motreanu and Panagiotopoulos [23] and Haslinger et al. [12]. We refer to Migorski [20–22]and the references therein for the results on the first order evolution and parabolic hemivariational inequalities. We mention that the hemivariational inequalities of hyperbolic type were firstly considered by Panagiotopoulos [29, 28] who studied models involving the one dimensional reaction-velocity laws. The hyperbolic hemivariational inequalities with a multivalued relation depending on the first order derivative of the unknown function were treated by Goeleven et al. [10], Haslinger et al. [12], Gasiński [8], Migorski [19], while the hemivariational inequalities with a subdifferential term which depends on the unknown function were studied by Panagiotopoulos and Pop [30], Haslinger et al. [12], Gasiński and Smołka [9], Ochal [25] and Migorski [18]. The contact problems for viscoelastic bodies have been recently investigated in several papers, see e.g. Chau et al. [3], Jarusek [14], Kuttler and Shillor [16], Rochdi et al. [31], Han and Sofonea [11] and the literature therein. The problem (1) with $A(t, \cdot)$ being multivalued maximal monotone, B coercive and $J \equiv 0$ was considered by Banks et al. [1] in the connection with identification of the damping term in the forced wave equation.

The goal of this paper is to provide the existence and uniqueness results for the problem (1). The main existence result can be proved in two steps (cf. [5]). First we assume regular initial data and reduce this problem to an evolution inclusion of the first order. The latter is solved by using a surjectivity result for multivalued operators of pseudomonotone type. In the second step we remove the restriction on the initial data and we are able to show the result in its generality. The uniqueness of a solution to (1) is obtained in a case when the damping operator is strongly monotone and the subdifferential operator satisfies a relaxed monotonocity condition.

The paper is organized as follows. In Section 1 we present a contact problem of viscoelasticity which serves as a model for the problem (1). In Section 2 we recall some necessary notation and present a result on properties of the Nemitsky operator corresponding to the damping operator. The main results of this paper are delivered in Section 3.

1. Motivation

In this section we describe shortly the classical contact model of viscoelasticity and we present its variational form.

We consider a deformable viscoelastic body which occupies a bounded open subset $\Omega \subset \mathbb{R}^d$, $d = 2, 3$. We suppose that the boundary $\Gamma = \partial\Omega$ is Lipschitz continuous and Γ is divided into three mutually disjoint measurable parts Γ_D, Γ_N and Γ_C such that $meas(\Gamma_D) > 0$. The body is clamped on Γ_D, so the displacement field vanishes there. Volume forces of density f_1 act in Ω and surface tractions of density f_2 are applied on Γ_N. The body may come in contact with a foundation over the potential contact surface Γ_C. We put $Q = \Omega \times (0, T)$ for $0 < T < \infty$. We denote by $u \colon Q \to \mathbb{R}^d$ the displacement field, by $\sigma \colon Q \to \mathcal{S}_d$ the stress tensor and by $\varepsilon(u) = (\varepsilon_{ij}(u))$, $\varepsilon_{ij}(u) = \frac{1}{2}(u_{i,j} + u_{j,i})$ the strain tensor, where i, $j = 1, \ldots, d$ and $\mathcal{S}_d$ denotes the space $\mathbb{R}^{d\times d}_s$ of symmetric matrices of order d).

We suppose the following multivalued counterpart of the Kelvin-Voigt viscoelastic constitutive relation

$$\sigma(u, u') \in \mathcal{C}(\varepsilon(u')) + \mathcal{G}(\varepsilon(u)),$$

where $\mathcal{C}$ and $\mathcal{G}$ are prescribed multivalued nonlinear and single valued linear constitutive maps, respectively. We remark that in the classical linear viscoelasticity the above law takes the form $\sigma_{ij} = c_{ijkl}\varepsilon_{kl}(u') + g_{ijkl}\varepsilon_{kl}(u)$, where $\mathcal{C} = \{c_{ijkl}\}$ and $\mathcal{G} = \{g_{ijkl}\}$, $i, j, k, l = 1, \ldots, d$ are the viscosity and elasticity tensors, respectively.

We denote by u_N and u_T the normal and the tangential components of the displacement u on Γ, $u_N = u \cdot n$, $u_T = u - u_N n$, where n is the outward unit vector to Γ. Similarly, the normal and the tangential components of the stress field on Γ are given by $\sigma_N = (\sigma n) \cdot n$ and $\sigma_T = \sigma n - \sigma_N n$, respectively. On the contact surface Γ_C we consider the following subdifferential boundary conditions. The normal stress σ_N and the normal displacement u_N satisfy the nonmonotone normal compliance response condition of the form

$$-\sigma_N \in \partial j_N(x, t, u_N) \quad \text{on} \quad \Gamma_C \times (0, T). \tag{2}$$

The friction law between the friction force σ_T and the tangential displacement u_T on Γ_C is given by

$$-\sigma_T \in \partial j_T(x, t, u_T) \quad \text{on} \quad \Gamma_C \times (0, T). \tag{3}$$

Here $j_N \colon \Gamma_C \times (0, T) \times \mathbb{R} \to \mathbb{R}$ and $j_T \colon \Gamma_C \times (0, T) \times \mathbb{R}^d \to \mathbb{R}$ are locally Lipschitz functions in their last variables and ∂j_N, ∂j_T represent the Clarke subdifferentials of $j_N(x, t, \cdot)$ and $j_T(x, t, \cdot)$, respectively. These boundary conditions include as special cases the classical boundary conditions of (see e.g. Panagiotopoulos [27], Chapter 2.3 and Naniewicz and Panagiotopoulos [24]).

Let us denote by u_0 and u_1 the initial displacement and the initial velocity. The classical formulation of the contact problem is stated as follows: find $u \colon Q \to \mathbb{R}^d$ and $\sigma \colon Q \to \mathcal{S}_d$ such that

$$\begin{cases} u'' - \mathrm{div}\sigma = f_1 & \textit{in } Q \\ \sigma \in \mathcal{C}(\varepsilon(u')) + \mathcal{G}(\varepsilon(u)) & \textit{in } Q \\ u = 0 & \textit{on } \Gamma_D \times (0, T) \\ \sigma n = f_2 & \textit{on } \Gamma_N \times (0, T) \\ -\sigma_N \in \partial j_N(x, t, u_N), \ -\sigma_T \in \partial j_T(x, t, u_T) & \textit{on } \Gamma_C \times (0, T) \\ u(0) = u_0, \quad u'(0) = u_1 & \textit{in } \Omega. \end{cases} \tag{4}$$

In order to give a variational formulation of this problem let $H = L^2(\Omega; \mathbb{R}^d)$, $\mathcal{H} = L^2(\Omega; \mathcal{S}_d)$, $H_1 = \{u \in H : \varepsilon(u) \in \mathcal{H}\} = H^1(\Omega; \mathbb{R}^d)$ and $V = \{v \in H_1 : v = 0 \text{ on } \Gamma_D\}$. Using the Green formula, the definition of the Clarke subdifferential and assuming the suitable regularity of the data (cf. Denkowski and Migórski [5] for details), we obtain the following variational formulation of (4): find $u\colon (0,T) \to V$ and $\sigma\colon (0,T) \to \mathcal{H}$ such that

$$\begin{cases} \langle u''(t), v\rangle_{V^*\times V} + (\sigma(t), \varepsilon(v))_{\mathcal{H}} + \\ \qquad + \displaystyle\int_{\Gamma_C} \left(j_N^0(x,t,u_N; v_N) + j_T^0(x,t,u_T; v_T)\right) d\Gamma(x) \geq \\ \qquad \geq \langle f(t), v\rangle_{V^*\times V} \quad \text{for all } \ v \in V \text{ and a.e. } t \in (0,T) \\ \sigma(t) \in \mathcal{C}(\varepsilon(u'(t))) + \mathcal{G}(\varepsilon(u(t))) \ \text{ for a.e. } t \in (0,T) \\ u(0) = u_0, \quad u'(0) = u_1, \end{cases} \tag{5}$$

where

$$\langle f(t), v\rangle_{V^*\times V} = (f_1(t), v)_H + (f_2(t), v)_{L^2(\Gamma_N;\mathbb{R}^d)} \quad \text{for } \ v \in V \text{ and a.e. } t.$$

Let $\mathcal{V} = L^2(0,T;V)$, $\mathcal{W} = \{w \in \mathcal{V} : w' \in \mathcal{V}^*\}$ and let $\overline{\gamma}\colon H^\delta(\Omega;\mathbb{R}^d) \to H^{1/2}(\Gamma;\mathbb{R}^d) \subset L^2(\Gamma;\mathbb{R}^d)$ be the trace operator where $\delta \in (1/2, 1)$. We define the operators $A\colon (0,T)\times V \to 2^{V^*}$ and $B\colon V \to V^*$ by

$$\langle A(t,u), v\rangle_{V^*\times V} = \left(\mathcal{C}(x,t,\varepsilon(u)), \varepsilon(v)\right)_{\mathcal{H}} \quad \text{for } u,v \in V \text{ and } t \in (0,T),$$

$$\langle Bu, v\rangle_{V^*\times V} = \left(\mathcal{G}(x,t,\varepsilon(u)), \varepsilon(v)\right)_{\mathcal{H}} \quad \text{for } u,v \in V \text{ and } t \in (0,T)$$

and the functional $J\colon (0,T)\times L^2(\Gamma_C;\mathbb{R}^d) \to \mathbb{R}$ by

$$J(t,v) = \int_{\Gamma_C} \Big(j_N(x,t,v_N(x)) + j_T(x,t,v_T(x)) \Big)\, d\Gamma(x)$$

for $t \in (0,T)$ and $v \in L^2(\Gamma_C;\mathbb{R}^d)$. Consider now the following inclusion

$$\begin{cases} \text{find } u \in \mathcal{V} \text{ with } u' \in \mathcal{W} \text{ such that} \\ u''(t) + A(t, u'(t)) + Bu(t) + \overline{\gamma}^*\left(\partial J(t, \overline{\gamma} u(t))\right) \ni f(t) \quad \text{a.e. } t \\ u(0) = u_0, \ u'(0) = u_1 \end{cases} \tag{6}$$

where $\overline{\gamma}^*$ denotes the adjoint operator to $\overline{\gamma}$. It can be shown (cf. Denkowski and Migórski [5] for details) that every solution to the inclusion (6) is also a solution to the problem (5). Therefore in what follows we are interested in the existence result for a problem of type (6).

2. Preliminaries

In this section we recall some definitions needed in the sequel and state a result that shows that certain properties of the damping mapping can be lifted to its Nemitsky operator.

Let V be a reflexive separable Banach space. We denote by $\langle \cdot, \cdot \rangle$ the pairing between V and its dual V^*.

DEFINITION 1 *A multivalued operator $T: V \to 2^{V^*}$ is said to be pseudomonotone if the following conditions hold:*

(j) the set Tv is nonempty, bounded, closed and convex for all $v \in V$;

(jj) T is usc from each finite dimensional subspace of V into V^* endowed with the weak topology;

(jjj) if $v_n \in V$, $v_n \to v$ weakly in V and $v_n^* \in Tv_n$ is such that $\limsup \langle v_n^*, v_n - v \rangle \leq 0$, then to each $y \in V$, there exists $v^*(y) \in Tv$ such that $\langle v^*(y), v - y \rangle \leq \liminf \langle v_n^*, v_n - y \rangle$.

LEMMA 2 *Assume that a multivalued operator $T: V \to 2^{V^*}$ satisfies conditions (j) and (jj) of the definition of pseudomonotonicity and T is bounded (i.e. it maps bounded sets into bounded sets). Then T is usc with respect to the strong topology in V and the weak topology in V^*.*

For the proof we refer to Lemma 1.4 in Kuttler [15].

We now comment on a measurability condition for multivalued mappings. For the following definitions, see Section 1.0 of Hu and Papageorgiou [13].

DEFINITION 3 *A multifunction $F: (0,T) \to 2^{V^*}$ is said to be measurable, if for every $U \subset V^*$ open, we have $F^-(U) = \{t \in (0,T) : F(t) \cap U \neq \emptyset\}$ is measurable.*

DEFINITION 4 *A multifunction $S: (0,T) \times V \to 2^{V^*}$ is said to be (strongly) measurable, if for every $C \subset V^*$ closed, we have $\{(t,v) \in (0,T) \times V : S(t,v) \cap C \neq \emptyset\}$ is a Borel set in $(0,T) \times V$.*

The following result due to Kuttler [15], Lemma 5.3 shows that the strong measurability condition implies a kind of measurability condition for multivalued operators which is useful in our setting.

LEMMA 5 *Suppose $S: (0,T) \times V \to 2^{V^*}$ has nonempty, closed, convex values and it satisfies the measurability condition of Definition 4 for every $C \subset V^*$ closed convex set. Then*

(*) for every $\alpha: (0,T) \to \mathbb{R}$ measurable and $x, y: (0,T) \to V$ measurable, the multifunction $F: (0,T) \to 2^{V^*}$ defined by

$$F(t) = \{w \in S(t, x(t)) : \langle w, x(t) - y(t) \rangle \leq \alpha(t)\}$$

is measurable (in the sense of Definition 3).

In what follows (see condition $H(A)(ii)$ below) instead of the standard definition (Definition 4) of measurability of a multivalued operator, we assume it satisfies condition $(*)$ of Lemma 5, which will be sufficient for our purposes.

Given $2 \le p < \infty$ we introduce the spaces $\mathcal{V} = L^p(0,T;V)$, $\mathcal{V}^* = L^q(0,T;V^*)$, $1/p + 1/q = 1$ and we denote by $\langle\langle \cdot, \cdot \rangle\rangle$ the duality between $\mathcal{V}$ and $\mathcal{V}^*$.

DEFINITION 6 *Let* $A\colon (0,T) \times V \to 2^{V^*}$ *be a multivalued operator. The operator* $\mathcal{A}\colon \mathcal{V} \to 2^{\mathcal{V}^*}$ *given by* $\mathcal{A}v = \{z \in \mathcal{V}^* : z(t) \in A(t, v(t))$ *a.e.* $t \in (0,T)\}$ *for* $v \in \mathcal{V}$ *is called the Nemitsky operator corresponding to* A.

We recall also the notion of L-pseudomonotonicity (see e.g. [6]). Let $L\colon D(L) \subset \mathcal{V} \to \mathcal{V}^*$ be a linear maximal monotone operator.

DEFINITION 7 *We say that the operator* $\mathcal{A}\colon \mathcal{V} \to 2^{\mathcal{V}^*}$ *is* L*-pseudomonotone, if the following conditions hold:*

(k) the set $\mathcal{A}v$ *is nonempty, weakly compact and convex for all* $v \in \mathcal{V}$;

(kk) $\mathcal{A}$ *is usc from each finite dimensional subspace of* $\mathcal{V}$ *into* $\mathcal{V}^*$ *furnished with the weak topology;*

(kkk) if $\{v_n\} \subset D(L)$, $v_n \to v$ *weakly in* $\mathcal{V}$, $Lv_n \to Lv$ *weakly in* $\mathcal{V}^*$, $v_n^* \in \mathcal{A}v_n$, $v_n^* \to v^*$ *weakly in* $\mathcal{V}^*$ *and* $\limsup \langle\langle v_n^*, v_n - v \rangle\rangle \le 0$, *then* $v^* \in \mathcal{A}v$ *and* $\langle\langle v_n^*, v_n \rangle\rangle \to \langle\langle v^*, v \rangle\rangle$.

The following result generalizes Theorem 2(b) of Berkovits and Mustonen [2].

THEOREM 8 *Assume that a multivalued operator satisfies the following hypothesis:*

$\underline{H(A)}$: $A\colon (0,T) \times V \to 2^{V^*}$ *is a multivalued operator such that*

(i) $A(t,\cdot)$ *satisfies conditions (j) and (jjj) in the definition of pseudomonotone operator;*

(ii) A *is measurable in the sense of condition* $(*)$;

(iii) there are $a_1 \in L^q(0,T)$ *and* $b_1 > 0$ *such that* $\|v^*\|_{V^*} \le a_1(t) + b_1 \|v\|_V^{p-1}$ *for all* $v^* \in A(t,v)$, $v \in V$ *and a.e.* $t \in (0,T)$;

(iv) there are constants $\beta_1 > 0$, $\beta_2 \ge 0$, $r \in (0,p)$ *and a function* $a \in L^1(0,T)$ *such that* $\langle v^*, v \rangle \ge \beta_1 \|v\|_V^p - \beta_2 \|v\|_V^r - a(t)$ *for all* $v^* \in A(t,v)$, $v \in V$ *and a.e.* $t \in (0,T)$.

Then the Nemitsky operator $\mathcal{A}$ corresponding to A has the following properties:

(1) there are constants $\bar{a}_1 \geq 0$ and $\bar{b}_1 > 0$ such that $\|v^*\|_{\mathcal{V}^*} \leq \bar{a}_1 + \bar{b}_1 \|v\|_{\mathcal{V}}^{p-1}$ for all $v^* \in \mathcal{A}v$ and $v \in \mathcal{V}$;

(2) there are constants $\overline{\beta}_2 \geq 0$ and $\bar{a} > 0$ such that $\langle\langle v^*, v \rangle\rangle \geq \beta_1 \|v\|_{\mathcal{V}}^p - \overline{\beta}_2 \|v\|_{\mathcal{V}}^r - \bar{a}$ for all $v^* \in \mathcal{A}v$ and $v \in \mathcal{V}$;

(3) $\mathcal{A}$ is L-pseudomonotone, where $L \colon D(L) \subset \mathcal{V} \to \mathcal{V}^*$ is given by $Lv = v'$ for all $v \in D(L) = \{v \in \mathcal{V} : v' \in \mathcal{V}^*, v(0) = 0\}$.

The detailed proof of this theorem can be found in Denkowski and Migórski [5]. We conclude this section by recalling (cf. Clarke [4]) the definitions of the generalized directional derivative and the generalized gradient of Clarke for a locally Lipschitz function $h \colon E \to \mathbb{R}$, where E is a Banach space. The generalized directional derivative of h at $x \in E$ in the direction $v \in E$, denoted by $h^0(x; v)$, is defined by

$$h^0(x; v) = \limsup_{y \to x,\ t \downarrow 0} \frac{h(y + tv) - h(y)}{t}.$$

The generalized gradient of h at x, denoted by $\partial h(x)$, is a subset of a dual space E^* given by $\partial h(x) = \{\zeta \in E^* : h^0(x; v) \geq \langle \zeta, v \rangle_{E^* \times E}$ for all $v \in E\}$.

3. Existence Theorem

In this section we deliver the main result of the paper on the existence of solutions to dynamic hemivariational inequalities.

Let V and Z be two reflexive, separable Banach spaces and let H be a Hilbert space. Suppose that $V \subset Z \subset H \approx H^* \subset Z^* \subset V^*$, where H^*, Z^* and V^* denote dual spaces to H, Z and V, respectively. We assume that all embeddings are dense and continuous, and $V \subset Z$ compactly. We denote by $\langle \cdot, \cdot \rangle$ the duality of V and V^* and the pairing between Z and Z^* as well. We introduce the following spaces:

$$\mathcal{V} = L^p(0, T; V), \quad \mathcal{Z} = L^p(0, T; Z), \quad \mathcal{H} = L^2(0, T; H),$$

$$\mathcal{Z}^* = L^q(0, T; Z^*), \quad \mathcal{V}^* = L^q(0, T; V^*) \text{ with } 1/p + 1/q = 1$$

with some $2 \leq p < \infty$ and $\mathcal{W} = \{v \in \mathcal{V} : v' \in \mathcal{V}^*\}$, where the time derivative involved in the definition of $\mathcal{W}$ is understood in the sense of vector valued distributions. We have $\mathcal{W} \subset \mathcal{V} \subset \mathcal{Z} \subset \mathcal{H} \subset \mathcal{Z}^* \subset \mathcal{V}^*$ with dense and continuous embeddings. Since we have assumed $V \subset Z$

compactly, we also know (cf. Theorem 5.1, p.58, Lions [17]) that $\mathcal{W} \subset \mathcal{Z}$ compactly. Moreover, $\mathcal{W} \subset C(0,T;H)$ is continuous. The pairing of $\mathcal{V}$ and $\mathcal{V}^*$ and also the duality between $\mathcal{Z}$ and $\mathcal{Z}^*$ are denoted by $\langle\langle f, g\rangle\rangle = \int_0^T \langle f(t), g(t)\rangle\, dt$.

Consider the following initial value problem for evolution hemivariational inequality of second order:

$$\begin{cases} \text{find } y \in \mathcal{V} \text{ such that } y' \in \mathcal{W} \text{ and} \\ y''(t) + A(t, y'(t)) + By(t) + \partial J(t, y(t)) \ni f(t) \quad \text{a.e. } t \in (0,T) \\ y(0) = y_0, \ y'(0) = y_1. \end{cases} \tag{7}$$

The problem (7) is called hemivariational inequality since it is equivalent to the following one:

$$\begin{cases} \text{find } y \in \mathcal{V} \text{ with } y' \in \mathcal{W} \text{ such that there is } \eta \in \mathcal{V}^* \text{ satisfying} \\ \langle y''(t) + \eta(t) + By(t) - f(t), v\rangle + J^0(t, y(t); v) \geq 0 \\ \qquad\qquad\qquad\qquad \text{for all } v \in V \text{ and a.e. } t \in (0,T) \\ \eta(t) \in A(t, y'(t)) \quad \text{a.e. } t \in (0,T) \\ y(0) = y_0, \ y'(0) = y_1, \end{cases}$$

where $J^0(t, v; w)$ is the generalized directional derivative of $J(t, \cdot)$ at a point $v \in Z$ in the direction $w \in Z$.

DEFINITION 9 *An element $y \in \mathcal{V}$ solves (7) if and only if $y' \in \mathcal{W}$ and there exist $\eta \in \mathcal{V}^*$ and $\zeta \in \mathcal{Z}^*$ such that*

$$\begin{cases} y''(t) + \eta(t) + By(t) + \zeta(t) = f(t) \quad a.e. \ t \in (0,T) \\ \eta(t) \in A(t, y'(t)), \quad \zeta(t) \in \partial J(t, y(t)) \quad a.e. \ t \in (0,T) \\ y(0) = y_0, \ y'(0) = y_1. \end{cases}$$

We admit the following hypotheses:

<u>$H(B)$</u>: $B \colon V \to V^*$ is a bounded, linear, positive and symmetric operator;

<u>$H(J)$</u>: $J \colon (0,T) \times Z \to \mathbb{R}$ is a function such that

(i) for each $z \in Z$, the map $J(\cdot, z)$ is measurable and $J(\cdot, 0) \in L^1(0,T)$;

(ii) for each $t \in (0,T)$, the function $J(t, \cdot)$ is locally Lipschitz;

(iii) there exists $\bar{c} > 0$ such that for all $\zeta \in \partial J(t, z)$, $z \in Z$ and $t \in (0,T)$, we have $\|\zeta\|_{Z^*} \leq \bar{c}\left(1 + \|z\|_Z^{2/q}\right)$;

$\underline{(H_0)}$: $f \in \mathcal{V}^*$, $y_0 \in V$, $y_1 \in H$;

$\underline{(H_1)}$: If $p = 2$, then $\beta_1 > \overline{c}\beta^2 T$, where $\beta > 0$ is an embedding constant of V into Z.

We start the study of (7) with the a priori estimates for the solutions.

LEMMA 10 *Assume that $H(A)$, $H(B)$, $H(J)$ and (H_0) hold and y is a solution to (7). If $p > 2$, then there is a constant $C > 0$ such that*

$$\|y\|_{C(0,T;V)} + \|y'\|_{\mathcal{W}} \leq C\left(1 + \|y_0\|_V^{2/q} + \|y_1\|_H^{2/q} + \|f\|_{\mathcal{V}^*}^{2/q}\right). \tag{8}$$

Moreover, the estimate (8) still holds for $p = 2$ provided (H_1) is satisfied.

If $Z = H$, then the estimate (8) holds for $p \geq 2$ without the hypothesis (H_1). Namely, we have

LEMMA 11 *If $H(A)$, $H(B)$, $H(J)$ and (H_0) hold, $p \geq 2$ and $Z = H$, then for every y solution to (7), the estimate (8) holds.*

The main result of this paper is the following

THEOREM 12 *If hypotheses $H(A)$, $H(B)$, $H(J)$, (H_0) and (H_1) hold, then the problem (7) has at least one solution.*

The idea of the proof is as follows. First, we consider the operator $K\colon \mathcal{V} \to C(0,T;V)$ given by $Kv(t) = \int_0^t v(s)\,ds + y_0$. Using K we rewrite (7) in the form: find $z \in \mathcal{W}$ such that

$$\begin{cases} z'(t) + A(t, z(t)) + B(Kz(t)) + \partial J(t, Kz(t)) \ni f(t) \text{ a.e. } t \\ z(0) = y_1. \end{cases} \tag{9}$$

We observe that $z \in \mathcal{W}$ is a solution to (9) iff $y = Kz$ solves (7). We deal now with the problem (9) under the additional hypothesis $y_1 \in V$. We define the following operators $\mathcal{A}_1\colon \mathcal{V} \to 2^{\mathcal{V}^*}$, $\mathcal{B}_1\colon \mathcal{V} \to \mathcal{V}^*$ and $\mathcal{N}_1\colon \mathcal{V} \to 2^{\mathcal{V}^*}$ by

$$\mathcal{A}_1 v = \{v^* \in \mathcal{V}^* \,:\, v^*(t) \in A(t, v(t) + y_1) \text{ a.e. t}\},$$

$$(\mathcal{B}_1 v)(t) = B(K(v(t) + y_1))$$

and

$$\mathcal{N}_1 v = \{z \in \mathcal{Z}^* \,:\, z(t) \in \partial J(t, K(v(t) + y_1)) \text{ a.e. t}\}$$

for all $v \in \mathcal{V}$, respectively. Using these operators, from (9) we have

$$\begin{cases} z' + \mathcal{A}_1 z + \mathcal{B}_1 z + \mathcal{N}_1 z \ni f \quad \text{a.e. } t \in (0,T) \\ z(0) = 0. \end{cases} \tag{10}$$

We note that $z \in \mathcal{W}$ solves (9) iff $z - y_1 \in \mathcal{W}$ solves (10). Next, defining $L\colon D(L) \subset \mathcal{V} \to \mathcal{V}^*$ and $\mathcal{T}\colon \mathcal{V} \to 2^{\mathcal{V}^*}$ by $Lz = z'$ with $D(L) = \{z \in \mathcal{W} : z(0) = 0\}$ and $\mathcal{T}z = (\mathcal{A}_1 + \mathcal{B}_1 + \mathcal{N}_1)z$, respectively, the problem (10) takes the form: find $z \in D(L)$ such that $(L + \mathcal{T})z \ni f$. In order to show the existence of solutions, we can prove that $\mathcal{T}$ is bounded, coercive and L-pseudomonotone, and apply a surjectivity result (cf. Theorem 1.3.73 in [6]). Finally, we suppose $y_1 \in H$ and we establish the existence of solutions in this case.

The existence of solution to (6) can be obtained analogously as for the model problem (7). This follows from the fact that the map $R\colon (0,T) \times Z \to 2^{Z^*}$ given by $R(t,z) = \overline{\gamma}^* (\partial J(t, \overline{\gamma} z(t)))$ has the same properties as $\partial J(t,z)$ (convex and weak compactness of the values, the strong-weak closedness of the graph and a growth condition).

References

[1] H.T. Banks, S. Reich, and I.G. Rosen. Estimation of nonlinear damping in second order distributed parameter systems. *Control - Theory and Advanced Techn.*, 6:395–415, 1990.

[2] J. Berkovits and V. Mustonen. Monotonicity methods for nonlinear evolution equations. *Nonlinear Analysis*, 27:1397–1405, 1996.

[3] O Chau, W. Han, and M. Sofonea. A dynamic frictional contact problem with normal damped response. *Acta Appl. Math.*, 71:159–178, 2002.

[4] F.H. Clarke. *Optimization and Nonsmooth Analysis.* Wiley - Interscience, New York, 1983.

[5] Z. Denkowski and S. Migórski. Existence of solutions to evolution second order hemivariational inequalities with multivalued damping, submitted.

[6] Z. Denkowski, S. Migórski, and N.S. Papageorgiou. *An Introduction to Nonlinear Analysis: Applications.* Kluwer Academic/Plenum Publishers, Boston, Dordrecht, London, New York, 2003.

[7] G. Duvaut and J.L. Lions. *Les Inéquations en Mécanique et en Physique.* Dunod, Paris, 1972.

[8] L. Gasiński. *Hyperbolic Hemivariational Inequalities and their Applications to Optimal Shape Design.* PhD thesis, Jagiellonian University, Cracow, Poland, 2000.

[9] L. Gasiński and M. Smołka. An existence theorem for wave-type hyperbolic hemivariational inequalities. *Math. Nachr.*, 242:1–12, 2002.

[10] D. Goeleven, M. Miettinen, and P.D. Panagiotopoulos. Dynamic hemivariational inequalities and their applications. *J. Optimiz. Theory and Appl.*, 103:567–601, 1999.

[11] W. Han and M. Sofonea. *Quasistatic Contact Problems in Viscoelasticity and Viscoplasticity.* AMS and International Press, 2002.

[12] J. Haslinger, M. Miettinen, and P.D. Panagiotopoulos. *Finite Element Method for Hemivariational Inequalities. Theory, Methods and Applications.* Kluwer Academic Publishers, Boston, Dordrecht, London, 1999.

[13] S. Hu and N.S. Papageorgiou. *Handbook of Multivalued Analysis, Volume I: Theory.* Kluwer, Dordrecht, 1997.

[14] J. Jarusek. Dynamic contact problems with given friction for viscoelastic bodies. *Czech. Math. J.*, 46:475–487, 1996.

[15] K. Kuttler. Non-degenerate implicit evolution inclusions. *Electronic J. Diff. Equations*, 34:1–20, 2000.

[16] K.L. Kuttler and M. Shillor. Set-valued pseudomonotone maps and degenerate evolution inclusions. *Comm. Contemp. Math.*, 1:87–123, 1999.

[17] J.L. Lions. *Quelques méthodes de résolution des problémes aux limites non linéaires.* Dunod, Paris, 1969.

[18] S. Migórski. Boundary hemivariational inequalities of hyperbolic type and applications, *J. Global Optim.*, in press, 2004.

[19] S. Migórski. Dynamic hemivariational inequality modeling viscoelastic contact problem with normal damped response and friction. *Applicable Analysis*, in press, 2004.

[20] S. Migórski. Existence and convergence results for evolution hemivariational inequalities. *Topological Methods Nonlinear Anal.*, 16:125–144, 2000.

[21] S. Migórski. Evolution hemivariational inequalities in infinite dimension and their control. *Nonlinear Analysis*, 47:101–112, 2001.

[22] S. Migórski. *Modeling, Analysis and Optimal Control of Systems Governed by Hemivariational Inequalities*, pp. 248–279, chapter in the book *"Industrial Mathematics and Statistics"* dedicated to commemorate the Golden Jubilee of Indian Institute of Technology, Kharagpur, India, 2002, J.C. Misra, ed., Narosa Publishing House. Delhi, 2003.

[23] D. Motreanu and P.D. Panagiotopoulos. *Minimax Theorems and Qualitative Properties of the Solutions of Hemivariational Inequalities and Applications.* Kluwer Academic Publishers, Boston, Dordrecht, London, 1999.

[24] Z. Naniewicz and P.D. Panagiotopoulos. *Mathematical Theory of Hemivariational Inequalities and Applications.* Marcel Dekker, Inc., New York, Basel, Hong Kong, 1995.

[25] A. Ochal. *Optimal Control of Evolution Hemivariational Inequalities.* PhD thesis, Jagiellonian University, Cracow, Poland, 2001.

[26] P.D. Panagiotopoulos. *Inequality Problems in Mechanics and Applications. Convex and Nonconvex Energy Functions.* Birkhäuser, Basel, 1985.

[27] P.D. Panagiotopoulos. *Hemivariational Inequalities, Applications in Mechanics and Engineering.* Springer-Verlag, Berlin, 1993.

[28] P.D. Panagiotopoulos. Hemivariational inequalities and fan-variational inequalities. new applications and results. *Atti Sem. Mat. Fis. Univ. Modena*, 43:159–191, 1995.

[29] P.D. Panagiotopoulos. Modelling of nonconvex nonsmooth energy problems: dynamic hemivariational inequalities with impact effects. *J. Comput. Appl. Math.*, 63:123–138, 1995.

[30] P.D. Panagiotopoulos and G. Pop. On a type of hyperbolic variational-hemivariational inequalities. *J. Applied Anal.*, 5 (1):95–112, 1999.

[31] M. Rochdi, M. Shillor, and M. Sofonea. A quasistatic contact problem with directional friction and damped response. *Applicable Analysis*, 68:409–422, 1998.

PROBABILISTIC INVESTIGATION ON DYNAMIC RESPONSE OF DECK SLABS OF HIGHWAY BRIDGES

Chul-Woo Kim
Research Scientist, Ph.D., Department of Civil Engineering, Kobe University, Japan
cwkim@kobe-u.ac.jp

Mitsuo Kawatani
Professor, Dr. Eng., Department of Civil Engineering, Kobe University, Japan
m-kawa@kobe-u.ac.jp

Abstract Probabilistic assessment on the code-specified impact factors of RC decks is investigated by means of a three-dimensional traffic-induced dynamic response analysis of bridges combined with the Monte Carlo simulation technique. The random variables considered in the simulation are the roadway roughness, bump height, traveling position of vehicles, vehicle running speed and axle load of three-axle vehicles. Statistical parameters of the random variables are taken from surveying data on Hanshin and Meishin Expressways in Japan. A simple span steel-girder bridge with RC decks that is experimentally verified is considered as a numerical example. This study demonstrates that the impact factor of the deck near expansion joints dominates the design impact factor due to the bump at the expansion joint of bridges.

Keywords: traffic-induced vibration, RC deck, bump, impact factor, reliability assessment

Introduction

The performance of reinforced concrete (RC) decks of highway bridges mainly depends on cracking damages. Thus the rational criterion for the performance level of RC decks provides useful assessment tool for decision making related to the inspection, repair, upgrading and replacement of existing steel plate girder bridges based on life-cycle costs, since the

RC deck, being directly subjected to wheel loads of vehicles, is more easily damaged than other structural members in steel highway bridges [5].

It is apparent that, except corrosion due to environmental factors, trucks or traffic loads play an important role in the deterioration of RC decks. Traffic loads are usually affected by the roadway roughness, dynamic properties of vehicles, vehicle speed, etc, and the dynamic effect is usually considered in design as the impact factor. For decks, moreover, a bump near expansion joints is another important factor because of the impulsive loading effect generated by vehicles passing over the bump [11] and [15].

Most of all the existing research topics related to the deck have been focused on static responses. Few research on dynamic responses of decks due to moving vehicles have been investigated, even though a fatigue problem of decks as a part of dynamic problems has been one of wide spreading research themes. Moreover, in civil infra-structures, the recent design concept trends a reliability-based design to consider many sources of uncertainties in structural design. However, researches on the impact factor of decks based on a probabilistic approach have not been advanced. Therefore, there is a need to fill this gap.

This paper reports a probabilistic assessment of code-specified impact factors for decks considering randomness of the influencing factors to dynamic responses of decks by means of a three-dimensional traffic-induced dynamic response analysis of bridges based on the modal analysis [9] combined with the Monte Carlo simulation (MCS) technique.

1. Governing Equations of Bridge-Vehicle Interaction System

The method called Lagrange equation of motion is adopted for the formulation of the governing equation of a bridge-vehicle interaction system as shown in Eq. (1).

$$\frac{d}{dt}\left(\frac{\partial T}{\partial \dot{q}_i}\right) - \frac{\partial T}{\partial q_i} + \frac{\partial V}{\partial q_i} + \frac{\partial U_d}{\partial \dot{q}_i} = 0 \tag{1}$$

where, T, V and U_d are the kinematic, potetial and dissipation energies of the interaction system, respectively. q_i is the *i-th* generalized co-ordinate.

The kinematic, potential including strain energy and dissipation energies due to the viscous damping of the bridge-vehicle interaction system are expressed in a set of generalized coordinates as follows [9]. It is noteworthy that the superscript dot on variables denotes differential with respect to time.

$$T = \frac{1}{2}[\dot{\mathbf{D}}^T \mathbf{M}_b \dot{\mathbf{D}} + \sum_{v=1}^{n_{veh}} T_v] \tag{2}$$

$$V = \frac{1}{2}[\mathbf{D}^T \mathbf{K}_b \mathbf{D} + \sum_{v=1}^{n_{veh}} \sum_{m=1}^{3} \sum_{u=1}^{2} V_v] \tag{3}$$

$$U_d = \frac{1}{2}[\dot{\mathbf{D}}^T \mathbf{C}_b \dot{\mathbf{D}} + \sum_{v=1}^{n_{veh}} \sum_{m=1}^{3} \sum_{u=1}^{2} U_{dv}] \tag{4}$$

where,

$$T_v = \sum_{k=1}^{2} \left(m_{v1k} \dot{Z}_{v1k}^2 + J_{xv1k} \dot{\theta}_{xv1k}^2 + J_{yvkk} \dot{\theta}_{yvkk}^2 \right) + m_{v22} \dot{Z}_{v22}^2 + J_{xv22} \dot{\theta}_{xv22}^2$$

$$V_v = K_{vm1u} R_{vm1u}^2 + K_{vm2u} \left(R_{vm2u} - Z_{0vmu} \right)^2 + 2 W_{vmu} Z_{0vmu}$$

$$U_{dv} = C_{vm1u} \dot{R}_{vm1u}^2 + C_{vm2u} \left(\dot{R}_{vm2u} - \dot{Z}_{0vmu} \right)^2$$

$$Z_{0vmu} = w(t, x_{vmu}) - Z_{rvmu}$$

$$R_{vmku} = \begin{cases} Z_{v11} - (-1)^m \lambda_{xvm} \theta_{yv11} & -(-1)^u \lambda_{yv1} \theta_{xv11} - Z_{vm2} \\ & +(-1)^u \lambda_{yv(m+1)} \theta_{xvm2} \\ & \text{for } m = 1, 2;\ k = 1;\ u = 1, 2 \\ Z_{v12} - (-1)^u \lambda_{yv2} \theta_{xv12} & \text{for } m = 1;\ k = 2;\ u = 1, 2 \\ 0 & \text{otherwise} \end{cases}$$

$$W_{vmu} = \begin{cases} \frac{1}{2} g \left[\left(1 - \frac{\lambda_{xv1}}{\lambda_{xv}} \right) m_{v11} + m_{v12} \right] & \text{for m=1; u=1,2} \\ \frac{1}{4} g \left[\left(1 - \frac{\lambda_{xv2}}{\lambda_{xv}} \right) m_{v11} + m_{v22} \right] & \text{for m=2, 3; u=1,2} \\ 0 & \text{otherwise} \end{cases}$$

In the equations, J and g indicate the mass moment of inertia of vehicles and gravity acceleration, respectively. $\mathbf{D}$ and $\dot{\mathbf{D}}$ indicate displacement and velocity vectors of a bridge, respectively; $\mathbf{M}_b$ and $\mathbf{K}_b$ respectively indicate mass and stiffness matrices of a bridge; $\mathbf{C}_b$, the damping matrix of a bridge derived from the assumption of a linear relation between the mass and stiffness matrices.

The symbols Z_{v11}, Z_{v12}, Z_{v22}, θ_{xv11}, θ_{xv12}, θ_{xv22}, θ_{yv11} and θ_{yv22} refer to vehicle motions in relation to the bounce, parallel hop of the front axle, parallel hop of the rear axle, rolling, axle tramp of the front axle, axle tramp of the rear axle, pitching and axle windup of the rear axle,

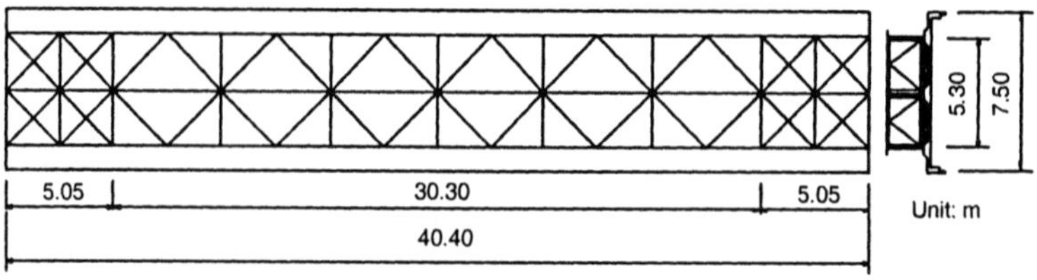

Figure 1. Plan view of bridge model

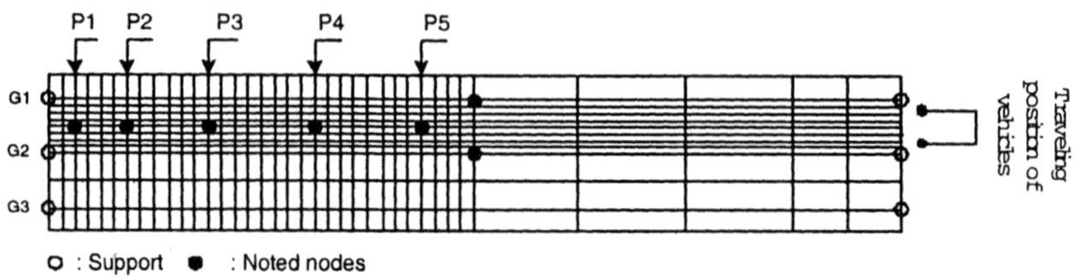

Figure 2. FE model of bridge

respectively. m_{v11}, m_{v12} and m_{v22} indicate the concentrated mass on the vehicle body, front axle and rear axle, respectively.

K_{vmku} and C_{vmku} are the spring constant and damping coefficient of the v-th vehicle; the subscript k is the index for indicating the vehicle body and axle ($k = 1$: vehicle body; $k = 2$: axle), m is the index for positions of axles or tires ($k = 1$ and $m = 1$: front axle; $k = 2$ and $m = 1$: rear axle; $k = 2$ and $m = 1$: wheel at the front axle; $k = 2$ and $m = 2$: front wheel at the tandem axle; $k = 2$ and $m = 3$: rear wheel at the tandem axle) and u is the index for indicating left and right sides of the v-th vehicle ($u = 1, 2$ indicating left and right side, respectively).

The symbol n_{veh} means numbers of the vehicles on a bridge. The variable z_{0vmu} denotes the vehicle displacement from the datum before deformation to a wheel after deformation of a bridge including roadway roughness. The longitudinal position of a wheel location x_{vmu} is relative to the bridge entrance. The pavement roughness of the bridge at a wheel is denoted by z_{rvmu}. The variable $w(t, x_{vmu})$ is the elastic deformation of the bridge at a location of x_{vmu} and a time of t. The subscript v indicates the v-th vehicle on the bridge.

The final formulation of governing differential equations for a bridge-vehicle interaction system is obtained from the relations in Eq. (1) to Eq. (4).

Table 1. Property of steel bridge

Mass per unit length (kg/m)		7550.000
Sectional area of girders (m^2)		0.142
Moment of inertia (m^4)		0.212
Torsional constant (m^4)		0.055
Damping constants for the 1st and 2nd modes		0.025
Fundamental frequency	1st: bending	2.340
(Experiment, Hz)	2nd: Torsion	3.810

2. Model Description

2.1 Bridge Model

A simple span bridge considered is a steel composite plate-girder bridge with span length of 40.4m, and composed with three girders. The span length and thickness of the RC deck are 2.65m and 17cm, respectively. Table 1 shows the properties of the bridge used in the dynamic response analysis. The fundamental frequencies for the bending and torsional modes taken from the eigenvalue analysis are calibrated to coincide with experimental values obtained from field-test data. Validity of the analytical responses is verified by comparing with field-test data [8].

The plan view and finite element model of the bridge are shown in Figure 1 and Figure 2, respectively. Analyzed panels are denoted as P1, P2, P3, P4 and P5 as shown in Figure 2. The FE model consists of 494 nodes, 444 flat shell elements and 223 beam elemts. The response of decks is calculated by superposing up to the 330th mode that coincides with the 5th bending mode of the deck with frequency of about 750Hz, since the dynamic responses are sufficiently converged within the 330th mode from a preliminary analysis.

2.2 Vehicle Model

Traffics with high percentage of heavy trucks on highway bridges usually occur at night, and the maximum traffic constitution among heavy trucks has been reported as the three-axle vehicle [14]. Moreover, Kim and Kawatani [12] demonstrate that the three-axle dump truck with a rear tandem axle gives rise to the maximum impact factor at the decks near expansion joints due to bumps. Thus, a dump truck with a tandem axle idealized as an eight-degree-of-freedom model is adopted as a vehicle model [9]. Properties of the vehicle model are summarized in Table 2.

Table 2. Property of vehicle

Geometry (m)	Tread	1.80
	Distance between front and rear axles	3.99
	Distance of tandem axle	1.32
	Distance between front axle and C.G.	2.99
Weight(kN)	Gross	191.00
	Sprung mass including payload	171.00
	Steer axle: unsprung mass	4.90
	Drive axle: unsprung mass	14.70
Spring constant (kN/m)	Front leaf spring	1577.00
	Rear leaf spring	4724.00
	Front tire	3146.00
	Rear tire	4724.00
Damping coefficient ($kN \cdot s/m$)	Front suspension	4.60
	Rear suspension	13.72
	Front tire	9.11
	Rear tire	27.34
Fundamental frequency (Hz)	Bounce	3.00
	Parallel hop	17.90

C.G: Centre of gravity

2.3 Random Variables

As random variables that effect on the dynamic response of decks, the roadway roughness on the bridge surface, bump height at the expansion joint of the bridge entrance, traveling position of vehicles, vehicle running speed and axle load at each axle of dump trucks are considered.

Roadway Profile. The fluctuations of roadway surface can be treated as a homogeneous, Gaussian random process with zero mean [4]. The simplest model describes the roadway surface as a cylindrical surface defined by a single longitudinal profile $z_r(x)$. Assuming $z_r(x)$ to be a zero mean, homogeneous, Gaussian random process as shown in Eq. (5), its probability structure can be defined by the auto-correlation function, or by the power spectral density(PSD).

$$z_r(x) = \sum_{k=1}^{M} a_k \sin(\omega_k x + \varphi_k) \tag{5}$$

where, α_k is Gaussian random variable with zero mean and variance $\sigma_k^2 = 4S(\omega_k)\Delta\omega$, φ_k is a random variable having uniform distribution between 0 and 2π, ω_k is the circular frequency of roadway surface roughness written as $\omega_k = \omega_L + (k - 1/2)\Delta\omega$, $\Delta\omega = (\omega_U - \omega_L)/M$, ω_U and

ω_L designate the upper and lower limit of the frequency, respectively, M means a large enough integer number and $S(\omega_k)$ is the PSD of a roadway profile.

The PSD can be obtained by a spectral analysis of the roadway profile measured along any longitudinal section. Following analytical description has been proposed to fit the measured PSD [7].

$$S(\Omega) = \frac{\alpha}{\Omega^n + \beta^n} \tag{6}$$

where, α is roughness coefficient, Ω $(= \omega/2\pi)$ is space frequency (cycle/m), β designates shape parameter and n means parameter to express the distribution of power of the PSD curve.

If a PSD for a roadway profile is defined, then, by means of the MCS method, samples of roadway profiles can be obtained using the sampling function shown in Eq. (5). As parameters in Eq. (6), $\alpha = 0.001$, $\beta = 0.05$ and $n = 2.0$ are used in this study based on measured data of Meishin Expressway in Japan [10]. The roadway roughness condition can be categorized as the road class "A" corresponds to a very good road according ISO 8086 code [2], which typically indicates a newly paved highway.

Bump Height near Expansion Joints. The extreme Type I distribution is assumed to describe bump heights at expansion joints of bridges based on the surveying results of national roadways in Japan [6]. Among the shapes of the measured bump profiles, the sine shaped bump profile that gives the most severe effect on the impact factors of decks from a preliminary study is adopted in the simulation. Although the mean value and standard deviation of the measured bump heights on the national roadways are $20.4mm$ and $7.0mm$, respectively, a half of the measured height is considered in the analysis for highway bridges [6]. The cumulative distribution function (CDF) and probability density function (PDF) for the random variables are

$$F_x(x) = e^{-e^{-\alpha(x-u)}} \tag{7}$$

$$f_x(x) = \alpha e^{-e^{-\alpha(x-u)}} e^{-\alpha(x-u)} \tag{8}$$

where, u and α are distribution parameters; $\alpha \approx 1.282/\sigma_x$, $u \approx \mu_x - 0.45\sigma_x$ [3]. The σ_x and μ_x indicate the standard deviation and mean value of a random variable x.

Traffic Data. The normal distribution is assumed for the running speed and traveling position of vehicles on highway bridges based on the database of Hanshin Expressway. The PDF of the normal distribution for a normal random variables x is shown in Eq. (9). The CDF of the normal random variables can be expressed as Eq. (10), even though there is no closed-form solution for the CDF of a normal random variable. The mean value and standard deviation of vehicle speeds are assumed as $70km/hr$ and $10km/hr$, respectively [14]. Those mean value and standard deviation for the traveling position of vehicles are $0.0m$ and $0.2m$ from a target passage [13].

$$f_x(x) = \frac{1}{\sigma_x\sqrt{2\pi}} exp\left[-\frac{1}{2}\left\{\frac{x-\mu_x}{\sigma_x}\right\}^2\right] \tag{9}$$

$$F_x(x) = \Phi\left(\frac{x-\mu_x}{\sigma_x}\right) \tag{10}$$

The lognormal distribution is assumed for the axle load of the three-axle vehicles based on the measured data of Hanshin Expressway. The mean value and standard deviation for the axle loads are $49.805kN$ and $12.056kN$ for the front axle, $90.507kN$ and $34.276kN$ for the front wheel of the tandem axle and $67.571kN$ and $31.637kN$ for rear wheel of the tandem axle [13]. The CDF and PDF for a lognormal random variable can be obtained by substituting $ln(x)$, $\mu_{ln(x)}$ and $\sigma_{ln(x)}$ into the Eq.(9) and Eq.(10) instead of x, μ_x and σ_x. It is noteworthy that the spring constants of vehicles are rearranged to have natural frequencies of 3.0Hz for the bounce and 17.9Hz for the axle hop motion according to each sample of axle loads.

3. Simulation of Impact Factor

3.1 Simulation and Probabilistic Feature

A number of sample roadway profiles, bump heights, vehicle speeds, traveling positions of vehicles and axle loads are generated by means of the MCS method. Impact factors of each deck are analyzed according to each sample of the random variables by means of traffic-induced dynamic response analysis of bridges [9]. In the simulation, no correlation among the considered random variables is assumed. A hundred samples of the simulated random variables are considered in the analysis, since the simulated impact factors tend to converge within 100 samples in a preliminary study.

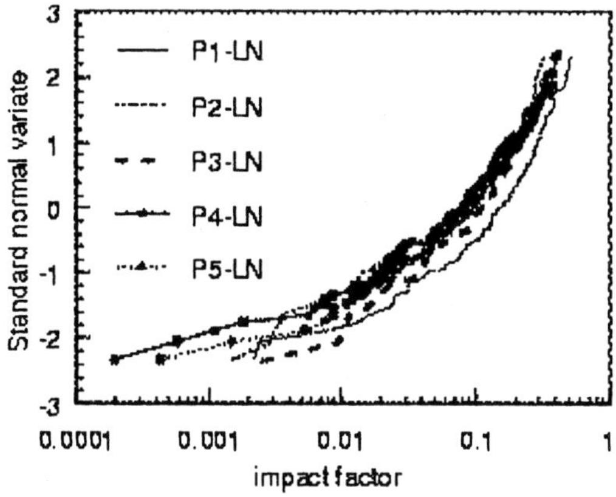

Figure 3. CDF of simulated impact factors of decks on lognormal probability paper

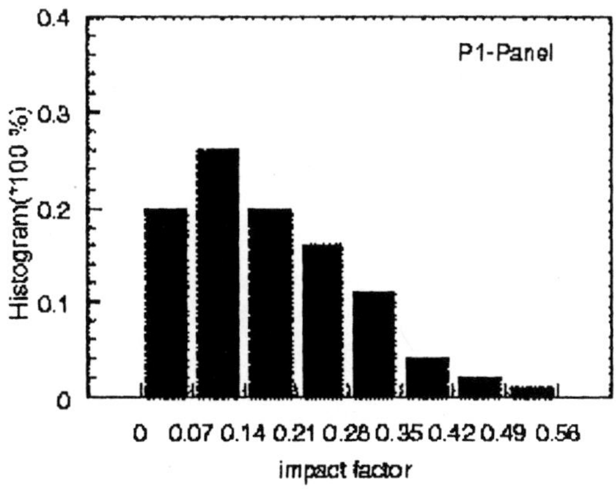

Figure 4. Distribution of simulated impact factors of P1-panel

Three types of distributions, such as normal, lognormal and extreme Type I distributions, are considered to investigate the probabilistic property of the simulated RC deck's impact factor. The CDF and histogram demonstrate that the impact factor can be concluded to follow lognormal distribution. Moreover the probability exceeding the code specified impact factors taken from the assumption of following the lognormal distribution gives the most frequent occurence among the three distributions. The CDF of the simulated impact factors plotted on lognormal probability paper is shown in Figure 3. The histogram of the simulated impact factors for the P1-panel is appeared in Figure 4.

3.2 Reliability of Code-Specified Impact Factors

Table 3. Code specified impact factors for deck slab

Code	impact factor	i_{code}
AASHTO (USA)	$i = 50/\,(3.3L + 125) \leq 0.3$	0.300
DIN1072 (Germany)	$i = 0.4 - 0.008L$	0.379
JSHB (Japan)	$i = 20/(L + 50)$	0.380
OHBDC (Ontario, Canada)	$i = 0.4$	0.400

$L = 2.65m$: Span length in meter

Table 4. Probability exceeding code specified impact factor (%)

	P1	P2	P3	P4	P5
AASHTO	11.15(3.77)	3.16(3.85)	4.25(5.20)	4.08(3.27)	3.58(3.35)
DIN1072	5.42(1.79)	1.42(1.88)	1.83(2.51)	2.03(1.48)	1.66(1.61)
JSHB	5.37(1.77)	1.40(1.87)	1.82(2.47)	2.01(1.47)	1.64(1.59)
OHBDC	4.50(1.49)	1.56(1.58)	1.49(2.09)	1.70(1.21)	1.37(1.34)

The reliability of the impact factors specified in AASHTO standard (USA), DIN1072 (Germany), Japanese Specifications of Highway Bridges (JSHB: Japan) and Ontario Highway Bridge Design Code (OHBDC Ontario, Canada) is investigated. The impact factors specified in the codes are summarized in Table 3.

The probability of exceeding the code-specified impact factors under the assumption of following lognormal distribution is summarized in Table 4. The value in the parenthesis indicates the exceeding probability against the code-specified impact factor without considering the bump at the expansion joint. Table 4 shows that the probability exceeding the code-specified impact factor for the deck near the bump is about three times greater than that of other decks.

The reliability index (RI) that calculated from the inverse of the exceeding probability is summarized in Figure 5 to compare the limit states defined in the EUROCODE [1]. The target reliability indices proposed in Eurocode are 1.5 for serviceability limit state (SLS), 3.8 for the ultimate limit state (ULS) and 1.5 to 3.8 for the fatigue limit state (FLS). The symbols -NB and -B indicate the results without considering bumps and with consdiering bumps at the expansion joint, respectively.

If the impact factor can be classified in the serviceability limit state then RI of the impact factor of AASHTO for the P1 panel is lower than that of the target reliability index for the SLS, although what kind of

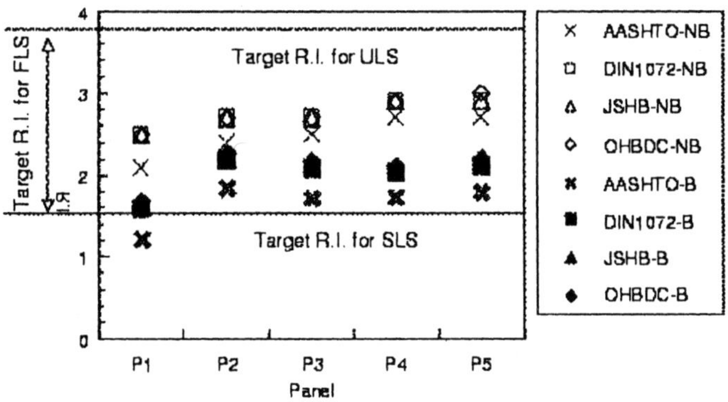

Figure 5. Reliability index of code-specified impact factors with target reliability index

limit state the impact factor is classified in has not been defined yet. On the other hand the reliability considering the condition of no bump at the expansion joint satisfies the SLS. It indicates that the bump is one of important factors for the impact factor of decks.

4. Concluding Remarks

The probabilistic feature of RC decks' simulated impact factors are examined. The reliability evaluation of code-specified impact factors is carried out considering randomness of the roadway roughness, bump height, vehicle speed, traveling position of vehicles and axle load. The study shows that the straight lines on the lognormal distribution paper can approximately represent probabilistic properties of the impact factor for the RC deck slab. The impact factor of the deck near expansion joints dominates the design impact factor. Therefore, if the impact factor of the deck near an expansion joint of an approaching side of bridges satisfies a given reliability due to a vehicle with tandem axle running on a bump, those reliabilities of other decks are satisfied automatically. In considering the impact factor of decks of bridges on roadways that have more severe bump condition than highway bridges, the reliability index against codes can decrease, therefore the use of code-specified impact factors of bridges on national roadways may overestimate the performance level of decks.

References

[1] Eurocode 1. Basis of design and action on structures, Part 1: Basis of design, 1993. Sixth draft.

[2] ISO 8606. Mechanical Vibration - Road Surface Profiles - Reporting of Measured Data, 1995; British Standard, BS 7853, 1996.

[3] J. R. Benjamin and C. A. Cornell. *Probability, Statistics and Decision for Civil Engineers.* McGraw-Hill, New York, 1970.

[4] C. J. Dodds and M. M. Robson. The description of road surface roughness. *Int. J. Sound and Vibrations*, 31(2):175–183, 1973.

[5] H. Furuta, I. Tsukiyama, M. Dogaki, and D. M. Frangopol. Maintenance support system of steel bridges based on life cycle cost and performance evaluation. *Proceedings of the 10th IFIP WG7.5 Working Conference on Reliability and Optimization of Structural Systems*, pages 205–213, 2002.

[6] H. Honda, Y. Kajikawa, and T. Kobori. Roughness characteristics at expansion joint on highway bridges. *Proceedings of JSCE, Note*, 328:173–176, 1982. (in Japanese).

[7] H. Honda, Y. Kajikawa, and T. Kobori. Spectra of road surface roughness on bridges. *ASCE Structural Division*, 1081(ST9):1956–1966, 1982.

[8] M. Kawatani and C. W. Kim. Effects of gap at expansion joint on traffic-induced vibration of highway bridge. *Proc. Int. Conference on Developments in Short and Medium Span Bridge Engineering '98*, CD-ROM, 1998.

[9] M. Kawatani and C. W. Kim. Computer simulation for dynamic wheel loads of heavy vehicles. *Int. J. Structural Engineering and Mechanics*, 12(4):409–428, 2001.

[10] M. Kawatani, Y. Kobayashi, and K. Takamori. Nonstationary random analysis with coupling vibration of bending and torsion of simple girder bridges under moving vehicles. *JSCE, J. Structural Eng. and Earthquake Eng.*, 15(1):107s–114s, 1998.

[11] C. W. Kim and M. Kawatani. A probabilistic investigation on impact factor of deck slabs of highway bridges. *Proc. of the 9th IFIP WG7.5 Working Conference on Reliability and Optimization of Structural Systems*, pages 125–133, 2000.

[12] C. W. Kim and M. Kawatani. Probabilistic investigation on impact factor of deck slabs due to truck configuration type. *Proceedings of the 10th IFIP WG7.5 Working Conference on Reliability and Optimization of Structural Systems*, pages 87–94, 2002.

[13] Comittee of Technology on Concrete of HEPC. Crack damage of RC deck slabs of highway bridges and its resistance. Technical report, Hanshin Expressway Management Technology Center, 1991. (in Japanese).

[14] M. Sakano, I. Mikami, and K. Miyagawa. Simultaneous loading effect of plural vehicles on fatigue damages of highway bridges. *IFIP Transactions B-12, Reliability and Optimization of Structural Systems, V*, pages 221–228, 1993.

[15] K. Yokoyama, J. Inoue, and T. Nagahara. Field test on the impact coefficient of steel deck and reinforced concrete slab of highway bridges. *JSCE, J. of Structural Engineering*, 35A:749–756, 1989. (in Japanese).

OPTIMAL MAINTENANCE PLANNING FOR BRIDGE STRUCTURES CONSIDERING EARTHQUAKE EFFECTS

Hitoshi Furuta
Kansai University, Informatics, Japan
furuta@res.kutc.kansai-u.ac.jp

Kazuhiro Koyama
Kansai University, Japan
kotaro-@sc.kutc.kansai-u.ac.jp

Abstract In the design of bridge structures, it is evident that the most important is to acquire their safety. However, there is a limitation to increase the safety because of the financial constraint. Especially, the reduction of construction cost is quite desirable under the severe economic condition in Japan.
Recently, Life-Cycle Cost (LCC) analysis has been paid attention as a possible and promising concept to achieve a rational maintenance program. In this study, a stochastic model of structural response is proposed, which accounts for the variation due to the uncertain characteristics of earthquakes, and the probability of failure is calculated based on the reliability theory. Using the failure probability, LCC can be calculated for the bridge structure with earthquake excitations.

Keywords: Bridge Structure, Failure Probability, Life-Cycle Cost, Reliability Theory, Seismic Analysis

Introduction

Many existing bridges in Japan are suffering from damage due to the deterioration of materials, heavy traffics and aging. In the future, it is evident that serious social problems will arise as the number of damaged bridges increases. Considering the present social and economic situation of Japan, it is urgent and important to establish an optimal maintenance strategy for such existing bridges so as to ensure their safety in satisfactory levels. Life-Cycle Cost (LCC) has been paid attention as a possible

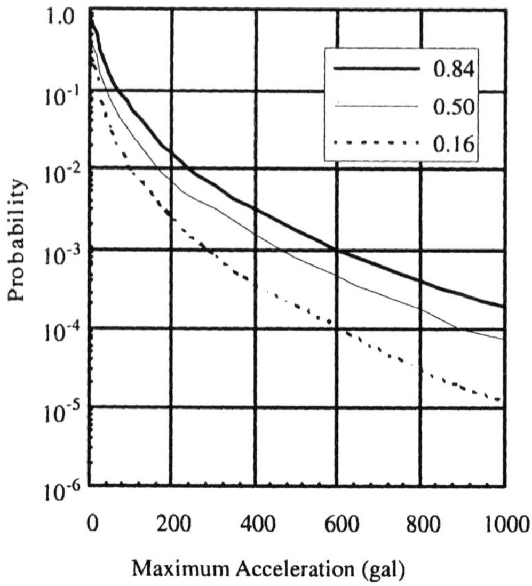

Figure 1. Seismic Hazard Curve

and promising method to achieve a rational maintenance program. Generally, LCC consists of initial cost, maintenance cost, and renewal cost. However, when considering Life-Cycle Cost in the region that frequent earthquakes occur, it is important to take into account the social and economical effects due to the collapse of structures as well as the minimization of maintenance cost. The loss by the collapse of structures due to the earthquakes can be defined in terms of an expected cost and introduced into the calculation of LCC. In this study, a stochastic model of structural response is proposed, which accounts for the variation due to the uncertain characteristics of earthquakes, and the probability of failure is calculated based on the reliability theory.

1. Earthquake Occurrence Probability in Service Time

In this study, the earthquake occurrence probability is evaluated by using seismic hazard curve. In the hazard curve, the annual exceedance probability of earthquake is calculated by considering the distribution of distance from epicenter, historical earthquake records, horizontal maximum acceleration and active fault. The hazard curve used here is shown in Figure 1.

2. Analysis of Required Yield Strength Spectrum

In this study, a probability model of yield strength is developed by using the yield strength spectrum. The yield strength spectrum shows a nonlinear relation between natural period and yield strength. This spectrum can be obtained for various ductility factors and damage indices. For natural period, target ductility factor, earthquake level, type of soil and number of seismic wave, the values presented in Table 1 are used.

Natural Period T	0.1‘5.0(sec)
Target Ductility Factor μ_T	1.0, 2.0
Earthquake Level	400, 800(gal)
Type of Soil	1,2,3
Seismic Wave	18

Table 1. Analysis Condition

Figure 2 shows the calculated results of the yield strength spectrum. Because there are many data, only representative values are shown in Figure 2,3. Figure 3 shows the analysis results using the data presented in Figure 2. There results are obtained through the regression analysis for the earthquake level of 800gal. In this study, the distribution of yield strength is obtained by using this spectrum.

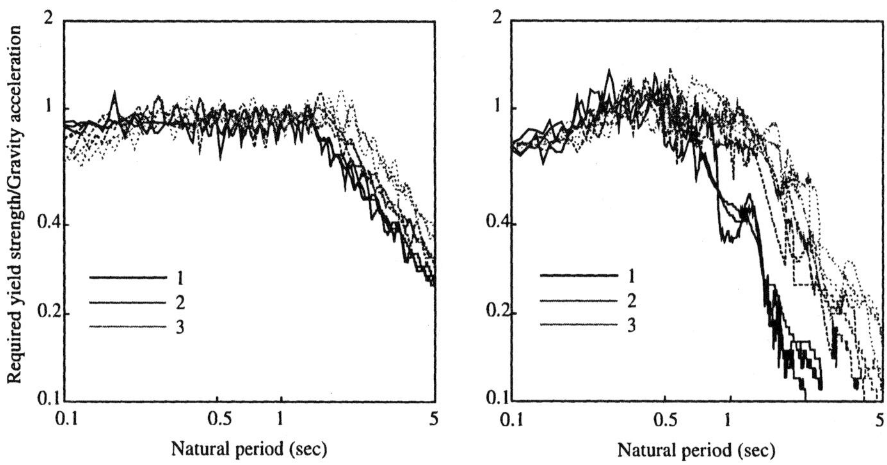

Figure 2. Required Yield Strength Spectra

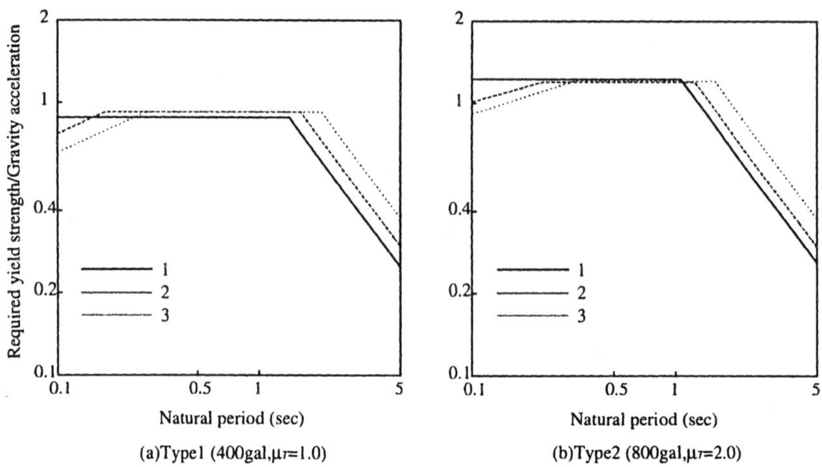

Figure 3. Standard Required Yield Strength Spectra

2.1 Probability Model of Required Yield Strength

When designing a structure, type of soil, ground maximum acceleration and target ductility factor should be given. The natural period of the structure is changed by the design. In order to make the design simple, it is not good that the probability model of the required yield strength changes depending on the design. In this study, a probability distribution model of yield strength is developed for various types of soil and target ductility factors. In this model the required yield strength can be constant regardless of the natural period. Here, the probability distribution model is assumed to be the lognormal distribution. The probability density function of the lognormal distribution is given in Figure 1

$$f_{P_{ny}}(x) = \frac{1}{\sigma_{ln(x)}\sqrt{2\pi}}\frac{1}{x}\left[-\frac{1}{2}\left(\frac{ln(x)-\mu_{ln(x)}}{\sigma_{ln(x)}}\right)^2\right] \tag{1}$$

$\mu_{ln(x)}$ is the standard value of required yield strength, $\sigma_{ln(x)}$ is standard deviation of every response ductility factor. In other words, if to understand a standard value of the probability function, the probability model of required yield strength can be calculated. We calculated a standard deviation about the type of soil and target ductility factor. (Refer to Table 2.)

Ductility Factor	Earthquake	Soil	Var_{ln}	σ_{ln}
1	1	1	0.0030	0.0551
		2	0.0029	0.0537
		3	0.0052	0.0722
	2	1	0.0049	0.0699
		2	0.0054	0.0736
		3	0.0044	0.0662
2	1	1	0.0222	0.1488
		2	0.0343	0.1852
		3	0.0287	0.1684
	2	1	0.0306	0.1749
		2	0.0361	0.1900
		3	0.0410	0.2025

Table 2. Result of Analysis

3. Reliability Analysis of Steel Bridge Pier

As an example, a steel bridge pier is employed, in which its failure probability and reliability index are calculated.

3.1 Analysis Model

As mentioned previously, a steel bridge pier is used for the analysis model. Figure 4 shows the detail of the steel bridge pier and its cross section. To simplify the design, it is assumed that the cross section is square without stiffening and the width of flange and web is equal. To avoid the local buckling, this pier is designed according to the gJapanese Specification of Highway Bridgeh. Then, the minimum thickness is determined so as to satisfy the following requirement: where the steel material is SM490Y.

In this study, the residual stress and the initial deflection are considered for the initial imperfection of the steel pier. This pier is used for a three-spanned continuous steel girder bridge with 40 m span length. The bridge has two main girders and Reinforced Concrete (RC) slabs. For the design condition, the vertical load P is calculated to be 10.84 MN, which corresponds to the reaction force to the superstructure. The height of the steel pier is 10m, and its width is 2,000 mm and its thickness is 42 to 60mm.

$$\frac{b_f}{t_f} = \frac{b_w}{t_w} \leq 48 \tag{2}$$

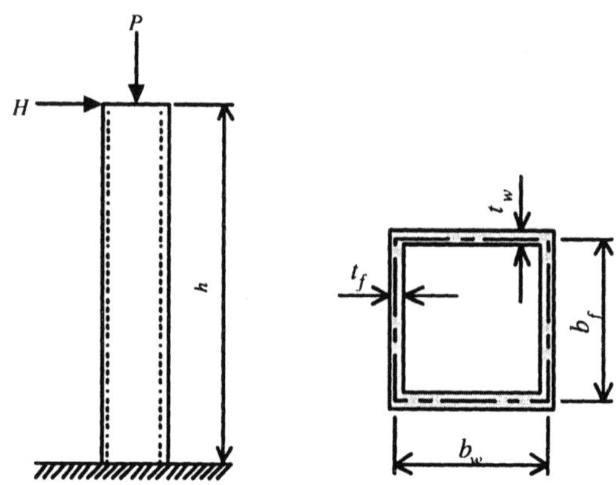

Figure 4. Analysis Model

3.2 Limit State Functions

In this study, both the serviceability limit condition and the ultimate limit condition are taken into account as the limit state.

Serviceability Limit Condition. The elastic limit of the structure is employed for the serviceability limit condition. The limit state function is given as follows:

$$Z = P_a - P_{yn} \tag{3}$$

where P_a means the yield strength of the structure and P_{yn} means the required yield strength of ductility factor μ=1.0.

Ultimate Limit Condition. In this study, the ductility factor μ is taken as the parameter which decides the ultimate limit condition. Considering the difference of structural type and structural material, the reliability analysis is performed for the ductility factor μ=1.0, 2.0, 3.0, 4.0, 5.0. Then, the limit state function is given as

$$Z = P_a - P_{yn} \tag{4}$$

where P_{yn} is required yield strength of the ductility factor μ=1.0, 2.0, 3.0, 4.0, 5.0.

3.3 Reliability Analysis Model

For the above ultimate and serviceability limit functions, the reliability analysis is performed, where the load combination is not considered.

Because the distribution of earthquake force becomes a non-normal distribution, the safety margin Z also becomes a non-normal distribution. Therefore, it is necessary to transfer Z to a normal distribution.

$$F_c(Z^*) = \Phi\left(\frac{z^* - \mu_c}{\sigma_c}\right) \tag{5}$$

$$f_c(z^*) = \frac{1}{\sigma_c}\varphi\left(\frac{z^* - \mu_c}{\sigma_c}\right) \tag{6}$$

where μ_c and σ_c are the mean value and the standard deviation of the normal distribution which is used for the approximation. The two values are unknown variables and to be obtained by solving the above equations. Then the reliability index is calculated as

$$\beta = \frac{\mu_c}{\sigma_c} \tag{7}$$

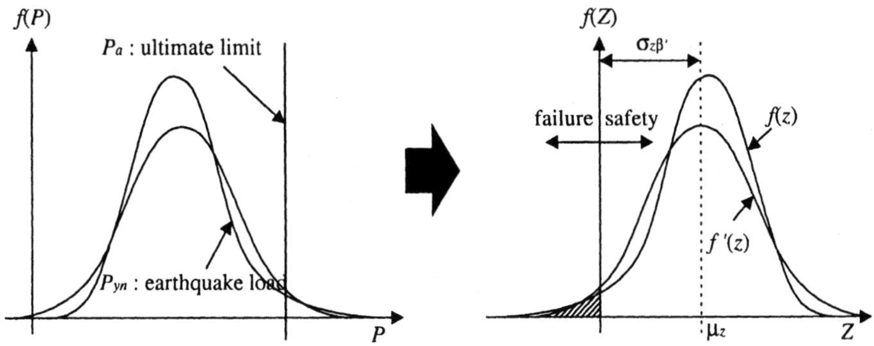

Figure 5. Reliability Index

3.4 Reliability Analysis of Steel Bridge Pier

Since it needs enormous task to analyze all the target ductility factor and type of soil, only types II soil is taken into consideration and the corresponding damage index to the earthquake is assumed as follow:
For type I earthquake, 400 gal is employed as the maximum acceleration. Type I earthquake is defined in the Japanese Specification of Highway Bridge. This type I earthquake is used to check the seismic design for middle earthquakes. Then, the damage index of the structure is assumed to be μ_T=1.0. On the contrary, Type II earthquake is used to check for strong earthquakes, in which the maximum acceleration is 800gal. Then, the damage index of the structure is μ_T=2.0.

Item	Unit	Cost(Yen)
Material Cost	t	97,585
Painting Cost	m^2	1,700
Transport Cost	t	9,000

Table 3. Initial Cost

4. Life-Cycle Cost Considering Earthquake Effects

Here, initial cost and loss of the steel bridge pier are described. Based on the result obtained previously, LCC is formulated and calculated for the steel bridge pier.

4.1 Initial Cost and Loss Cost

As the initial cost, only pier is considered, because the sufficient data for the whole bridge is not prepared. Then, the initial cost consists of material cost, painting cost and transportation cost. (Table 3) It is also assumed that type I earthquake will lose 80%of the initial cost and type II earthquake will lose 110%.

4.2 Formulation of LCC Considering Earthquake Effects

The failure probability of structure is calculated by (8).

$$P_{f,t=L} = P(t) \times P_f \tag{8}$$

Then, LCC is formulated as (9), where i is the social discount rate and assumed to be 2%.

$$LCC = C_I + P_{f,t=L}\frac{C_F}{(1+i)^T} \tag{9}$$

4.3 Result of Analysis

The relation between LCC and failure probability is shown in Figure 6, in which LCC is calculated with the same assumption as those used in the reliability analysis.

We calculated LCC about the order which was equal to the reliability analysis.

Case 1. For the case shown in Figure 6, LCC is minimized when the thickness $t_f=t_w$ is 50(mm). Then, the failure probability P_f is 0.03 and the reliability index β is 1.86.

Case 2. For the case shown in Figure 7, LCC is minimized. When the thickness $t_f=t_w$ is 56(mm), in which P_f is 0.12 and reliability index is 1.17.

From these results, it is confirmed that it is possible to propose the optimal design plan in service time including the loss cost, which has the minimum LCC from the standpoint of reliability-based design, by introducing the loss of the structure due to the earthquake and the repairing cost.

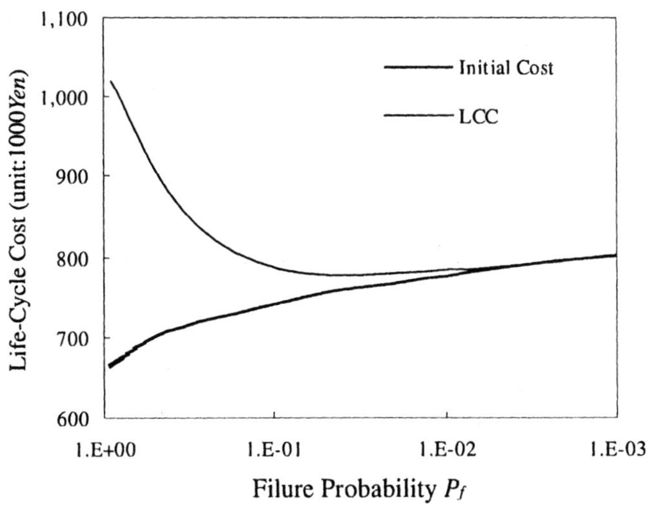

Figure 6. Type I eathquake,μ_T=1.0

5. Conclusion

In this study, an attempt was made to propose a calculation method of LCC considering earthquake effects based on reliability index. Through several numerical calculations, the following conclusions were derived:

1 The proposed method can calculate LCC and failure probability for the structure which shows simple behavior by the earthquake.

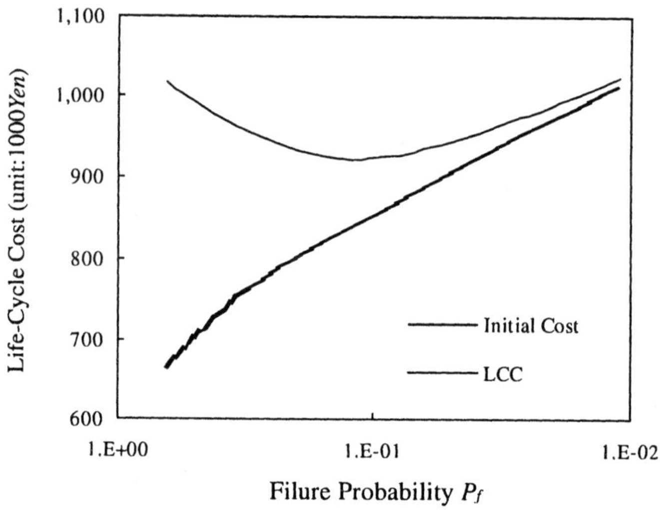

Figure 7. Type II eathquake, μ_T=2.0

2 Considering the collapse of the structure, LCC could be evaluated by failure probability.

3 Using the reliability-based design concept, it is possible to provide a design plan to make the reliability index maximum and LCC minimum.

4 In the calculation of LCC, many factors are interrelated and include various uncertainties.

Therefore, there still remain many issues to overcome in the future. For instance, in this study, the probability model of the maximum acceleration and required yield strength was calculated by the historical earthquake records and the standard input earthquake wave of "Japanese Specification of Highway Bridge". However, when comparing the return period of the earthquake, the period of the earthquake records was too short and the available data is insufficient to estimate the earthquake intensity and the seismography. Moreover, the data of initial cost and maintenance cost were insufficient to calculate LCC of the whole bridge with accuracy. In the calculation of LCC, user cost should be accounted for as well as the earthquake occurrence and the maintenance by the deterioration of the structure.

UNIFORM DECAY RATES OF SOLUTIONS TO A NONLINEAR WAVE EQUATION WITH BOUNDARY CONDITION OF MEMORY TYPE

Marcelo M. Cavalcanti*
Department of Mathematics - State University of Maringá
87020-900, Maringá - PR, Brazil
mmcavalcanti@uem.br

Valéria N. Domingos Cavalcanti†
Department of Mathematics - State University of Maringá
87020-900, Maringá - PR, Brazil
vndcavalcanti@uem.br

Mauro L. Santos
Departamento de Matemática, Universidade Federal do Pará
Campus Universitário do Guamá,
Rua Augusto Corrêa 01, Cep 66075-110, Pará, Brazil.
ls@ufpa.br

Abstract In this article we study the hyperbolic problem (1) where Ω is a bounded region in $\mathbf{R}^n$ whose boundary is partitioned into disjoint sets Γ_0, Γ_1. We prove that the dissipation given by the memory term is strong enough to assure exponential (or polynomial) decay provided the relaxation function also decays exponentially (or polynomially). In both cases the solution decays with the same rate of the relaxation function.

Keywords: wave equation, gradient nonlinearity, boundary memory term.

*Supported by CNPq(Brazil) - 301326/1996-7
†Supported by CNPq(Brazil) - 300567/1999-5

Introduction

In this work we study the existence of global solutions and the asymptotic behavior of the energy related to the following nonlinear wave equation with a boundary condition of memory type

$$\begin{aligned} u_{tt} - \Delta u + F(x,t,u,\nabla u) &= 0 \quad \text{in} \quad \Omega\times]0,+\infty[, && \text{(1a)}\\ u &= 0 \quad \text{on} \quad \Gamma_0\times]0,+\infty[, && \text{(1b)}\\ u + \int_0^t g(t-s)\frac{\partial u}{\partial \nu}(s)ds &= 0 \quad \text{on} \quad \Gamma_1\times]0,+\infty[, && \text{(1c)}\\ u(x,0) = u^0(x), \quad u_t(x,0) &= u^1(x) \quad \text{in} \quad \Omega, && \text{(1d)} \end{aligned}$$

where Ω is a bounded domain of $\mathbf{R}^n$, $n \geq 1$, with smooth boundary $\Gamma = \Gamma_0 \cup \Gamma_1$. Here, Γ_0 and Γ_1 are closed, disjoint, $\Gamma_0 \neq \emptyset$ and ν is the unit normal vector pointing towards the exterior of Ω. Equation (1c) is a nonlocal boundary condition responsible for the memory effect. Considering the history condition, we must add to conditions (1b)-(1c) the one given by

$$u = 0 \quad \text{on} \quad \Gamma_0\times]-\infty,0].$$

We observe that in problem (1a)-(1d), u represents the transverse displacement and the relaxation function g is a positive non-increasing function belonging to $W^{2,1}(0,+\infty)$. Furthermore, suppose that the function $F : \overline{\Omega} \times [0,+\infty[\times\mathbf{R}^{n+1} \to \mathbf{R}$ is of class C^1 and satisfies

$$|F(x,t,\xi,\zeta)| \leq C_0(1 + |\xi|^{\gamma+1} + |\zeta|) \tag{2}$$

where C_0 is a positive constant, and $\zeta = (\zeta_1, ..., \zeta_n)$.

Let γ be a constant such that $\gamma > 0$ for $n = 1, 2$, and $0 < \gamma \leq 2/(n-2)$ for $n \geq 3$. Assume that there is a non-negative function $\varphi(t)$ in the space $L^\infty(0,\infty) \cap L^1(0,\infty)$ such that

$$F(x,t,\xi,\zeta)\eta \geq |\xi|^\gamma \xi\eta - \varphi(t)(1 + |\eta||\zeta|), \quad \forall \eta \in \mathbf{R}, \tag{3}$$

and, particularly,

$$F(x,t,\xi,\zeta)(m\cdot\zeta) \geq |\xi|^\gamma \xi(m\cdot\zeta) - \varphi(t)(1 + |\zeta||m\cdot\zeta|). \tag{4}$$

Consider the existence of positive constants $C_0, \ldots, C_n$, which verify

$$|F_t(x,t,\xi,\zeta)| \leq C_0\left(1 + |\xi|^{\gamma+1} + |\zeta|\right), \tag{5}$$

$$|F_\xi(x,t,\xi,\zeta)| \leq C_0(1 + |\xi|^\gamma), \tag{6}$$

$$|F_{\zeta_i}(x,t,\xi,\zeta)| \leq C_i \quad \text{for } i = 1, 2, \ldots, n. \tag{7}$$

and also consider that there exist positive constants D_1, D_2, such that for all $\xi, \hat{\xi}, \eta, \hat{\eta} \in \mathbf{R}$ and for all $\zeta, \hat{\zeta} \in \mathbf{R}^n$,

$$\begin{aligned} &(F(x,t,\xi,\zeta) - F(x,t,\hat{\xi},\hat{\zeta}))(\eta - \hat{\eta}) \\ &\geq -D_1(|\xi|^\gamma + |\hat{\xi}|^\gamma)|\xi - \hat{\xi}||\eta - \hat{\eta}| - D_2|\eta - \hat{\eta}||\zeta - \hat{\zeta}| . \end{aligned} \tag{8}$$

Defining

$$F(x,t,u,\nabla u) = |u|^\gamma u + \varphi(t) \sum_{i=1}^{n} \sin\left(\frac{\partial u}{\partial x_i}\right),$$

where φ is a sufficiently regular function, we obtain an example of a function F which verifies the above hypotheses.

In order to obtain the decay rates stated in theorems (2.1)-(2.2), we will assume that the function $\varphi(t)$ considered in (4) satisfies one of the following assumptions

$$\varphi(t) \leq \theta_0 e^{-b_1 t} \quad \text{or} \quad \varphi(t) \leq \frac{\theta_1}{(1+t)^{p+1}}, \tag{9}$$

where θ_0, θ_1 and b_1 are positive constants and $p > 1$.

The integral equation (1c) describes the memory effect which can be caused, for example, by the interaction with another viscoelastic element. Indeed, from the physical point of view, condition (1c) means that Ω is composed of a material which is clamped in a rigid body in Γ_0 and is clamped in a body with viscoelastic properties in the complementary part of its boundary named Γ_1. So, it is expected that if the kernel of the memory decays (exponentially or polynomially) the same occurs to the solutions of problem (1a) - (1d).

In what follows we are going to assume that there exists $x_0 \in \mathbf{R}^n$ such that

$$\begin{aligned} \Gamma_0 &= \{x \in \Gamma \;:\; \nu(x)\cdot(x - x_0) \leq 0\}, \\ \Gamma_1 &= \{x \in \Gamma \;:\; \nu(x)\cdot(x - x_0) > 0\}. \end{aligned}$$

Defining $m(x) = x - x_0$, the compactness of Γ_1 implies that there exists a positive constant δ_0 such that

$$0 < \delta_0 \leq m(x)\cdot\nu(x), \quad \forall x \in \Gamma_1. \tag{10}$$

There is not much in literature regarding the existence and asymptotic behavior of evolution equations subject to memory conditions acting on the boundary. It is worth mentioning some papers in connection with viscoelastic effects on the boundary. In this direction we can cite the work by Aassila , Cavalcanti and Soriano [1]who considered the linear wave equation subject to nonlinear feedback and viscoelastic effects on the boundary, and proved uniform (exponential and algebraic) decay rates. Also, we can cite the article of Andrade and Munõz Rivera [2]where it was considered a one-dimensional nonlinear wave equation subject to a nonlocal and nonlinear boundary memory effect. In this work the authors showed that the dissipation occasioned by the memory term was strong enough to guarantee global estimates and, consequently, allowed them to prove existence of global smooth solution for small data and to obtain exponential(or polynomial) decay provided the kernel decays exponentially(or polynomially). In the same context we can mention the work of Santos [11], where decay rates were proved concerning the wave equation with coefficients depending on time and subject to a memory condition on the boundary.

A natural question that arises in this context is about the non-existence results for the wave equation in the presence of viscoelastic effects acting on the boundary. Concerning to this subject we can mention the work of Kirane and Tartar [6]who obtained non-existence results and Qin [9]who proved a blow up result for the nonlinear one dimensional wave equation with memory boundary condition.

In connection with the above discussion, regarding viscoelastic problems, it is important to cite the works of Ciarletta [4], Fabrizio and Morro [5]and Qin [8].

The main goal of the present paper is to complement the above mentioned works. The majority of the results are obtained in an one dimensional domain while our paper deals with a n-dimensional problem which brings up some additional difficulties, mainly in what concerns the geometric conditions. In addition, as we have a nonlinear problem whose nonlinearity $F = F(x, t, u, \nabla u)$ depends on the gradient, we do not have any information about the influence of the integral $\int_\Omega F(x, t, u, \nabla u)u_t dx$ on the energy $E(t)$ (see page 6) or about the sign of the derivative $E'(t)$. In other words, we cannot guarantee that $E'(t) \leq 0$ which plays an essential role in establishing the desired decay rates.

Note that condition (1b) implies that the solution of system (1a)-(1d) must belong to the following space

$$V := \{v \in H^1(\Omega) \ : \ v = 0 \quad \text{on} \quad \Gamma_0\}.$$

The notations we use in this paper are standard and can be found in Lion's book [7]. In the sequel, C (sometimes $C_1, C_2, \dots$) is going to denote various positive constants which do not depend on t and neither depend on the initial data. This paper is organized as follows. In section 2 we establish the existence and uniqueness for regular and weak solutions to the system(1a)-(1d). In section 3 we prove the uniform exponential decay and in section 4 we prove the uniform polynomial decay.

1. Notations and Main Results

In this section we present some notations and we are going to study the existence of regular and weak solutions to the system (1a)-(1d). First, we will use equation (1c) to estimate the term $\frac{\partial u}{\partial \nu}$.

Defining the convolution product operator by

$$(g * \varphi)(t) = \int_0^t g(t-s)\varphi(s)ds$$

and differentiating the equation (1c) with respect to t, we obtain the Volterra equation

$$\frac{\partial u}{\partial \nu} + \frac{1}{g(0)}\left(g' * \frac{\partial u}{\partial \nu}\right) = -\frac{1}{g(0)}u_t \quad \text{on} \quad \Gamma_1\times]0,+\infty[.$$

Applying the Volterra's inverse operator, we get

$$\frac{\partial u}{\partial \nu} = -\frac{1}{g(0)}(u_t + k * u_t) \quad \text{on} \quad \Gamma_1 \times (0,+\infty).$$

where the resolvent kernel satisfies

$$k + \frac{1}{g(0)}g' * k = -\frac{1}{g(0)}g'.$$

Defining $\eta = \frac{1}{g(0)}$, we get

$$\frac{\partial u}{\partial \nu} = -\eta\left(u_t + k(0)u - k(t)u^0 + k' * u\right) \quad \text{on} \quad \Gamma_1\times]0,+\infty[. \tag{11}$$

Reciprocally, considering that the initial data satisfies $u^0 = 0$ on Γ_1, (11) implies (1c). Since we are interested in relaxation functions of exponential or polynomial type and identity (11) involves the resolvent kernel k, we want to investigate if k has the same properties. The following Lemma answers this question.

Let h be a relaxation function and k its resolvent kernel, that is,

$$k(t) - k * h(t) = h(t). \tag{12}$$

LEMMA 1 *If h is a positive continuous function, then k is also a positive continuous function. Moreover,*

1 If there exist positive constants c_0 and γ with $c_0 < \gamma$ such that

$$h(t) \le c_0 e^{-\gamma t},$$

we conclude that the function k satisfies

$$k(t) \leq \frac{c_0(\gamma - \epsilon)}{\gamma - \epsilon - c_0} e^{-\epsilon t},$$

for all $0 < \epsilon < \gamma - c_0$.

2 Let us consider $p > 1$ and define by $c_p := \sup_{t \in \mathbf{R}_+} \int_0^t (1+t)^p (1+t-s)^{-p} (1+s)^{-p}\, ds$. Provided there exists a positive constant c_0 with $c_0 c_p < 1$ such that

$$h(t) \leq c_0 (1+t)^{-p},$$

the function k satisfies

$$k(t) \leq \frac{c_0}{1 - c_0 c_p} (1+t)^{-p}.$$

proof. Note that $k(0) = h(0) > 0$. If we take $t_0 = \inf\{t \in \mathbf{R}_+ : k(t) = 0\}$ we obtain that $k(t) > 0$ for all $t \in [0, t_0[$. If $t_0 < +\infty$, from equation (12) we get that $-k * h(t_0) = h(t_0)$ which is a contradiction. Therefore $k(t) > 0$ for all $t \in [0, +\infty[$. Now, fixing ϵ, such that $0 < \epsilon < \gamma - c_0$ and defining

$$k_\epsilon(t) := e^{\epsilon t} k(t), \quad h_\epsilon(t) := e^{\epsilon t} h(t),$$

we get from (12) that $k_\epsilon(t) = h_\epsilon(t) + k_\epsilon * h_\epsilon(t)$. Hence

$$\begin{aligned} \sup_{s \in [0,t]} k_\epsilon(s) &\leq \sup_{s \in [0,t]} h_\epsilon(s) + \left(\int_0^\infty c_0 e^{(\epsilon - \gamma)s}\, ds \right) \sup_{s \in [0,t]} k_\epsilon(s) \\ &\leq c_0 + \frac{c_0}{(\gamma - \epsilon)} \sup_{s \in [0,t]} k_\epsilon(s). \end{aligned}$$

Therefore

$$k_\epsilon(t) \leq \frac{c_0(\gamma - \epsilon)}{\gamma - \epsilon - c_0},$$

which proves our first assertion. To show the second part let us introduce the following definitions

$$k_p(t) := (1+t)^p k(t), \quad h_p(t) := (1+t)^p h(t).$$

Multiplying equation (12) by $(1+t)^p$ we obtain

$$k_p(t) = h_p(t) + \int_0^t k_p(t-s)(1+t)^p (1+t-s)^{-p} (1+s)^{-p} h_p(s)\, ds$$

and, consequently,

$$\sup_{s\in[0,t]} k_p(s) \leq \sup_{s\in[0,t]} h_p(s) + c_0 c_p \sup_{s\in[0,t]} k_p(s) \leq c_0 + c_0 c_p \sup_{s\in[0,t]} k_p(s),$$

which implies

$$k_p(t) \leq \frac{c_0}{1 - c_0 c_p}.$$

This concludes the proof of Lemma.

Remark: In Racke [10, Lemma 7.4], it is assured that c_p is a finite positive constant. Also, according to this Lemma, in what follows, we are going to use (11) instead of (1c).

In order to prove the following Lemma, let us define

$$(g\Diamond\varphi)(t) := \int_0^t g(t-s)|\varphi(t)-\varphi(s)|^2 ds.$$

LEMMA 2 *For real functions $g, \varphi \in C^1([0,\infty[)$ we have*

$$(g*\varphi)\varphi_t = -\frac{1}{2}g(t)|\varphi(t)|^2 + \frac{1}{2}g'\Diamond\varphi - \frac{1}{2}\frac{d}{dt}\left[g\Diamond\varphi - (\int_0^t g(s)ds)|\varphi|^2\right].$$

The proof of this lemma follows by differentiating the term $g \diamond \varphi$.

The first order energy of system (1a)-(1d) is defined by

$$\begin{aligned} E(t) : &= \frac{1}{2}\int_\Omega |u_t(x,t)|^2 dx + \frac{1}{2}\int_\Omega |\nabla u(x,t)|^2 dx \\ &+ \frac{1}{\gamma+2}\int_\Omega |u(x,t)|^{\gamma+2} dx - \frac{\eta}{2}(k'\Diamond u)(t) \\ &+ \frac{\eta}{2}k(t)\int_{\Gamma_1} |u(x,t)|^2 d\Gamma. \end{aligned}$$

The well-posedness of system (1a)-(1d) as well as the decay rates expected are presented in the following Theorem.

THEOREM 3 *Let $k \in W^{2,1}(\mathbf{R}_+)$, assume that assumptions (2)-(8) hold and suppose that $\{u^0, u^1\} \in \left(V \cap H^2(\Omega)\right)^2$, satisfying the compatibility condition*

$$\frac{\partial u^0}{\partial \nu} + \eta u^1 = 0 \quad on \quad \Gamma_1. \tag{13}$$

Then, problem (1a)-(1d) possesses a unique solution u such that

$$u \in L^\infty(0,\infty, V\cap H^2(\Omega)),\ u' \in L^\infty(0,\infty,V),\ u'' \in L^\infty(0,\infty,V). \tag{14}$$

In addition, assuming that there exist positive constants b_1, b_2 which verify one of the conditions below

$$k(0) > 0, \quad k'(t) \leq -b_1 k(t), \quad k''(t) \geq -b_2 k'(t), \tag{15}$$

or

$$k(0) > 0, \quad k'(t) \leq -b_1 k(t)^{1+\frac{1}{p}}, \quad k''(t) \geq b_2[-k'(t)]^{1+\frac{1}{p+1}}, \ p > 1 \tag{16}$$

and, moreover, that hypotheses (9)-(10) hold, we obtain that the energy $E(t)$ associated to problem (1a)-(1d) decays, respectively, with the following rates of decay

$$E(t) \leq \alpha_1 e^{-\alpha_2 t} E(0) \tag{17}$$

or

$$E(t) \leq \frac{C}{(1+t)^{p+1}} E(0), \tag{18}$$

where α_1, α_2 and C are positive constants.

THEOREM 4 *Let $k \in W^{2,1}(\mathbf{R}_+)$; suppose that $\{u^0, u^1\} \in V \times L^2(\Omega)$ and the assumptions (2)-(10) and (15)-(16) hold. Then, problem (1a)-(1d) has a unique weak solution u in the space*

$$C^0([0,\infty); V) \cap C^1([0,\infty); L^2(\Omega)).$$

Furthermore, the decay rates presented in (17)-(18) hold for the weak solution u.

proof. The proof of existence and uniqueness for regular and weak solutions can be obtained following exactly identical procedure as in the work [3]of the authors Cavalcanti, Domingos Cavalcanti and Soriano. Consequently it will be omitted.

2. Exponential Decay

In this section we shall study the asymptotic behavior of the solutions of system (1a)-(1d) when the resolvent kernels k is exponentially decreasing, that is, there exist positive constants b_1, b_2 such that

$$k(0) > 0, \quad k'(t) \leq -b_1 k(t), \quad k''(t) \geq -b_2 k'(t) \quad . \tag{19}$$

Note that this conditions implies that

$$k(t) \leq k(0) e^{-b_1 t} \quad \text{for} \quad t > 0.$$

Our point of departure will be to establish some inequalities for the solution of system (1a)-(1d).

LEMMA 5 *Any regular solution u of the system (*1a*)-(*1d*) satisfy*

$$\begin{aligned}\frac{d}{dt}E(t) &\leq -\frac{\eta}{2}\int_{\Gamma_1}|u_t|^2 d\Gamma_1+\frac{\eta}{2}k^2(t)\int_{\Gamma_1}|u_0|^2 d\Gamma_1\\ &+\frac{\eta}{2}k'(t)\int_{\Gamma_1}|u|^2 d\Gamma_1-\frac{\eta}{2}\int_{\Gamma_1}k''\Diamond u d\Gamma_1\\ &+\varphi(t)\int_{\Omega}(1+|u_t||\nabla u|)dx.\end{aligned}$$

proof. Multiplying the equation (1a) by u_t and integrating by parts over Ω we get

$$\frac{1}{2}\frac{d}{dt}\int_{\Omega}(|u_t|^2+|\nabla u|^2)dx=-\int_{\Omega}F(x,t,u,\nabla u)u_t dx+\int_{\Gamma_1}\frac{\partial u}{\partial\nu}u_t d\Gamma_1.$$

Taking (3), (11) into account and using Lemma 2 our conclusion follows.

Let us consider the following binary operator

$$(k\diamond\varphi)(t):=\int_0^t k(t-s)(\varphi(t)-\varphi(s))ds.$$

Then employing Hölder's inequality for $0\leq\mu\leq 1$ we have

$$|(k\diamond\varphi)(t)|^2\leq\left[\int_0^t|k(s)|^{2(1-\mu)}ds\right](|k|^{2\mu}\Diamond\varphi)(t). \tag{20}$$

Let us define the functionals

$$\mathcal{N}(t):=\int_{\Omega}(|u_t|^2+|\nabla u|^2+|u|^{\gamma+2})dx,$$

$$\psi(t)=2\int_{\Omega}(m\cdot\nabla u)u_t+\theta\int_{\Omega}uu_t dx, \tag{21}$$

where $\theta\in]n-2,n[$ and $\theta>\frac{2n}{\gamma+2}$. The following Lemma plays an important role for the construction of the Lyapunov functional.

LEMMA 6 *For any regular solution of the system (1a)-(1d) we get*

$$\begin{aligned}\frac{d}{dt}\psi(t) \;\leq\;& \int_{\Gamma_1}(m\cdot\nu)|u_t|^2 d\Gamma_1 + (\theta - n)\int_\Omega |u_t|^2 dx \\ &-(\theta-(n-2))\int_\Omega |\nabla u|^2 dx \\ &+\int_{\Gamma_1}\frac{\partial u}{\partial \nu}\{m\cdot\nabla u + \theta u\} d\Gamma_1 - \int_{\Gamma_1}(m\cdot\nu)|\nabla u|^2 d\Gamma_1 \\ &-(\theta - \frac{2n}{\gamma+2})\int_\Omega |u|^{\gamma+2}dx + \theta\varphi(t)\int_\Omega (1+|u||\nabla u|)dx \\ &+2\varphi(t)\int_\Omega (1+|\nabla u||m\cdot\nabla u|)dx.\end{aligned}$$

proof. Differentiating the equation (21) with respect to t and substituting the equation (1a) in the expression obtained we deduce

$$\begin{aligned}\frac{d}{dt}\psi(t) \;=\;& \int_{\Gamma_1}(m\cdot\nu)|u_t|^2 d\Gamma_1 + (\theta - n)\int_\Omega |u_t|^2 dx \\ &+\int_{\Gamma_1}\frac{\partial u}{\partial \nu}\{m\cdot\nabla u + \theta u\} d\Gamma_1 \\ &-\int_{\Gamma_1}(m\cdot\nu)|\nabla u|^2 d\Gamma_1 - (\theta-(n-2))\int_\Omega |\nabla u|^2 dx \\ &-2\int_\Omega F(x,t,u,\nabla u)m\cdot\nabla u dx - \theta\int_\Omega F(x,t,u,\nabla u)u dx.\end{aligned} \tag{22}$$

From the inequalities (3) and (4) we obtain

$$\begin{aligned}-\theta\int_\Omega F(x,t,u,\nabla u)u dx \;\leq\;& -\theta\int_\Omega |u|^{\gamma+2}dx \\ &+\theta\varphi(t)\int_\Omega(1+|u||\nabla u|)dx,\end{aligned} \tag{23}$$

$$\begin{aligned}-2\int_\Omega F(x,t,u,\nabla u)m\cdot\nabla u dx \;\leq\;& -2\int_\Omega |u|^\gamma u(m\cdot\nabla u)dx \\ &+2\varphi(t)\int_\Omega(1+|\nabla u||m\cdot\nabla u|)dx.\end{aligned} \tag{24}$$

Substituting the inequalities (23)-(24) into (22) and noting that

$$-\int_{\Gamma_1}(m\cdot\nu)|u|^{\gamma+2}d\Gamma_1 \leq 0$$

our conclusion follows.

To show that the energy decay exponentially we shall need of the following Lemma.

LEMMA 7 *Let f be a real positive function of class C^1. If there exists positive constants γ_0, γ_1 and c_0 such that*

$$f'(t) \leq -\gamma_0 f(t) + c_0 e^{-\gamma_1 t},$$

then there exist positive constants γ and c such that

$$f(t) \leq (f(0) + c)e^{-\gamma t}.$$

proof. First, let us suppose that $\gamma_0 < \gamma_1$. Define $F(t)$ by

$$F(t) := f(t) + \frac{c_0}{\gamma_1 - \gamma_0} e^{-\gamma_1 t}.$$

Then

$$F'(t) = f'(t) - \frac{\gamma_1 c_0}{\gamma_1 - \gamma_0} e^{-\gamma_1 t} \leq -\gamma_0 F(t).$$

Integrating above inequality over $]0, t[$ we arrive to

$$F(t) \leq F(0)e^{-\gamma_0 t} \quad \Rightarrow \quad f(t) \leq \left(f(0) + \frac{c_0}{\gamma_1 - \gamma_0}\right) e^{-\gamma_0 t}.$$

Now, we shall assume that $\gamma_0 \geq \gamma_1$. In this conditions we obtain

$$f'(t) \leq -\gamma_1 f(t) + c_0 e^{-\gamma_1 t} \quad \Rightarrow \quad \left[e^{\gamma_1 t} f(t)\right]' \leq c_0.$$

Integrating last expression over $]0, t[$ we obtain

$$f(t) \leq [f(0) + c_0 t]\, e^{-\gamma_1 t}.$$

Since $t \leq (\gamma_1 - \epsilon)e^{(\gamma_1 - \epsilon)t}$ for any $0 < \epsilon < \gamma_1$ we conclude that

$$f(t) \leq [f(0) + c_0(\gamma_1 - \epsilon)]\, e^{-\epsilon t}.$$

This completes the proof.

Finally, we shall show the inequality (17) of the Teo 3. Using hypothesis (15) and Young inequality in Lemma 5 we get

$$\begin{aligned} \frac{d}{dt}E(t) \quad &\leq \quad -\frac{\eta}{2}\int_{\Gamma_1} (|u_t|^2 - b_2 k' \Diamond u + b_1 k(t)|u|^2 - |k(t)u_0|^2) d\Gamma_1 \\ &\quad + C_\epsilon \varphi^2(t) + \epsilon C \int_\Omega (|u_t|^2 + |\nabla u|^2) dx, \end{aligned} \tag{25}$$

where ϵ is an arbitrary positive constant. Applying Young and Poincaré's inequalities in Lemma 6 we obtain

$$\begin{aligned}\frac{d}{dt}\psi(t) \leq & \int_{\Gamma_1}(m\cdot\nu)|u_t|^2 d\gamma_1 + (\theta - n)\int_\Omega |u_t|^2 dx \\ & -(\theta-(n-2))\int_\Omega |\nabla u|^2 dx \\ & -(\theta - \frac{2n}{\gamma+2})\int_\Omega |u|^{\gamma+2}dx + C_\epsilon\varphi^2(t) \\ & +\epsilon C\{\int_{\Gamma_1}(m\cdot\nu)|\nabla u|^2 d\Gamma_1 + \mathcal{N}(t)\} + C_\epsilon\int_{\Gamma_1}|\frac{\partial u}{\partial\nu}|^2 d\Gamma_1 \\ & -\int_{\Gamma_1}(m\cdot\nu)|\nabla u|^2 d\Gamma_1.\end{aligned}$$

Noting that the boundary condition (11) can be written as $\frac{\partial u}{\partial \nu} = -\eta\{u_t + k(t)u - k'\diamond u - k(t)u_0\}$ we arrive at

$$\begin{aligned}\frac{d}{dt}\psi(t) \leq & (\theta - n)\int_\Omega |u_t|^2 dx - (\theta-(n-2))\int_\Omega |\nabla u|^2 dx \\ & -(\theta - \frac{2n}{\gamma+2})\int_\Omega |u|^{\gamma+2}dx \\ & -\int_{\Gamma_1}(m\cdot\nu)|\nabla u|^2 d\Gamma_1 + C_\epsilon\varphi^2(t) \\ & +\epsilon C\{\int_{\Gamma_1}(m\cdot\nu)|\nabla u|^2 d\Gamma_1 + \mathcal{N}(t)\} \\ & +C_\epsilon\int_{\Gamma_1}(|u_t|^2 + |k(t)u|^2 + |k'\diamond u|^2 + |k(t)u_0|^2)d\Gamma_1.\end{aligned} \tag{26}$$

On the other hand applying the inequality (20) with $\mu = \frac{1}{2}$ in inequality (26) we obtain

$$\begin{aligned}\frac{d}{dt}\psi(t) \leq & (\theta - n)\int_\Omega |u_t|^2 dx - (\theta-(n-2))\int_\Omega |\nabla u|^2 dx \\ & -(\theta - \frac{2n}{\gamma+2})\int_\Omega |u|^{\gamma+2}dx \\ & -\int_{\Gamma_1}(m\cdot\nu)|\nabla u|^2 d\Gamma_1 + C_\epsilon\varphi^2(t) \\ & +\epsilon C\{\int_{\Gamma_1}(m\cdot\nu)|\nabla u|^2 d\Gamma_1 + \mathcal{N}(t)\} \\ & +C_\epsilon\int_{\Gamma_1}(|u_t|^2 + k(t)|u|^2 - k'\Diamond u + |k(t)u_0|^2)d\Gamma_1.\end{aligned} \tag{27}$$

Let us introduce the Lyapunov functional

$$\mathcal{L}(t) := NE(t) + \psi(t), \tag{28}$$

with $N > 0$. Taking N large and ϵ small enough, the previous inequalities imply that

$$\frac{d}{dt}\mathcal{L}(t) \quad \leq -C_0 E(t) + C_1 R^2(t) E(0),$$

where $R(t) = k(t) + \varphi(t)$. Moreover, using Young's inequality and taking N sufficiently large we find that

$$q_0 E(t) \leq \mathcal{L}(t) \leq q_1 E(t), \tag{29}$$

for some positive constants q_0 and q_1. From this inequality we conclude that

$$\frac{d}{dt}\mathcal{L}(t) \quad \leq -\tfrac{C_0}{q_1}\mathcal{L}(t) + C_1 R^2(t) E(0),$$

which implies, in view of Lemma 7 and from the exponential decay of k, φ, that

$$\mathcal{L}(t) \leq \{\mathcal{L}(0) + C\} e^{-\alpha_2 t},$$

for some positive constants C, α_2. From the inequality (29) our conclusion follows.

3. Polynomial Rate of Decay

Here our attention will be focused on the uniform decay rate when the resolvent kernel k decays polynomially like $(1+t)^{-p}$. In this case we will show that the solution also decays polynomially with the same rate. Therefore, we will assume that the resolvent kernel k satisfies

$$k(0) > 0, \quad k'(t) \leq -b_1 k(t)^{1+\frac{1}{p}}, \quad k''(t) \geq b_2[-k'(t)]^{1+\frac{1}{p+1}} \tag{30}$$

for some $p > 1$ and some positive constants b_1 and b_2.

The lemmas below will play an important role in the sequel.

LEMMA 8 *Let u be a solution of system* (1a)-(1d)*. Then, for $p > 1$, $0 < r < 1$ and $t \geq 0$, we have*

$$\left(\int_{\Gamma_1} |k'| \Diamond u d\Gamma_1\right)^{\frac{1+(1-r)(p+1)}{(1-r)(p+1)}}$$

$$\leq \left(2\int_0^t |k'(s)|^r ds \|u\|^2_{L^\infty(0,t;L^2(\Gamma_1))}\right)^{\frac{1}{(1-r)(p+1)}} \int_{\Gamma_1} |k'|^{1+\frac{1}{p+1}} \Diamond u d\Gamma_1$$

while for $r = 0$ we get

$$\left(\int_{\Gamma 1} |k'| \diamondsuit u d\Gamma_1\right)^{\frac{p+2}{p+1}}$$
$$\leq 2\left(\int_0^t ||u(s,.)||^2_{L^2(\Gamma_1)} ds + t||u(s,.)||^2_{L^2(\Gamma_1)}\right)^{p+1} \int_{\Gamma_1} |k'|^{1+\frac{1}{p+1}} \diamondsuit u d\Gamma_1.$$

proof. See e. g. [12]

LEMMA 9 *Let $f \geq 0$ be a differentiable function satisfying*

$$f'(t) \leq -\frac{c_1}{f(0)^{\frac{1}{\alpha}}} f(t)^{1+\frac{1}{\alpha}} + \frac{c_2}{(1+t)^\beta} f(0) \quad \textit{for} \quad t \geq 0,$$

for some positive constants c_1, c_2, α and β such that

$$\beta \geq \alpha + 1.$$

Then there exists a constant $c > 0$ such that

$$f(t) \leq \frac{c}{(1+t)^\alpha} f(0) \quad \textit{for} \quad t \geq 0.$$

proof. See e. g. [11]

Finally, we shall prove the inequality (18). Using hypothesis (30) in Lemma 5 yields

$$E'(t) \leq -\frac{\eta}{2} \int_{\Gamma_1} \left(|u_t|^2 + b_2[-k']^{1+\frac{1}{p+1}} \diamondsuit u + b_1 k^{1+\frac{1}{p}}(t)|u|^2 - |k(t)u_0|^2\right) d\Gamma_1$$
$$+ \quad C_\epsilon \varphi^2(t) + \epsilon C \int_\Omega (|u_t|^2 + |\nabla u|^2) dx.$$

Considering inequality (20) with $\mu = \frac{p+2}{2(p+1)}$ and taking hypothesis (30) into account we obtain the estimate

$$|k' \diamond u|^2 \leq C[-k']^{1+\frac{1}{p+1}} \diamondsuit u.$$

Using the above inequalities in Lemma 6, yields

$$\begin{aligned}
\frac{d}{dt}\psi(t) \leq\ & +(\theta - n) \int_\Omega |u_t|^2 dx - (\theta - (n-2)) \int_\Omega |\nabla u|^2 dx \\
& -(\theta - \frac{2n}{\gamma+2}) \int_\Omega |u|^{\gamma+2} dx \\
& - \int_{\Gamma_1} (m \cdot \nu) |\nabla u|^2 d\Gamma_1 \\
& + C_\epsilon \varphi^2(t) + \epsilon C \{ \int_{\Gamma_1} (m \cdot \nu) |\nabla u|^2 d\Gamma_1 + \mathcal{N}(t) \} \\
& + C \int_{\Gamma_1} \left(|u_t|^2 + k^{1+\frac{1}{p}}(t)|u|^2 + [-k']^{1+\frac{1}{p+1}} \diamondsuit u + |k(t)u_0|^2\right) d\Gamma_1
\end{aligned}$$

In this conditions, taking N sufficiently large and ϵ small enough the Lyapunov functional defined in (28) satisfies

$$\begin{aligned}\frac{d}{dt}\mathcal{L}(t) \quad &\leq \quad -C_0\mathcal{N}(t) + C_1 R^2(t)E(0) \\ &\quad -C_2\int_{\Gamma_1}[-k']^{1+\frac{1}{p+1}}\Diamond u d\Gamma_1. \qquad (31)\end{aligned}$$

Let us fix $0 < r < 1$ such that $\frac{1}{p+1} < r < \frac{p}{p+1}$. From (30) we have that

$$\int_0^\infty |k'|^r \leq C\int_0^\infty \frac{1}{(1+t)^{r(p+1)}} < \infty.$$

Using this estimate in Lemma 9 we get

$$\begin{aligned}&\int_{\Gamma_1}[-k']^{1+\frac{1}{p+1}}\Diamond u d\Gamma_1 \qquad (32)\\ &\geq CE(0)^{-\frac{1}{(1-r)(p+1)}}\left(\int_{\Gamma_1}[-k']\Diamond u d\Gamma_1\right)^{1+\frac{1}{(1-r)(p+1)}}.\end{aligned}$$

On the other hand, from the Trace theorem we deduce

$$E(t)^{1+\frac{1}{(1-r)(p+1)}} \leq CE(0)^{\frac{1}{(1-r)(p+1)}}\mathcal{N}(t). \qquad (33)$$

Substituting (32)-(33) into (30) we obtain

$$\begin{aligned}\frac{d}{dt}\mathcal{L}(t) \quad &\leq -CE(0)^{-\frac{1}{(1-r)(p+1)}}E(t)^{1+\frac{1}{(1-r)(p+1)}} + C_1R^2(t)E(0)\\ &\quad -CE(0)^{-\frac{1}{(1-r)(p+1)}}\left(\int_{\Gamma_1}[-k']\Diamond u d\Gamma_1\right)^{1+\frac{1}{(1-r)(p+1)}}.\end{aligned}$$

Taking into account the inequality (29) we conclude that

$$\frac{d}{dt}\mathcal{L}(t) \leq -\frac{C}{\mathcal{L}(0)^{\frac{1}{(1-r)(p+1)}}}\mathcal{L}(t)^{1+\frac{1}{(1-r)(p+1)}} + C_1R^2(t)E(0),$$

which implies, applying Lemma 9, that

$$\mathcal{L}(t) \leq \frac{C}{(1+t)^{(1-r)(p+1)}}\mathcal{L}(0).$$

Since $(1-r)(p+1) > 1$ we get, for $t \geq 0$, the following bounds

$$\begin{aligned} t\|u\|^2_{L^2(\Gamma_1)} \leq t\mathcal{L}(t) \quad &< \quad \infty,\\ \int_0^t \|u\|^2_{L^2(\Gamma_1)}ds \leq C\int_0^t \mathcal{L}(s)ds \quad &< \quad \infty.\end{aligned}$$

Considering the above estimates in Lemma 9 with $r = 0$ it holds that

$$\int_{\Gamma_1} [-k']^{1+\frac{1}{p+1}} \Diamond u d\Gamma_1 \geq \frac{c}{E(0)^{\frac{1}{p+1}}} \left(\int_{\Gamma_1} [-k'] \Diamond u d\Gamma \right)^{1+\frac{1}{p+1}}.$$

Using the last inequality instead of (32) and reasoning in the same way as above we conclude that

$$\frac{d}{dt}\mathcal{L}(t) \leq -\frac{c}{\mathcal{L}(0)^{\frac{1}{p+1}}} \mathcal{L}(t)^{1+\frac{1}{p+1}} + C_1 R^2(t) E(0).$$

Applying Lemma 9 again, we obtain

$$\mathcal{L}(t) \leq \frac{c}{(1+t)^{p+1}} \mathcal{L}(0).$$

Finally, from (29) we conclude

$$E(t) \leq \frac{c}{(1+t)^{p+1}} E(0),$$

which completes the proof.

Acknowledgments

The authors wish to thank Irena Lasiecka and all the organizers for their kind attention and the nice moments during the 21 st IFIP.

References

[1] M. Aassila, M. M. Cavalcanti, and J. A. Soriano. Asymptotic stability and energy decay rates for solutions of the wave equation with memory in a star-shaped domain. *SIAM J. Control Optim.*, 38(5):1581–1602, 2000.

[2] D. Andrade and J. E. Muñoz Rivera. Exponential decay of non-linear wave equation with viscoelastic boundary condition. *Math. Meth. Appl. Sci*, 23:41–61, 2000.

[3] M. M. Cavalcanti, V. N. Domingos Cavalcanti, and J. A. Soriano. Existence and boundary stabilization of a nonlinear hyperbolic equation with time-dependent coefficients. *Electron. J. Differential Equations*, 1998(08):1–21, 1998.

[4] M. Ciarletta. A differential problem for the heat equation with a boundary condition with memory. *Appl. Math. Lett.*, 10(1):95–101, 1997.

[5] M. Fabrizio and A. Morro. A boundary condition with memory in electromagnetism. *Arch. Rational Mech. Anal.*, 136:359–381, 1996.

[6] M. Kirane and N. Tartar. Non-existence results for a semilinear hyperbolic problem with boundary condition of memory type. *Journal for Analysis and Its Applications*, 19(2):453–468, 2000.

[7] J. L. Lions. *Quelques Mèthodes de resolution de problèmes aux limites non lineaires.* Dunod Gauthiers Villars, Paris, 1969.

[8] T. Qin. Global solvability of nonlinear wave equation with a viscoelastic boundary condition. *Chin. Ann. Math.*, 14B(3):335–346, 1993.

[9] T. Qin. Breakdown of solutions to nonlinear wave equations with a viscoelastic boundary condition. *Arab. J. Sci. Engng.*, 19(2A):195–201, 1994.

[10] R. Racke. *Lectures on nonlinear evolution equations. Initial value problems. Aspect of Mathematics E19.* Friedr. Vieweg & Sohn, Braunschweig, Wiesbaden, 1992.

[11] M. L. Santos. Decay rates for solutions of a system of wave equations with memory. *E. J. Diff. Eqs.*, 2002(38):1–17.

[12] M. L. Santos. Asymptotic behavior of solutions to wave equations with a memory condition at the boundary. *E. J. Diff. Eqs.*, 2001(73):1–11.

BAYESIAN DECONVOLUTION OF FUNCTIONS IN RKHS USING MCMC TECHNIQUES

Gianluigi Pillonetto
Department of Information Engineering, University of Padova, Via Gradenigo, 6/B, Padova, Italy
giapi@dei.unipd.it

Bradley M. Bell
Applied Physics Laboratory, University of Washington, Seattle, Washington, USA
brad@apl.washington.edu

Abstract We propose a novel stochastic approach to reconstruct the unknown input of a partly known dynamical system from noisy output data. We assume that the unknown function belongs to a Reproducing Kernel Hilbert Space (RKHS). We then design an algorithm based on the Markov chain Monte Carlo (MCMC) framework which is able to recover the minimum variance estimate of the input given the output data.

Keywords: Bayesian regularization; stochastic processes; stochastic simulation

1. Introduction

Deconvolution is the process of reconstructing the input of a dynamical linear system starting from sparse and noisy output data. This problem is important and encountered in many domains of applied science (see e.g. [2, 7]). It is also often difficult to solve since it is subject to ill-posedness and ill-conditioning [2, 14]. Usually, the system designer does not have sufficient information to overcome these difficulties by restricting the unknown function to a finite dimensional model.[1] The most attractive and employed technique to effectively solve the deconvolution problem is instead nonparametric regularization, which does not restrict the unknown function to a particular parameteric form (see e.g. [9, 14]). Many nonparametric deconvolution algorithms select a function from an infinite dimensional Reproducing Kernel Hilbert Space (RKHS)

[1, 5]. The key feature of such spaces is their capability to approximate arbitrarily well a very rich class of functions [8]. The unknown function is then estimated as the solution of a Tychonov-type variational problem [13], containing a quadratic term related to the adherence of experimental data and another one which penalizes unlikely solutions, i.e. functions whose norm amplitude is large. The resulting estimator has an interpretation in stochastic terms [7, 14]. In fact, under certain Gaussian assumptions, it provides the minimum variance estimate of the unknown function given the data.[2] In real applications, a Tykonov-type estimator has unknown parameters that must be included in the estimation procedure. For example, the regularization parameter, which is a key one since it establishes the right amount of regularization to include in the estimation process, is almost always unknown [12]. Other unknown variables can be present in the linear relationship between the function and the measurements. If we employ a stochastic framework where we model all these additional unknown parameters as random variables, it turns out that the minimum variance estimate of the function requires the evaluation of analytically intractable integrals (see Section 1 in [10]).

In this paper we present a new approach to face this problem together with an efficient algorithm based on a stochastic simulation technique known in literature as Markov chain Monte Carlo (MCMC). Our technique is able to reconstruct a function belonging to a *generic* RKHS together with all the other unknown parameters present in the problem. In contrast to the approach in [10], our computational scheme avoids any kind of discretization of the domain where the function of interest is defined. The paper is organized as follows. In Section 2 we describe the measurement model and recall some properties of RKHSs (that are used in the other sections). In Section 3 we describe our stochastic deconvolution model and the resulting estimation problem. In Section 4 the algorithm which implements the model introduced in Section 3 is illustrated. The performance of the new approach is then tested in Section 5 by one simulated case study. Conclusions are finally offered in Section 6.

2. Preliminaries

2.1 The Measurement Model

For any vector w, we use w_i to refer to the $i-th$ component of w. Moreover, all the vectors are column vectors. We define our problem in mathematical terms. We are given a vector of measurements $y \in \Re^n$. The measurement values depend on an unknown function $f : X \to$

$\Re$, where X is a compact domain on $\Re^q$, and on an unknown random vector $\nu \in \Re^n$. The dependence between the function and the $i-th$ measurement is

$$y_i = L_i(f, \theta) + \nu_i \tag{1}$$

where the mapping $f \to L_i(f, \theta)$ is a linear and continuous functional between a space containing continuous functions and $\Re$. Moreover, $\theta \in \Re^d$ is an unknown random vector whose probability density function, prior to making the measurements, is $p_\theta(\theta)$. We are also given a model for the statistics of the measurement noise. To be specific, we assume ν is a zero-mean Gaussian random vector whose positive definite covariance matrix is $\Sigma_\nu(\theta)$. We assume that the functions L, p_θ, and Σ_ν are known. In addition, given θ, ν is independent from f. The problem of estimating θ and f without additional information is ill-posed (ill-posed problems are described on page 7 of [13]).

2.2 Reproducing Kernel Hilbert Spaces

Our approach to this ill-posed inverse problem is to place a Bayesian prior on a special type of function space H called Reproducing Kernel Hilbert Space (RKHS). We briefly sketch some properties of these spaces which are relevant in the context of the present work. We use $L^2(X)$ to denote the classical Lebesque space of square integrable functions on X, equipped with the inner product $< .,. >_2$.

DEFINITION 1 *We say that $M : X \times X \to \Re$ is positive definite if for all finite sets $\{x_1, x_2, ..., x_k\} \subset X$ the $k \times k$ matrix whose (i,j) entry is $M(x_i, x_j)$ is positive semi-definite. Moreover, we say that M is a Mercer kernel if it is continuous, symmetric and positive definite.*

The following theorem can be obtained by combining the Spectral Theorem for compact operators and Mercer's theorem (see [4]).

THEOREM 2 *If M is a Mercer Kernel, there exist a sequence $\{\lambda_j \geq 0 : \lambda_{j+1} \geq \lambda_j, j = 1, \ldots, \infty\}$ and a basis in $L^2(X)$ of continuous functions $\{\phi_j : j = 1, \ldots, \infty\}$ such that*

$$\begin{aligned}
< \phi_j, \phi_k >_2 &= \begin{cases} 1 & \text{if } j = k \\ 0 & \text{otherwise} \end{cases} \\
\int_X M(s,t)\phi_j(t)dt &= \lambda_j \phi_j(s) \\
M(s,t) &= \sum_{j=1}^{\infty} \lambda_j \phi_j(s) \phi_j(t)
\end{aligned}$$

where the above convergence is uniform in $X \times X$.

The following proposition can be derived from [1, 4].

THEOREM 3 *For each Mercer kernel* M *there exists a unique Hilbert space* H *such that*

- *for each* $x \in X$, $M(x, \cdot) \in H$
- *the span of the set* $\{M(x, .),\ x \in X\}$ *is dense in* H
- *for each* $f \in H$ *and* $x \in X$, $f(x) = < f(\cdot), M(x, \cdot) >_H$

If $\lambda_j > 0$ *for every* j, *the associated RKHS* H *has the representation*

$$H = \left\{ f \in L^2(X) \;\middle|\; f = \sum_{j=1}^{\infty} a_j \phi_j \quad \text{where} \quad \sum_{j=1}^{\infty} \frac{a_j^2}{\lambda_j} < \infty \right\}$$

equipped with the inner product $< .,. >_H$ *where, given* $f, g \in H$ *with* $f = \sum_{j=1}^{\infty} a_j \phi_j$ *and* $g = \sum_{j=1}^{\infty} b_j \phi_j$, *we have*

$$< f, g >_H = \sum_{j=1}^{\infty} \frac{a_j b_j}{\lambda_j}$$

The theorem above enables us to interpret H as a certain subset of smooth functions in $L^2(X)$ 3 generated by the eigenvectors $\{\phi_j\}$. The smoothness condition does in particular concern the behavior of the generalized Fourier coefficients $\{a_k\}$ and is regulated by the eigenvalues of M

We conclude this section by defining the notation Θ to be a subset of $\Re^d$ and $K : \Theta \times X \times X \to \Re$ to be a parameterized Mercer Kernel; i.e., for each $\theta \in \Theta$, $K(\theta, \cdot, \cdot)$ is a Mercer Kernel. In addition $\phi(\cdot, \theta)$ and $\lambda_i(\theta)$ are the eigen-functions and eigen-values corresponding to $K(\theta, \cdot, \cdot)$.

DEFINITION 4 *We use* $L^N(\theta)$ *to denote the* $n \times N$ *matrix whose* (i, j) *entry is* $L_i(\phi_j, \theta)$. *In addition, we denote with* $\Lambda^N(\theta)$ *the* $N \times N$ *diagonal matrix whose* $i-th$ *entry of the diagonal is equal to* $\lambda_i(\theta)$.

DEFINITION 5 *Let* $\boldsymbol{N}$ *the set of natural numbers. Given* $A \subset \boldsymbol{N}$, *we define the following notation:*

$$K_{-A}(\theta, x, y) = K(\theta, x, y) - \sum_{j \in A} \lambda_j(\theta) \phi_j(x, \theta) \phi_j(y, \theta)$$

2.3 An Example of RKHS

In this section we review some properties of an example RKHS space parameterized by the integer $m > 0$. We define the Green's function G_m and the reproducing kernel K_m on $[0, T] \times [0, T]$, $T \in \Re$ as

$$
\begin{aligned}
G_m(x, y) &= \begin{cases} 0 & \text{if } x \leq y \\ 1 & \text{if } x > y \text{ and } m = 1 \\ (x - y)^{m-1}/(m-1)! & \text{otherwise} \end{cases} \\
K_m(x, y) &= \int_0^T G_m(x, \tau) G_m(y, \tau) d\tau
\end{aligned}
$$

Given a function $f : [0, T] \to \Re$, we use $f^{(i)}$ to denote the $i-th$ derivative of f. The RKHS associated to K_m is then

$$
W_m = \left\{ f : [0, T] \to \Re \;\middle|\; \begin{array}{l} f^{(m)} \in L^2[0, T] \\ \text{and for } j = 0, ..., m-1, \quad f^{(j)}(0) = 0 \\ \text{and } f^{(j)} \text{ is absolutely continuous} \end{array} \right\}
$$

equipped with the inner product

$$
< f, g >_{W_m} = < f^{(m)}, f^{(m)} >_2
$$

For the special case $m = 1$, the following closed forms for ϕ_j and λ_j are available (see [15])

$$
\begin{aligned}
\lambda_{W_1, j} &= T^2/[(j - 1)\pi + \pi/2]^2 \\
\phi_{W_1, j}(x) &= \sqrt{2/T} \sin[(x/T)(j\pi - \pi/2)]
\end{aligned}
\tag{2}
$$

3. Statement of the Estimation Problem

We now define our estimation problem in a Bayesian framework. We start by defining a Bayesian prior for the unknown function f on the RKHS corresponding to $K(\theta, \cdot, \cdot)$.

Assumption: *Given θ, the function f is a random field of the form $f(x) = \sum_{j=1}^{\infty} a_i \phi_i(x)$, where $\{a_i\}$ are Gaussian and independent random variables and the variance of a_i is $\lambda_i(\theta)$. In addition, each a_i is independent of the measurement noise ν.*

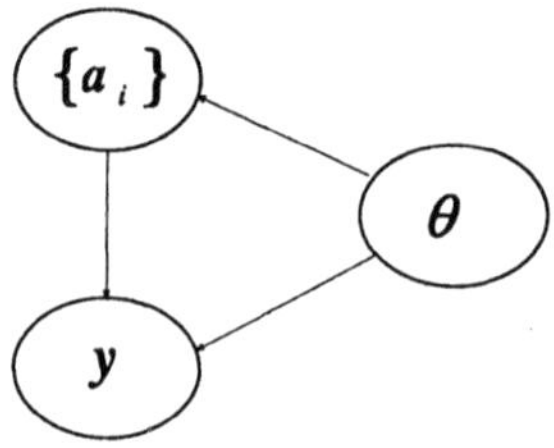

Figure 1. Bayesian network describing the stochastic deconvolution problem

A graphical description of the joint probability density function of y, f and θ is given by the Bayesian network in Figure 1. It is important to note that the node θ is connected via a direct link with both f and y. We also note that given the other two nodes in the network, y depends on f and θ through $\Sigma_\nu(\theta)$ and $L(f,\theta)$. In addition, the probability density function of f given θ depends on θ through the sequence $\{\lambda_i(\theta)\}$. *i.e.*, the smoothness parameters contained in θ parameterize the norm in the RKHS H.

THEOREM 6 *The minimum variance estimate of the function f, given θ and y, is*

$$\hat{f} = \underset{f \in H}{\arg\min} \; [y - L(f,\theta)]^T \Sigma_\nu^{-1}(\theta)[y - L(f,\theta)] + \|f,\theta\|_H^2 \tag{3}$$

where H is the RKHS corresponding to $K(\theta,\cdot,\cdot)$.
(see Appendix for the proof)

Equation (3) shows that if f is the only unknown in the model depicted in Figure 1, its optimal estimate is provided by a Tikhonov-type variational problem. The solution of such problem is linear in the data y and admits a closed form which is well known in literature (see e.g. [14]). However, in real applications θ is seldom completely known. The estimation problem we aim to solve when θ is uncertain is described below.

Problem: *Let $p(y|a_i,\theta)$ and $p(a_i|\theta)$ the probability density functions of y given (a_i,θ) and of a_i given θ, respectively. Let also $p(y)$ the marginal probability density function of y. Given the Bayesian network of Figure 1 and known the data y, determine the minimum variance estimate of $f(x)$, i.e. compute $\hat{f}(x) = \sum_{i=1}^{\infty} \hat{a}_i \phi_i(x)$ where*

$$\hat{a}_i = \frac{\int_{\Re^{d+1}} a_i p(y|a_i,\theta) p(a_i|\theta) p(\theta) da_i d\theta}{p(y)}$$

The function $\hat{f}$ takes into account all the possible sources of uncertainty present in the problem and represents our ideal estimate of the random field f. However, its determination turns out difficult since the computation of $\hat{a}_i$ will in general require the solution of an analytically intractable integral. We describe the strategies developed in order to circumvent these problems in the next Section.

4. MCMC Deconvolution Algorithms in RKHS

We solve our stochastic deconvolution problem by reducing it to the reconstruction (in sampled form) of two finite-dimensional probability density functions. The numerical procedure relies on the MCMC framework (for an overview on MCMC theory see e.g. [6]). To simplify our notation below, dependence of some operators on θ is implicit.

4.1 Step 1: Reconstruction of $\mathbf{p}(\theta|y)$

The first goal is to reconstruct the probability density function of θ given y after integrating out the unknown random field f from the probabilistic model of Figure 1. For this aim, the following proposition is useful. It can be proved by employing the linearity of the operator L and the fact that ν is independent from f.

THEOREM 7 *We have*

$$p(y|\theta) = \frac{1}{[det(2\pi\Sigma_y)]^{0.5}} exp(-\frac{1}{2}y^T\Sigma_y^{-1}y)$$

where Σ_y *is an* $n \times n$ *matrix, such that*

$$\Sigma_y(i,j) = L_i[L_j[K(s,t)]] + \Sigma_\nu(i,j)$$

We then have that $p(\theta|y) \propto p(y|\theta)p(\theta)$ and a MCMC strategy can be employed in order to recover in sampled form this marginal posterior. In particular, we firstly obtain an approximated covariance matrix Σ of the random vector θ given y as the inverse of the Hessian of the minus log of $p(y|\theta)p(\theta)$ computed at its mode (in θ). We then resort to a random-walk Metropolis scheme to reconstruct $p(\theta|y)$. In other words, we use a proposal density which consists of a Gaussian distribution centered at the current point of the Markov chain with covariance matrix proportional to Σ.

4.2 Step 2: Determination of the Minimum Variance Estimate of f given y

THEOREM 8 *Let $A = \{x \in \boldsymbol{N}; x \leq N\}$. Also, let $\Sigma_{\nu,-A}$ the $n \times n$ matrix such that $\Sigma_{\nu,-A}(i,j) = \Sigma_\nu(i,j) + L_i L_j[K_{-A}(x,y)]$. Then, by denoting with $p(y|a^N, \theta)$ the probability density function of y given a^N and θ, we have:*

$$p(y|a^N, \theta) = \frac{\exp\left(-\frac{1}{2}(y - L^N a^N)^T (\Sigma_{\nu,-A})^{-1}(y - L^N a^N)\right)}{[det(2\pi\Sigma_{\nu,-A})]^{0.5}}$$

Proof: The model of measurements can be rewritten as follows

$$y = L^N a^N + \sum_{j=N+1}^{\infty} L[a_j \phi_j(s)] + \nu \doteq L^N a^N + \xi$$

where ξ is zero-mean normal random vector independent from a^N and having covariance matrix equal to $\Sigma_{\nu,-A}$. This completes the proof.

The following result can be obtained using Bayes formula and Theorem 8.

THEOREM 9 *Given y and θ, the random vector a^N is Gaussian having covariance matrix $\hat{\Sigma}_{a^N}$ and mean $\hat{\mu}_{a^N}$ where:*

$$\hat{\Sigma}_{a^N} = [(\Lambda^N)^{-1} + (L^N)^T \Sigma_{\nu,-A}^{-1}(L^N)]^{-1} \tag{4}$$

$$\hat{\mu}_{a^N} = \hat{\Sigma}_{a^N}(L^N)^T \Sigma_{\nu,-A}^{-1} y \tag{5}$$

We remark that Λ^N, L^N and $\Sigma_{\nu,-A}$ may depend on θ. Thus, $\hat{\Sigma}_{a^N}$ can be computed for some of those values of θ located in high probability regions in accordance with the marginal posterior obtained (in sampled form) at step 1 of the proposed algorithm. This analysis obtains crucial information regarding how the a posteriori probability density function of a component a_i differs from its a priori probability density function. The spectrum of many physical transformations L is located at low frequencies. If one also has that the higher i, the lower is the spectral content of ϕ_i, as e.g. in eq. (2), many of the amplitudes in the set $\{a_i\}$ may be insensitive to the output data. This means that from a certain index i the minimum variance estimate of a_i will be close to the mean of the prior, i.e. close to zero. This makes it possible to find a value of N so that, for every t, $\sum_{i=1}^N \hat{a}_i\phi_i(t) \approx \sum_{i=1}^\infty \hat{a}_i\phi_i(t)$.
Finally, after determining the number of amplitudes a_i which is worth reconstructing, the marginal posterior of a^N given y can be recovered.

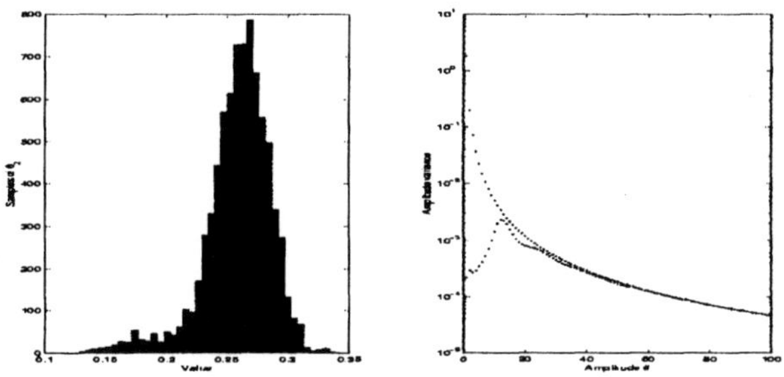

Figure 2. Simulation *Left* A posteriori probability density function of θ_2 obtained in sampled form by MCMC. *Right* A posteriori and a priori variance of a_i as function of i having set θ_1 and θ_2 to their minimum variance estimates

For this aim, let θ_k be a sample drawn from the distribution of θ given y as obtained in step 1. Then, it suffices drawing samples from a Gaussian distribution having mean $\hat{\mu}_{a^N}(\theta_k)$ and covariance matrix $\hat{\Sigma}_{a^N}(\theta_k)$ by using a sufficiently large set of realizations θ_k.

5. Numerical Experiments

We consider a semi-blind deconvolution problem, i.e. a deconvolution problem where the relationship between the unknown input and the output data is only partly known. Let

$$\beta_{pq}(t) = \frac{\Gamma(p+q)}{\Gamma(p)\Gamma(q)} t^{p-1}(1-t)^{q-1}$$

Then, the simulated function f to reconstruct, taken from [3], is $f(x) = \sum_{j=1}^{2} w_j \beta_{p_j,q_j}(x)$ where $0 \leq x \leq 1$ and $w_1 = 0.3$, $w_2 = 0.6$, $p_1 = 12$, $p_2 = 4$, $q_1 = 7$, $q_2 = 11$. f is modeled as the unknown input of a shift-invariant linear system with impulse response equal to $\chi([0, \theta_2])$, where $\chi(A)$ is the indicator function of a set A and θ_2 is equal to 0.27, a value drawn from a uniform random variable between 0 and 1. The function has to be reconstructed from 50 output observations, collected by using a uniform sampling grid and corrupted by a white Gaussian process with a constant CV% equal to 10. We model the unknown function as $\sum_{i=1}^{\infty} a_i \phi_{W_1,i}$ where a_i are independent Gaussian random variables of variance $\lambda_i = \theta_1 \lambda_{W_1,i}$ (note that θ_1 represents the regularization parameter). Moreover, we model θ_1 and θ_2 as uniform and independent random variables on $[0, \rho]$, with $\rho \to +\infty$, and $[0, 1]$, respectively. In Figure 2 (left panel) we report the a posteriori probability density function of θ_2 as reconstructed in sampled form in the first step of our algorithm by a MCMC run where 5500 samples were generated.[4] The minimum vari-

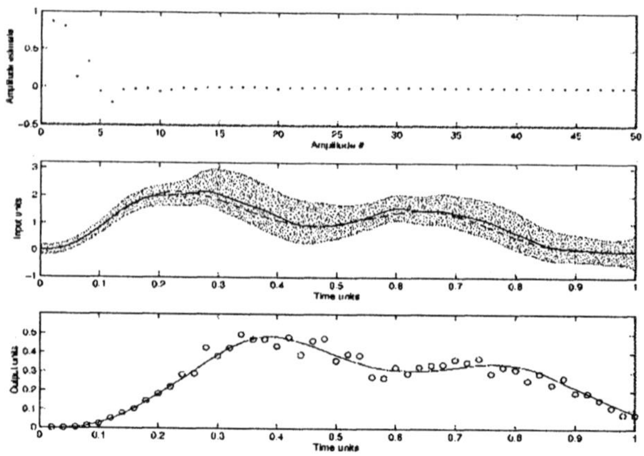

Figure 3. Simulation *Top* Minimum variance estimate of a_i as function of i. *Middle* minimum variance estimate of f (continuous line) with 95% confidence interval (shaded area) and true function (dashed line). *Bottom panel* Reconvolution (continuous line) vs noisy samples (bullets)

ance estimate of this parameter turns out to be 0.258, a value close to truth. Such estimate together with that of θ_1 has been then employed in order to compute the covariance matrix of eq.(4) when N is set to 100. In Figure 2 (right panel) we plot the diagonal elements of such matrix and of the matrix $\Lambda^N(\hat{\theta}_1)$. This plot suggest that only few components a_i are identifiable from the data. The second step of the algorithm has been then performed by setting N to 50 and generating 1500 samples of a^N. In Figure 3 we plot the minimum variance estimates of the components of a^N (top panel) and of f (solid line, middle panel), which appears close to the true function (dashed line, middle panel), together with the 95% confidence interval (shaded area, middle panel). Finally, in the bottom panel of Figure 3 the reconvolution against the noisy samples is depicted.

6. Conclusions

We have proposed a new Bayesian deconvolution algorithm based on the MCMC framework. The technique we have introduced improves on the existing deconvolution algorithms proposed in literature in some important aspects. In particular, differently from the approach developed in [10], our approach is well suited for the reconstruction of a function belonging to a generic RKHS and avoids any kind of discretization of the domain X where the unknown function is defined.

Future developments of this work could consist in extending the estimation technique here presented for reconstructing functions from non-linearly related output data.

Appendix: Proof of Theorem 6

In the sequel, the dependence of Σ_ν, L and $\{\lambda_j\}$ on θ will be implicit. Given a function $f \in H$, where $f = \sum_{j=1}^{\infty} a_j \phi_j(x)$, let a^N the vector containing the first N components of $\{a_j\}$. Let also $f_a^N(x) = \sum_{j=1}^{N} a_j^N \phi_j(x)$, where $x \in X$. We define the following prior distribution for a^N

$$\frac{1}{(2\pi)^{N/2}\sqrt{\lambda_1 \cdots \lambda_N}} \exp\left(-\frac{1}{2}\|f_a^N(x)\|_H^2\right)$$

The conditional density for y given a^N and θ is

$$p^N(y|a^N, \theta) = \frac{1}{\sqrt{\det[2\pi\Sigma_\nu]}} \exp\left(-\frac{1}{2}(y - L[f_a^N])^T \Sigma_\nu^{-1}(y - L[f_a^N])\right)$$

The corresponding negative log of the likelihood for a certain $y \in \Re^n$ and a certain a^N is

$$\begin{aligned} l^N(y, a^N|\theta) &= \frac{\|f_a^N\|_H^2}{2} + \frac{1}{2}\sum_{j=1}^{N}\log(2\pi\lambda_j) + \frac{1}{2}\log\det[2\pi\Sigma_\nu] \\ &+ \frac{1}{2}(y - L[f_a^N])^T \Sigma_\nu^{-1}(y - L[f_a^N]) \end{aligned} \tag{1}$$

We point out that H, being a RKHS, is a subset of the space of continuous functions and convergence in the topology induced by $\|\|_H$ implies uniform convergence (see [4]). Then, as $N \to \infty$, the following pointwise convergence holds

$$l^N(y, f_a^N|\theta) - \frac{1}{2}\sum_{j=1}^{N}\log(2\pi\lambda_j) \to l(y, f|\theta)$$

where $l(y, f|\theta)$ is defined by

$$l(y, f|\theta) = \frac{1}{2}\|f\|_H^2 + \frac{1}{2}\log\det[2\pi\Sigma_\nu] + \frac{1}{2}(y - L[f])^T\Sigma_\nu^{-1}(y - L[f])$$

Thus, given the model of Figure 1, maximizing $l(y, f|\theta)$ with respect of f corresponds to recovering the maximum a posteriori estimate of f given y and θ. This, combined with the linearity of L and the gaussianity of f, completes the proof.

Notes

1. Otherwise, this would reduce deconvolution to a standard parametric estimation problem, easily solvable by traditional methods as nonlinear least squares
2. There is a strong relationship between RKHSs and Gaussian processes, see e.g. Section 1.4 in [14].
3. all the functions in H are in fact continuous, see Proposition 3 on pag.36 of [4].
4. Results displayed in the sequel have been obtained by assessing the convergence of the generated Markov chains through the binary control of Raftery and Lewis [11]. In particular, we have always required to estimate quantiles 0.025, 0.25, 0.5, 0.75, 0.975 of all the unknown parameters with precision respectively 0.005, 0.01, 0.01, 0.01, 0.005 and with probability 0.95

References

[1] N. Aronszajn. Theory of reproducing kernels. *Transactions of the American Mathematical Society*, 68:337–404, 1950.

[2] M. Bertero. Linear inverse and ill-posed problems. *Advances in Electronics and Electron Physics*, 75:1–120, 1989.

[3] P. Craven and G. Wahba. Estimating the correct degree of smoothing by the method of generalized cross-validation. *Numerische Mathematik*, 31:377–403, 1979.

[4] F. Cucker and S. Smale. On the mathematical foundations of learning. *Bulletin of the American mathematical society*, 39:1–49, 2001.

[5] T. Evgeniou, M. Pontil, and T. Poggio. A unified framework for regularization networks and support vector machines. CBCL Paper 171, Massachusetts Institute of Technology, Cambridge, MA, March 1999.

[6] W.R. Gilks, S. Richardson, and D.J. Spiegelhalter. *Markov chain Monte Carlo in Practice.* London: Chapman and Hall, 1996.

[7] G. De Nicolao, G. Sparacino, and C. Cobelli. Nonparametric input estimation in physiological systems: problems, methods and case studies. *Automatica*, 33:851–870, 1997.

[8] G. De Nicolao and G. Ferrari Trecate. Consistent identification of narx models via regluarization networks. *IEEE Transactions on Automatic Control.*

[9] G. De Nicolao and G. Ferrari Trecate. Regularization networks: fast weight calculation via kalman filtering. *IEEE Transactions on Neural Networks*, 12:228–235, 2001.

[10] P.Magni, R.Bellazzi, and G.De Nicolao. Bayesian function learning using mcmc methods. *IEEE Transactions on Pattern Analysis and Machine Intelligence*, 20:1219–1331, 1998.

[11] A.E. Raftery and S.M. Lewis. *Implementing MCMC*, pages 115–130. Markov Chain Monte Carlo in Practice. W.R. Gilks, S.Richardson, and D.J. Spiegelhalter, eds. London: Chapman and Hall, 1996.

[12] J.A. Rice. Choice of smoothing parameter in deconvolution problems. *Contemp. Math.*, 59:137–151, 1986.

[13] A.N. Tychonov and V.Y. Arsenin. *Solutions of Ill-Posed Problems.* Washington, D.C.: Winston/Wiley, 1977.

[14] G. Wahba. *Spline models for observational data.* SIAM, Philadelphia, 1990.

[15] A.M. Yaglom. *Correlation theory of stationary and related random functions*, volume 1. Springer-Verlag, New York, 1987.

MODELING STOCHASTIC HYBRID SYSTEMS

Mrinal K. Ghosh *
Department of Mathematics, Indian Institute of Science, Bangalore - 560 012, India
mkg@math.iisc.ernet.in

Arunabha Bagchi
Department of Applied Mathematics, University of Twente, Post Box 217, 7500 AE Enschede, The Netherlands
a.bagchi@math.utwente.nl

Abstract Stochastic hybrid systems arise in numerous applications of systems with multiple models; e.g., air traffc management, flexible manufacturing systems, fault tolerant control systems etc. In a typical hybrid system, the state space is hybrid in the sense that some components take values in a Euclidean space, while some other components are discrete. In this paper we propose two stochastic hybrid models, both of which permit diffusion and hybrid jump. Such models are essential for studying air traffic management in a stochastic framework.

Keywords: Stochastic hybrid systems, Markov processes, Ito-Skorohod type stochastic differential equations, hybrid jumps.

Introduction

In this article we study some classes of stochastic hybrid models. Stochastic hybrid systems arise in numerous applications of systems with multiple modes, e.g., flexible manufacturing systems, air traffic management, fault tolerant control systems etc. For various applications of stochastic hybrid systems we refer to [3], [8], [1], [11]and the references therein. In a typical hybrid system, the state space is hybrid in the sense that some components take values in a Euclidean space while some other

*Research supported by the IST project "HYBRIDGE", IST-2001-32460, of the European Commission

components are discrete. The evolution of continuous and discrete components are intertwined in an intricate manner. This makes the analysis of a hybrid system quite involved and challenging. Several classes of stochastic hybrid systems have been studied in the literature, e.g., counting processes with diffusion intensity [10], [13], diffusion processes with Markovian switching parameters [11], [16], switching diffusions [8], [9], piecewise deterministic processes [5], [15], Markov decision drift processes [1]etc. All these stochastic hybrid systems arise in different kinds of applications.

Here we address two kinds of stochastic hybrid models. In the first model we construct a Markov process $(X(t), \theta(t))$, where $X(t) \in \mathbb{R}^d$ and $\theta(t) \in \Theta = \{1, 2, \ldots, N\}$. Here $X(t)$ is governed by a stochastic differential equation of Ito-Skorohod type with drift coefficient, diffusion matrix and the 'jump' function depending on the discrete component $\theta(t)$. Thus $X(t)$ switches from one jump diffusion path to another as the discrete component $\theta(t)$ moves from one state to another. On the other hand, the discrete component $\theta(t)$ is a "controlled Markov chain" with a transition matrix that depends on the continuous component $X(t)$. A change in the discrete state $\theta(t)$ makes a switching in the continuous state. This apart the continuous state does jump at random times. At times this may lead to a situation where a switching triggers a jump and vice-versa. This model is discussed in the next section. Section 3 is devoted to the study of a very general stochastic hybrid system. The state of the system at time t, denoted by $(X(t), \theta(t))$ takes values in $\cup_n (S_n \times \Theta_n)$, where $\Theta_n = \{1, 2, \ldots, N_n\}$ and S_n is a subset of $\mathbb{R}^{d_n}$. Between the jumps $(X(t), \theta(t))$ is a switching diffusion. That is $\theta(\cdot)$ is a pure jump process taking values in Θ_n; between successive jumps of $\theta(t)$, $X(t)$ is a diffusion process. On the other hand, the infinitesimal jump rates of $\theta(t)$ depends on $X(t)$. Let A_n be a subset of S_n. If $X(t)$ starting from some point in S_n, hits A_n then it executes an instantaneous jump to some S_m. The destination of $X(t)$ at this moment is determined by a pre-determined map. The discrete component at this moment is also reset by a given map. We investigate a Markovian structure of this system by introducing another switching component in the systems.

A typical construction of a hybrid systems is based on stochastic differential equations driven by Wiener processes and Poisson random measures. For a comprehensive treatment of stochastic differential equation driven by Wiener processes and Poisson random measure we refer to [6], [7]and [12].

1. Stochastic Hybrid Model I

In this section we construct a Markov process $(X(t), \theta(t))$ taking values in $\mathbb{R}^d \times \Theta$ where $\Theta = \{1, 2, \ldots, N\}$. The evolution of the process is governed by equations of the following form:

$$\left.\begin{aligned} dX(t) &= b(X(t),\theta(t))dt + \sigma(X(t),\theta(t))dW(t) + \\ &\quad \textstyle\int_{\mathbb{R}} g(X(t),\theta(t),u)p(dt,du) \\ P(\theta(t+\delta t &= j),\theta(t) = i, X(s),\theta(s), s \le t) \\ &= \lambda_{ij}(X(t))\delta t + 0(\delta t), i \neq j \\ X(0) &= X_0,\ \ \theta(0) = \theta_0. \end{aligned}\right\} \qquad (1)$$

Here b, σ, g, λ are suitable functions, $\lambda_{ij} \geq 0, i \neq j, \sum_{j=1}^N \lambda_{ij} = 0, W(\cdot)$ is a standard Wiener process and $p(\cdot,\cdot)$ is a certain Poisson random measure on $\mathbb{R}_+ \times \mathbb{R}$ to be specified shortly. Under certain conditions we establish the existence of a pathwise unique solution of (1). We make certain assumptions on b, σ, g, λ. Let

$$\begin{aligned} b &: \mathbb{R}^d \times \Theta \to \mathbb{R}^d \\ \sigma &: \mathbb{R}^d \times \Theta \to \mathbb{R}^{d\times d} \\ g &: \mathbb{R}^d \times \Theta \times \mathbb{R} \to \mathbb{R} \\ \lambda_{ij} &: \mathbb{R}^d \to \mathbb{R},\ \ i,j = 1,2,\ldots,N. \end{aligned}$$

We make the following assumptions on the above functions.

(A1) For each $i = 1, 2, \ldots, N$, $b(\cdot, i)$ is bounded and Lipschitz continuous.

(A2) For each $i = 1, 2, \ldots, N$, $\sigma(\cdot, i)$ is bounded and Lipschitz continuous.

(A3) For $i, j = 1, 2, \ldots, N$, $\lambda_{ij}(\cdot)$ are bounded and measurable, $\lambda_{ij}(\cdot) \geq 0$ for $i \neq j$, and $\sum_{j=1}^N \lambda_{ij}(\cdot) = 0$.

(A4) Let K_1 be the support of $g(\cdot,\cdot,\cdot)$ and let U_1 be the projection of K_1 on $\mathbb{R}$. We assume that U_1 is bounded.

Note that in (1), the process $\theta(t)$ is a pure jump process. Thus by the results of [6], $\theta(t)$ may be represented by an integral with respect to a Poisson random measure. Following [4], [8], [9]we proceed to obtain this representation explicitly. To this end, we first embed Θ into $\mathbb{R}^N$ by identifying i with e_i, the ith unit vector in $\mathbb{R}^N$. For $i, j \in \Theta, : x \in \mathbb{R}^d$, let $\Delta_{ij}(x)$ be consecutive (with respect to the lexicographic ordering on $\Theta \times \Theta$) left closed, right open intervals on the real line, each having length $\lambda_{ij}(x)$. Define a function

$$h : \mathbb{R}^d \times \Theta \times \mathbb{R} \to \mathbb{R}^N$$

by

$$h(x,i,u) = \begin{cases} j-i & \text{if} \quad u \in \Delta_{ij}(x) \\ 0 & \text{otherwise.} \end{cases} \tag{2}$$

Let $(X(t), \theta(t))$ be an $\mathbb{R}^d \times \Theta$-valued process given by the following stochastic differential equation of Ito-Skorohod type.

$$\left.\begin{aligned} dX(t) &= b(X(t),\theta(t))dt + \sigma(X(t),\theta(t))dW(t) + \\ &\quad \textstyle\int_{\mathbb{R}} g(X(t-),\theta(t-),u)p(dt,du), \\ d\theta(t) &= \textstyle\int_{\mathbb{R}} h(X(t-),\theta(t-),u)p(dt,du) \\ &\text{for } t \geq 0,\ X(0) = X_0,\ \theta(0) = \theta_0. \end{aligned}\right\} \tag{3}$$

Here:

(i) X_0 is a prescribed $\mathbb{R}^d$-valued random variable.

(ii) θ_0 is a given Θ-valued random variable.

(iii) $W(\cdot)$ is a d-dimensional standard Wiener process.

(iv) $p(dt, du)$ is a Poisson random measure with intensity $dt \times l(du)$, where l is the Lebesgue measure on $\mathbb{R}$.

By the construction of the function h in (2), it is clear that a solution of (3) is also a solution of (1). Thus we prove the existence of an a.s. unique strong solution of (3). To achieve this we use the method in [6].

THEOREM 1 *Assume (A1)-(A4). Let $p(\cdot,\cdot), W(\cdot), X_0, \theta_0$ be independent. Then the equation (3) has an a.s. unique strong solution.*

PROOF: Let $(\Omega, \mathcal{F}, P)$ be the underlying (complete) probability space on which $p(\cdot,\cdot), W(\cdot), X_0, \theta_0$ are defined. Let $\tilde{p}(\cdot)$ be the Poisson process on $(\Omega, \mathcal{F}, P)$ corresponding to the given Poisson random measure $p(\cdot,\cdot)$. Let $K_2 =$ support of $h(\cdot,\cdot,\cdot)$ and U_2 the projection of K_2 on $\mathbb{R}$. By (A3), U_2 is a bounded set. Let $U = U_1 \cup U_2$. Then U is also bounded. Let $D_{\tilde{p}}$ denote the domain of the Poisson process $\tilde{p}(\cdot)$. Let

$$D = \{t \in D_{\tilde{p}} \ : \ \tilde{p}(t) \in U\}.$$

Since $l(U) < \infty$, D is a discrete set in $(0,\infty)$ a.s. Let $\tau_1 < \tau_2 < \ldots < \tau_n < \ldots$ be the enumeration of all elements in D. Let

$$\mathcal{F}_t = \sigma\{W(s),\ p(A,B) \mid s \leq t,\ A \in \mathcal{B}([0,t]),\ B \in \mathcal{B}(\mathbb{R})\}.$$

Then it is easy to see that τ_n is an $\mathcal{F}_t$- stopping time for each n and $\tau_n \uparrow \infty$ a.s. First we establish the existence and uniqueness of the solution in the time interval $[0,\tau_1]$. To achieve this consider the following stochastic differential equation:

$$Y(t) = X_0 + \int_0^t b(Y(s),\theta_0)dt + \sigma(Y(s),\theta_0)dW(s). \tag{4}$$

First assume $X_0 = x \in \mathbb{R}^d$ and $\theta_0 = i \in \Theta$ for some x, i. Under (A1), (A2), the equation (4) has an a.s. unique strong solution which depends measurably on x i,: and : $W(\cdot)$. The solution for the initial condition X_0, θ_0 is obtained by replacing (x,i) by (X_0,θ_0). Now set

$$X_1(t) = \begin{cases} Y(t) & \text{if } 0 \leq t < \tau_1 \\ Y(\tau_1-) + g(Y(\tau_1-),\theta_0,\tilde{p}(\tau_1)) & \text{if } t = \tau_1 \end{cases}$$

$$\theta_1(t) = \begin{cases} \theta_0 & \text{if } 0 \leq t < \tau_1 \\ \theta_0 + h(Y(\tau_1-),\theta_0,\tilde{p}(\tau_1)) & \text{if } t = \tau_1. \end{cases}$$

The process $\{X_1(t),\theta_1(t)\}_{t\in[0,\tau_1]}$ is clearly the unique solution of (3) in the time interval $[0,\tau_1]$. Next, let $\tilde{X} = X_1(\tau_1), \tilde{\theta} = \theta(\tau_1)$,: $\tilde{W}(\cdot) = \tilde{W}(\cdot + \tau_1) - W(\tau_1)$, and $\hat{p} = (\hat{p}(t))$, where $D_{\hat{p}} = \{s : s + \tau_1 \in D_{\tilde{p}}\}$ and $\hat{p}(s) = \tilde{p}(s+\tau_1)$. Proceeding as before we can determine the process $(\tilde{X}_2(t),\tilde{\theta}_2(t))$ on $[0,\hat{\tau}_1]$ with respect to $\tilde{X}, \tilde{\theta}, \tilde{W}$ and $\hat{p}$. Clearly $\hat{\tau}_1 = \tau_2 - \tau_1$. Define $(X(t),\theta(t))$ by

$$(X(t),\theta(t)) = \begin{cases} (X_1(t),\theta_1(t)) & \text{if } t \in [0,\tau_1] \\ (\tilde{X}_2(t-\tau_1),\tilde{\theta}_2(t-\tau_1)) & \text{if } t \in [\tau_1,\tau_2]. \end{cases}$$

It is now clear that $(X(t),\theta(t))$ is the unique solution of (3) in the interval $[0,\tau_2]$. Proceedings this way $(X(t),\theta(t))$ is determined uniquely in $[0,\tau_n]$ for every n. Hence a.s. $(X(t),\theta(t))$ is determined uniquely for all time. ■

Some comments are in order.

Remark 2.1

(i) The boundedness assumption on b and σ in (A1), (A2) may be relaxed. It may be replaced by a growth condition of the following type: there exists a constant C such

$$||b(x,i)||^2 + ||\sigma(x,i)||^2 \leq C(1+||x||^2).$$

Similarly the Lipschitz continuity assumption in (A1), (A2) may be replaced by locally Lipschitz continuity.

(ii) If for each $i = 1, 2, \ldots, N$, $\sigma(\cdot, i)\sigma^*(\cdot, i)$ is uniformly elliptic, i.e., the least eigenvalue of $\sigma(\cdot, i)\sigma^*(\cdot, i)$ is uniformly bounded away from zero, then we can drop any kind of continuity assumption on $b(\cdot, i)$. In fact if $b(\cdot, i)$ is bounded and measurable and $\sigma(\cdot, i)$ is bounded and Lipschitz and (A3), (A4) hold, then under the uniform ellipticity condition it can be shown as in [8], [9], (3) has an a.s. unique strong solution.

(iii) It is clear from the construction that the process $(X(t), \theta(t))$ is Markov. Let $\mathcal{L}$ denote the extended generator of $(X(t), \theta(t))$. Then for $f \in C^2(\mathbb{R}^d \times \Theta) \subset D(\mathcal{L})$, it can be shown that

$$\mathcal{L} : f(x,i) = L_i f(x,i) + \sum_{j=1}^{N} \int_{\mathbb{R}^d} [f(y,j) - f(x,i)] \nu_{x,i}(dy \times \{j\}) \qquad (5)$$

where

$$L_i f(x,i) = \sum_{k=1}^{d} b_k(x,i) \frac{\partial f(x,i)}{\partial x_k} + \frac{1}{2} \sum_{j,k=1}^{d} \sum_{l=1}^{d} \sigma_{jl}(x,i)\sigma_{kl}(x,i) \frac{\partial^2 f(x,i)}{\partial x_j \partial x_k} \qquad (6)$$

and

$$\nu_{x,i}(A \times \{j\}) = \int_{\mathbb{R}} I_{A \times \{j\}}(x + g(x,i,u),\ i + h(x,i,u))du \text{ for } A \in \mathcal{B}(\mathbb{R}^d).$$

(iv) Note that the times at which jumps or switchings occur are determined by the stopping times $\tau_n, n = 1, 2, \ldots$. But at every τ_n, a jump or a switching may not occur. For example if at $t = \tau_1$, $g(Y(\tau_1-), \theta_0, \tilde{p}(\tau_1)) = 0$, there is no jump in the trajectory of $X(t)$ at this time. Similarly, if $h(Y(\tau_1-), \theta_0, \tilde{p}(\tau_1)) = 0$, then $\theta(t)$ remains at θ_0 and thus there is no switching in the trajectory of $X(t)$ at this time. If $g(Y(\tau_1-), \theta_0, \tilde{p}(\tau_1)) \neq 0$ but $h(Y(\tau_1-), \theta_0, \tilde{p}(\tau_1) = 0$, then there will be a jump at $t = \tau_1$, but no switching at $t = \tau_1$. Similarly if $g(Y(\tau_1-), \theta_0, \tilde{p}(\tau_1)) = 0$, but $h(Y(\tau_1-), \theta_0, \tilde{p}(\tau_1)) \neq 0$, there is no jump but only a switching occurs at $t = \tau_1$. On the other hand if $g(Y(\tau_1-), \theta_0, \tilde{p}(\tau_1) \neq 0$ and

$h(Y(\tau_1-), \theta_0, \tilde{p}(\tau_1)) \neq 0$ there is a simultaneous jump and switching at $t = \tau_1$. This kind of mechanism goes on for $t > \tau_1$.

(v) We now focus our attention to a specific case where jumps and switching always occur simultaneously. Let

$$\tilde{g} : \mathbb{R}^d \times \Theta \to \mathbb{R}^d$$

be a function which is bounded and measurable. Let the function $g(\cdot, \cdot, \cdot)$ be given by

$$g(x, i, u) = \begin{cases} \tilde{g}(x, j) & \text{if} \quad u \in \Delta_{ij}(x) \\ 0 & \text{otherwise.} \end{cases} \tag{7}$$

If $g(\cdot, \cdot, \cdot)$ is of the above form, then from (2) and (7), it is clear that the jumps and switchings always occur together. In this specific case the extended generator of $(X(t), \theta(t))$ can be expressed explicitly in terms of $b, \sigma, \tilde{g}$ and λ_{ij}. Let $\mathcal{L}$ denote the extended generator of $(X(t), \theta(t))$. Let $f : \mathbb{R}^d \times \Theta \to \mathbb{R}$ be a smooth function. Then using Ito's formula one can show that

$$\mathcal{L} : f(x, i) = L_i f(x, i) + \sum_{j=1}^{N} \lambda_{i,j}(x)(f(x + g(x, i)) - f(x, i)$$

where $L_i f(x, i)$ is as in (6).

(vi) Consider the non-degenerate case, i.e., when for each i

$$\sigma(\cdot, i)\sigma^*(\cdot, i)$$

is uniformly elliptic. In this case if for each $i, b(\cdot, i)$ is bounded and measurable, and $\sigma(\cdot, i)$ is bounded and Lipschitz continuous, (A3) holds and $g(\cdot, \cdot, \cdot)$ is of the form (7), then one can show as in [9]that the process $(X(t), \theta(t))$ is strong Feller.

2. Stochastic Hybrid Model II

In this section we study a very general stochastic hybrid system. We refer to [2], [14]for analogous controlled stochastic hybrid systems. The state of the system at time t, denoted by $(X(t), \theta(t))$, takes values in $\cup_{n=1}^{\infty}(S_n \times \Theta_n)$, where $\Theta_n = \{1, 2, \ldots, M_n\}$ and S_n is a subset of $\mathbb{R}^{d_n}$.

Between the jumps of $X(t)$ the state equations are of the form

$$\left.\begin{array}{l} dX(t) = b^n(X(t),\theta(t))dt + \sigma^n(X(t),\theta(t))dW^n(t) \\ P(\theta(t+\delta t) = j|\theta(t) = i, X(s), \theta(s), s \le t) \\ = \lambda_{ij}^n(X(t))\delta t + 0(\delta t), i \neq j, \\ X(0) = X_0, \;\; \theta(0) = \theta_0, \end{array}\right\} \quad (8)$$

where for each $n \in \mathbb{N}$

$$\begin{aligned} b^n &: S_n \times \Theta_n \to \mathbb{R}^{d_n} \\ \sigma^n &: S_n \times \Theta_n \to \mathbb{R}^{d_n \times d_n} \\ \lambda_{ij}^n &: S_n \to \mathbb{R} \end{aligned}$$

are suitable functions, $\lambda_{ij}^n(\cdot) \geq 0, i \neq j,: \sum_{j=1}^{M_n} \lambda_{ij}^n(\cdot) = 0,: X_0, \theta_0$ are S_n- and Θ_n- valued random variables, and $W^n(\cdot)$ is a standard d_n-dimensional Wiener process. For each $n \in \mathbb{N}$, let $A_n \subset S_n,: D_n \subset S_n$. The set A_n is the set of instantaneous jump, whereas D_n is the destination set. If at some random time $X(t)$ hits A_n, then it executes an instantaneous jump. The destination of $(X(t), \theta(t))$ at this juncture is determined by a map

$$g_n : A_n \times \Theta_n \to \cup_m (D_m \times \Theta_m).$$

After reaching the destination, the process $(X(t), \theta(t))$ follows the same evolutionary mechanism over and over again.

To ensure the existence of such a pair of processes we need to make certain assumptions.

For each $n \in \mathbb{N}$, let S_n be the closure of a connected open subset of some Euclidean space $\mathbb{R}^{d_n}$. For each $n \in \mathbb{N}$, A_n and D_n are closed, and $A_n \cap D_n = \phi$.

We now make the following assumptions.

(A5) For each $n \in \mathbb{N}$ and $i \in \Theta_n$, $b^n(\cdot, i)$ is Lipschitz continuous.

(A6) For each $n \in \mathbb{N}$ and $i \in \Theta_n$, $\sigma^n(\cdot, i)$ is Lipschitz continuous.

(A7) For each $n \in \mathbb{N}$, $i, j \in \Theta_n$, $\lambda_{ij}^n(\cdot)$ are bounded and measurable.

(A8) The maps $g_n, n \in \mathbb{N}$, are bounded and uniformly continuous.

(A9) $\inf_n d(A_n, D_n) > 0$.

Let $(\Omega, \mathcal{F}, P)$ be the underlying (complete) probability space on which $W^n(\cdot), X_0, \theta_0$ are defined. As in the previous section $\theta(t)$ can be expressed as an integral with respect to a Poisson random measure. Let $p(\cdot,\cdot)$ be $\mathbb{R}_+ \times \mathbb{R}$-valued Poisson random measure with the intensity $dt \times l(du)$ as in the previous section. Construct the maps

$$h^n : \mathbb{R}^{d_n} \times \Theta_n \times \mathbb{R} \to \mathbb{R}^{M_n}$$

as in the previous section such that

$$d\theta(t) = \int_{\mathbb{R}} h^n(X(t-), \theta(t-), u) p(dt, du). \tag{9}$$

Let

$$\mathcal{F}_t^n = \sigma\{W^n(s), p(A, B) | s \le t, A \in \mathcal{B}([0,t]), B \in \mathcal{B}(\mathbb{R})\}.$$

Let $X_0, \theta_0, W^n(\cdot), p(\cdot,\cdot)$ be independent. Then as in the previous section, we can show that under (A5), (A6) and (A7), the equation (8) has an a.s. unique strong solution, denoted by $(X^n(t), \theta^n(t))$ which takes values in $\mathbb{R}^{d_n} \times \Theta_n$. Let

$$\tau_1 = \inf\{t \ge 0| : X(t) \in A_n\}. \tag{10}$$

Then τ_1 is an $\mathcal{F}_t^n$ stopping time. Now define the process $(X(t), \theta(t))$ by

$$(X(t), \theta(t)) = \begin{cases} (X^n(t), \theta^n(t)), & 0 \le t < \tau_1 \\ g_n(X^n(\tau_1-), \theta^n(\tau_1-)), & t = \tau_1. \end{cases} \tag{11}$$

Note that $(X(\tau_1), \theta(\tau_1)) \in D_m \times \Theta_m$, for some $m \in \mathbb{N}$. From τ_1 on the system continues with the same mechanism from the state $(X(\tau_1), \theta(\tau_1))$. Let

$$\mathcal{F}_t = \vee_n \mathcal{F}_t^n.$$

Thus there is a sequence of $\mathcal{F}_t$ stopping times $0 = \tau_0 \le \tau_1 < \tau_2 < \tau_3 < \ldots < \tau_m < \ldots$ such that $\tau_m \uparrow \infty$ a.s. and in the interval $[\tau_m, \tau_{m+1}),:$ $m = 0, 1, \ldots$ the process $(X(t), \theta(t))$ evolves according to (8) for some index $n \in \mathbb{N}$. At times $\tau_m, m \ge 1$, there is an instantaneous jump determined by the map g_m.

Note that, though in each interval of the type $[\tau_m, \tau_{m+1})$, the evolution of $(X(t), \theta(t))$ follows a Markovian type dynamics, the process $(X(t), \theta(t)),: t \in [0, \infty)$, is not a Markov process. This is because we have not thus far accounted for a dynamical variable $\eta(t)$, to be introduced shortly, which is intricately linked with the evolution of $(X(t), \theta(t))$. Let $\eta(t)$ be an $\mathbb{N}$ valued process defined by

$$\eta(t) = n \text{ if } (X(t), \theta(t)) \in S_n \times \Theta_n. \tag{12}$$

The process $\eta(t)$ is a piecewise constant process, it changes from n to m when $(X(t), \theta(t))$ jumps from the regime $S_n \times \Theta_n$ to the regime $S_m \times \Theta_m$. Thus $\eta(t)$ is an indicator of a regime and a change in $\eta(t)$ means a switching in the regimes in which $(X(t), \theta(t))$ evolves. One can show that the process $(X(t), \theta(t), \eta(t))$ is Markov. To see this more clearly we investigate the equations governing the process $(X(t), \theta(t), \eta(t))$. To this end, let

$$\tilde{S} = \{(x, i, n) | x \in S_n, \;\; i \in \Theta_n\}$$
$$\tilde{A} = \{(x, i, n) | x \in A_n, \;\; i \in \Theta_n\}$$
$$\tilde{D} = \{(x, i, n) | x \in D_n, \;\; i \in \Theta_n\}$$

Clearly $(X(t), \theta(t), \eta(t))$ is an $\tilde{S}$-valued process. The set $\tilde{A}$ is the set where jumps occur and $\tilde{D}$ is the destination set for this process. The sets $\cup_n(S_n \times \Theta_n)$, $\cup_n(A_n \times \Theta_n)$ and $\cup_n(D_n \times \Theta_n)$ can be embedded in $\tilde{S}, \tilde{A}$ and $\tilde{D}$ respectively.

Let d^0 denote the injection map of $\bigcup_n(D_n \times \Theta_n)$ into $\tilde{D}$. Define three maps $\tilde{g}_i : \tilde{A} \to \tilde{D}, : i = 1, 2, :: \tilde{h} : \tilde{A} \to \mathbb{N}$

$$\left.\begin{array}{lcl} \tilde{g}_1(x, i, n) & = & \text{the first component in } d^0(g_n(x, i)) \\ \tilde{g}_2(x, i, n) & = & \text{the second component in } d^0(g_n(x, i)) \\ \tilde{h}(x, i, n) & = & \text{the third argument in } d^0(g_n(x, i)). \end{array}\right\} \quad (13)$$

To describe the evolution of $(X(t), \theta(t), \eta(t))$ there is a sequence of $\mathcal{F}_t$ stopping times

$$\tau_1 < \tau_2 < \tau_3 < \ldots < \tau_m < \ldots$$

$\tau_m \uparrow \infty$ a.s. which are the successive hitting times of $\tilde{A}$, such that for $t = \tau_m$

$$\left.\begin{array}{lcl} (X(\tau_m), \theta(\tau_m)) & = & (\tilde{g}_1(X(\tau_m-), \theta(\tau_m-), \eta(\tau_m-)), \\ & & \quad \tilde{g}_2(X(\tau_m-), \theta(\tau_m-), \eta(\tau_m-))) \\ \eta(\tau_m) & = & \tilde{h}(X(\tau_m-), \theta(\tau_m-), \eta(\tau_m-)), \end{array}\right\} \quad (14)$$

where $\tilde{g}_i, \tilde{h}$ are defined in (13). For $\tau_m < t < \tau_{m+1}$

$$\left.\begin{array}{lcl} dX(t) & = & b(X(t), \theta(t), \eta(t))dt + \sigma(X(t), \theta(t)), \eta(t))dW^{\eta(t)}(t) \\ d\theta(t) & = & \int_{\mathbb{R}} h(X(t-), \theta(t-), \eta(t-)u)p(dt, du) \end{array}\right\} \quad (15)$$

where $b(x, i, n) = b^n(x, i)$, $\sigma(x, i, n) = \sigma^n(x, i)$, $h(x, i, n, u) = h^n(x, i, u)$.

The stopping time τ_{m+1} is defined by

$$\tau_{m+1} = \inf\{t > \tau_m | (X(t-), \theta(t-), \eta(t-)) \in \tilde{A}\}.$$

The equations for $(X(t), \theta(t), \eta(t))$ may thus be summarized as follows:

$$\begin{aligned} dX(t) = & [b(X(t), \theta(t), \eta(t)) \\ & + \sum_{m=0}^{\infty} [\tilde{g}_1(X(\tau_m-), \theta(\tau_m-), \eta(\tau_m-)) - X(\tau_m-)]\delta(t - \tau_m)]dt \\ & + \sigma(X(t), \theta(t), \eta(t))dW^{\eta(t)}(t), \end{aligned}$$

$$\begin{aligned} d\theta(t) = & \int_{\mathbb{R}} h(X(t-), \theta(t-), \eta(t-)u)p(dt, du) \\ & + \sum_{m=0}^{\infty} [\tilde{g}_2(X(\tau_m-), \theta(\tau_m-), \eta(\tau_m-)) - \theta(\tau_m-)]\delta(t - \tau_m)]dt \end{aligned}$$

$$d\eta(t) = \sum_{m=0}^{\infty} [\tilde{h}(X(\tau_m-), \theta(\tau_m-), \eta(\tau_m-)) - \eta(\tau_m-)]I_{\{\tau_m \leq t\}}$$

where δ is the Dirac measure.

From the above equation it is clear that the $\tilde{S}$-valued process

$$(X(t), \theta(t), \eta(t))$$

is a Markov process. Note that the stochastic hybrid model constructed in this section generalizes the stochastic hybrid models studied in [2], [14]. In the stochastic hybrid model studied in [2], there is no discrete component like $\theta(t)$. In [14]the discrete component $\theta(t)$ is included, but this component remains unchanged when the continuous component $X(t)$ makes an instantaneous jump. In our model we have removed this restriction on the dynamics of $\theta(t)$, and allow it to change when $X(t)$ changes. Thus we automatically have simultaneous jumps and switchings. Moreover in [14], the same functions b, σ, λ_{ij} are used in every component of the state space $S_n \times \Theta_n$, whereas in our model, these functions depend on the index n. Thus our dynamics are more general than the one treated in [14]. Hence we have constructed a stochastic hybrid model which is more general than the models in [2]and [14].

3. Conclusion

In this paper we have explicitly constructed two stochastic hybrid systems. We established the existance and uniqueness of a strong solution in both cases, and showed that both solutions are Markov processes. The important point is that both models allow for simultaneous jumps in the trajectory and model parameters, the "so-called" case of a hybrid jump.

Acknowlegments

We are grateful to Dr. H. Blom of NLR (Amsterdam) for stimulating discussions during the course of this research.

References

[1] F. A. van der Duyn Schouten and A. Hordijk. Average optimal policies in markov decision drift processes with applications to a queueing and a replacement model. *Adv. Appl. Prob.*, 15:274–303, 1983.

[2] A. Bensoussan and J. L. Menadi. Stochastic hybrid control. *J.Math. Anal. Appl.*, 249:261–268, 2000.

[3] H. A. P. Blom. *Bayesian estimation for decision-directed stochastic control.* PhD thesis, Delft Univ. of Technology, 1990.

[4] R. W. Brockett and G. L. Blankenship. A representation theorem for linear differential equations with markovian switching. *Proc. Allerton Conf. Circ. Syst. Th.*, pages 671–679, 1977.

[5] M. H. A. Davis. *Markov Moldels and Optimisation.* Chapman and Hall, 1999.

[6] N. Ikeda and S. Watanabe. *Stochastic Differential Equations and Diffusion Processes, Second Edition.* North-Holland, Kodansha, 1989.

[7] J. Jacod and A. N. Shiryayev. *Limit Theorems for Stochastic Processes.* Springer-Verlag, 1980.

[8] A. Arapostathis, M. K. Ghosh and S. I. Marcus. Optimal control of switching diffusions with application to flexible manufacturing systems. *SIAM J. Control Optim.*, 31:1183–1204, 1993.

[9] A. Arapostathis, M. K. Ghosh and S. I. Marcus. Ergodic control of switching diffusions. *SIAM J. Control Optim.*, 35:1952–1988, 1997.

[10] S. I. Marcus. Average optimal policies in markov decision drift processes with applications to a queueing and a replacement model. *Adv. Appl. Prob.*, 15:274–303, 1983.

[11] M. Mariton. *Jump Linear Systems in Automatic Control.* Marcel Dekker, 1990.

[12] A. V. Skorohod. *Asymptotic Methods in the Theory of Stochastic Differential Equations.* AMS, 1989.

[13] D. L. Snyder. *Random Point Processes.* Wiley, 1975.

[14] M. K. Ghosh, V. S. Borkar and P. Sahay. Optimal control of a stochastic hybrid system with discounted cost. *J. Optim. Theory Appl.*, 101:557–580, 1991.

[15] D. Vermes. Optimal control of piecewise deterministic processes. *Stochastics*, 14:165–207, 1985.

[16] W. M. Wonham. *Random differential equations in control theory.* Probability Analysis in Applied Mathematics, vol. 2, A. T. Bharucha-Reid (Editor). Academic Press, 1970.

MATHEMATICAL MODELS AND STATE OBSERVATION OF THE GLUCOSE-INSULIN HOMEOSTASIS

Andrea De Gaetano
IASI-CNR BioMatLab, Università Cattolica del Sacro Cuore, Largo A. Gemelli 8, 00168 Roma, Italy
andrea.degaetano@biomatematica.it

Domenico Di Martino
Istituto di Analisi dei Sistemi ed Informatica (IASI-CNR), Viale Manzoni 30, 00185 Roma, Italy
dimartin@iasi.rm.cnr.it, domenico@ing.univaq.it

Alfredo Germani
Dipartimento di Ingegneria Elettrica, Università degli Studi dell'Aquila, Monteluco di Roio, 67040 L'Aquila, Italy
germani@ing.univaq.it

Costanzo Manes
Dipartimento di Ingegneria Elettrica, Università degli Studi dell'Aquila, Monteluco di Roio, 67040 L'Aquila, Italy
manes@ing.univaq.it

Abstract This paper explores the possibility of using an asymptotic state observer for the real-time reconstruction of insulin blood concentration in an individual by using only measurements of the glucose blood concentration. The interest in this topic relies on the fact that the glucose measurements are much more economical and faster then the insulin measurements. An algorithm providing reliable insulin concentrations in real-time is essential for the realization of an "artificial pancreas", an automatic device aimed to infuse the required amount of insulin into the circulatory system of a diabetic patient. An important issue for a good observer design is the determination of satisfactory models of the glucose-insulin homeostasis. Different models have been considered and discussed in this paper. For all models presented an asymptotic state observer has been constructed and numerical simulations have been successfully carried out.

Keywords: Homeostasis, nonlinear systems, state observer, drift-observability

Introduction

It is well-known that glucose and insulin blood concentrations are two extremely important variables in a diabetic individual. Unfortunately, only glucose blood concentration can be easily monitored in real-time, while the measurement of the insulin concentration is expensive and not immediate. In many control applications, when not all the variables of a system can be directly measured, an "asymptotic state observer" is used, which is an algorithm that processes the available measurements and provides estimates of all the system variables, with an error that asymptotically converges to zero. This fact suggests the use of a state observer for the reconstruction of the insulin blood concentration using only glucose concentration data. This paper investigates the use of the observers presented in [13] and [6], that under some conditions guarantee exponential decay of the estimation error. The design of a good observer requires the knowledge of a good model of the system under investigation. The problem of developing satisfactory models of the glucose-insulin homeostasis has been widely investigated by many authors in the last two decades (see [7], [16], [10], [9], [2], [5], [11], [15], [4]). At today the most used model in physiological research on glucose metabolism is the so called *Minimal Model* [2], originally proposed for the interpretation of the glucose and insulin plasma concentrations following the intra-venous glucose tolerance test (IVGTT). In the Minimal Model two dynamic subsystems can be singled out. The parameters of each subsystem can be evaluated using glucose and insulin data in a separate identification procedures. In [7] the authors showed that in some situations the coupling of the two subsystems does not admit an equilibrium and the concentration of active insulin in the "distant" compartment increases without bounds. For this reason, in this paper two modifications of the Minimal Model are considered. For each model an asymptotic state observer is computed and verified through numerical simulations.

1. Asymptotic State Observers

Consider a dynamic system described, for $t \geq 0$, by nonlinear differential equations of the form

$$\dot{x}(t) = f\big(x(t)\big) + g\big(x(t)\big)u(t), \tag{1}$$
$$y(t) = h\big(x(t)\big), \tag{2}$$

where $x(t) \in \mathcal{X} \subseteq \mathbb{R}^n$ is the system state, $u(t) \in \mathcal{U} \subseteq \mathbb{R}$ is the input function and $y(t) \in \mathbb{R}$ is the output. $f(x)$ and $g(x)$ are $C^k(\mathcal{X})$ vector fields, with k an integer allowing all differentiations needed.

In most applications the input $u(t)$ and the output $y(t)$ of the system are the only quantities available through measurements, while the system state $x(t)$ remains unaccessible. An important issue in systems and control theory is the problem of state reconstruction through on line processing of the measured input and output signals. An algorithm that asymptotically reconstructs the state is called an *asymptotic state observer*, and is usually described by differential equations. The existence of an asymptotic observer depends on the *observability* properties of the system. A system is said *drift-observable* if it is such that, when the input is identically zero, different states produce different outputs (see [6]). Such property depends only on the pair $\big(f(x), h(x)\big)$, and claims the theoretical possibility of the state reconstruction from the measured output data, when $u(t) \equiv 0$. A system is said *uniformly observable* when different states produce different outputs, for any input function $u(t)$ (see [12]). This is a rather strong property for nonlinear systems, and depends on the triple $\big(f(x), g(x), h(x)\big)$. A weaker property is the *almost-uniform observability*, that characterizes systems such that different states produce different outputs, for any *constant* input (see [13]). The study of the drift-observability and of the almost-uniform observability of nonlinear systems is made through the construction of suitable vector functions that map the system state at a given time t into the output and its derivatives at the same time t.

The following two square maps can be defined for system (1)–(2)

$$\Phi(x) \stackrel{\text{def}}{=} \begin{bmatrix} h(x) \\ L_f h(x) \\ \vdots \\ L_f^{n-1} h(x) \end{bmatrix}, \quad \Psi(x,u) \stackrel{\text{def}}{=} \begin{bmatrix} h(x) \\ L_{f+gu} h(x) \\ \vdots \\ L_{f+gu}^{n-1} h(x) \end{bmatrix}, \tag{3}$$

where $L_f^k h(x)$ denotes the k-th order repeated Lie derivative of the function $h(x)$ along the field $f(x)$, and in the same way, $L_{f+gu}^k h(x)$ denotes repeated Lie derivative of $h(x)$ along $f(x)+g(x)u$, with u constant parameter. Formally

$$L_f^0 h(x) = h(x), \qquad L_{f+gu}^0 h(x) = h(x),$$
$$L_f^{k+1} h(x) = \frac{\partial L_f^k h}{\partial x} f(x), \qquad L_{f+gu}^{k+1} h(x) = \frac{\partial L_{f+gu}^k h}{\partial x}\big(f(x)+g(x)u\big), \tag{4}$$

Recall that the observation relative degree of a triple $\big(f(x), g(x), h(x)\big)$ in a set $\Omega \subseteq \mathbb{R}^n$ is an integer $r \leq n$ such that

$$\begin{aligned} &\forall x \in \Omega,\ L_g L_f^k h(x) \equiv 0, \quad k = 0, 1, \ldots, r-2, \\ &\exists x \in \Omega:\ L_g L_f^{r-1} h(x) \neq 0. \end{aligned} \tag{5}$$

The observation relative degree is said *full* or *maximal* if it is equal to n. Quite obviously, when $u = 0$ the two maps (3) coincide, so that $\varPhi(x) = \varPsi(x, 0)$. Moreover, it is not difficult to show that if the relative degree is maximal it follows that $\varPsi(x, u) = \varPhi(x)$. Let $z(t)$ be the vector made of the output and its derivatives up to order $n-1$, i.e.

$$z(t) = \begin{bmatrix} y(t) \\ \dot{y}(t) \\ \vdots \\ y^{(n-1)}(t) \end{bmatrix}. \tag{6}$$

It can be easily checked that when $u(t) \equiv 0$ or the relative degree is maximal it is $z(t) = \varPhi(x(t))$, while if $u(t) \equiv \bar{u}$ it is $z(t) = \varPsi(x(t), \bar{u})$. The map $\Phi(x)$ is named *drift-observability map.* The drift-observability property in an open set $\Omega \subseteq \mathbb{R}^n$ implies that in Ω there exists the inverse map $x = \Phi^{-1}(z)$. This means that when $u(t) \equiv 0$ or when the relative degree is full the state can be reconstructed from the knowledge of the output, at least from a theoretical point of view. In the same way, the uniform-observability in $\Omega \times \mathcal{U}$ implies the existence of the inverse $x = \varPsi^{-1}(z, \bar{u})$, forall $x \in \Omega$ and $\bar{u} \in \mathcal{U}$. This means that the state can be reconstructed from the knowledge of the output and of the constant input $\bar{u}$.

Let $Q(x)$ and $\bar{Q}(x, u)$ denote the Jacobians

$$Q(x) \stackrel{\text{def}}{=} \frac{\partial \varPhi(x)}{\partial x}, \qquad \bar{Q}(x, u) \stackrel{\text{def}}{=} \frac{\partial \varPsi(x, u)}{\partial x}. \tag{7}$$

Drift-observability of the system (1)–(2) in a set Ω implies nonsingularity of $Q(x)$ in Ω, and allows the construction of the asymptotic observer presented in [6], described by the differential equation

$$\dot{\hat{x}}(t) = f\big(\hat{x}(t)\big) + g\big(\hat{x}(t)\big)u(t) + Q^{-1}\big(\hat{x}(t)\big)K\Big(y(t) - h\big(\hat{x}(t)\big)\Big). \tag{8}$$

Almost-uniform observability in a set $\Omega \times \mathcal{U}$ implies invertibility of $\bar{Q}(x,u)$ for all $(x,u) \in \Omega \times \mathcal{U}$, and allows the construction of the following asymptotic observer

$$\dot{\hat{x}}(t) = f\big(\hat{x}(t)\big) + g\big(\hat{x}(t)\big)u(t) + \bar{Q}^{-1}\big(\hat{x}(t), u(t)\big)K\Big(y(t) - h\big(\hat{x}(t)\big)\Big), \quad (9)$$

presented in [13]. In both observers (8)-(9) the constant vector K is the *observer gain*, and is a design parameter. In [6, 13] it is shown that, under suitable assumptions, there exists a choice for K such that (8) or (9) are exponential observers for system (1)-(2), i.e. there exist positive constants μ and α such that

$$\|x(t) - \hat{x}(t)\| \leq \mu\, e^{-\alpha t}\|x(0) - \hat{x}(0)\|, \qquad t \geq 0. \quad (10)$$

In the case of observer (8) the convergence is guaranteed if the system has relative degree n and the input $u(t)$ is bounded. If the system relative degree is smaller than n, (8) is still an exponential observer for (1)-(2), provided that the amplitude of the input $u(t)$ satisfies a specific bound (for more details see [6]). For systems with generic relative degree (9) is an exponential observer provided that the input derivative satisfies a specific bound (*slowly varying input*, for more details see [13]). A strategy for the choice of the gain vector K is described in the convergence proofs reported in [6, 13]. In particular, K should be chosen such to assign eigenvalues to the matrix $A_b - KC_b$, where (A_b, C_b) define a Brunowski pair of dimension n.

In this paper the state observers (8) and (9) are applied for the reconstruction of the blood insulin concentration through on-line processing of the measured blood glucose concentration (the system output $y(t)$).

2. The Minimal Model

There are two main experimental procedures currently in use for the estimation of the insulin sensitivity in a subject: the euglycemic hyperinsulinemic clamp (EHC) [8] and the intra venous glucose tolerance test (IVGTT) [2]. With respect to the EHC, the IVGTT is easier to execute and provides more informations. The test consists of injecting I.V. a bolus of glucose and frequently sampling the glucose and insulin plasma concentrations afterwards, for a period of about three hours. The physiological model most used in the interpretation of the IVGTT is known as the *Minimal Model* [2]:

$$\dot{G}(t) = -\big(p_1 + X(t)\big)G(t) + p_1 G_b, \qquad G(0) = p_0 \quad (11)$$

$$\dot{X}(t) = -p_2 X(t) + p_3\big(I(t) - I_b\big), \qquad X(0) = 0 \quad (12)$$

$$\dot{I}(t) = p_4\big[G(t) - p_5\big]^+ t - p_6\big(I(t) - I_b\big), \qquad I(0) = p_7 + I_b \quad (13)$$

where $[\cdot]^+$ denotes the positive part of its argument, and

- $G(t)$ $[mg/dl]$ is the blood glucose concentration at time t $[min]$;
- $I(t)$ $[\mu UI/ml]$ is the blood insulin concentration;
- $X(t)$ $[min^{-1}]$ is an auxiliary function representing insulin-excitable tissue glucose uptake activity, proportional to insulin concentration in a "distant" compartment;
- G_b $[mg/dl]$ is the subject's baseline glycemia;
- I_b $[\mu UI/ml]$ is the subject's baseline insulinemia;
- p_0 $[mg/dl]$ is the theoretical glycemia at time 0 after the instantaneous glucose bolus;
- p_1 $[min^{-1}]$ is the glucose "mass action" rate constant, i.e. the insulin-independent rate constant of tissue glucose uptake, "glucose effectiveness";
- p_2 $[min^{-1}]$ is the rate constant expressing the spontaneous decrease of tissue glucose uptake ability;
- p_3 $[min^{-2}(\mu UI/ml)^{-1}]$ is the insulin-dependent rate of increase in tissue glucose uptake ability, per unit of insulin concentration excess over baseline insulin;
- p_4 $[(\mu UI/ml)^{-1}(mg/dl)^{-1}min^{-2}]$ is the rate of pancreatic release of insulin after the bolus, per minute and per mg/dl of glucose concentration above the "target" glycemia;
- p_5 $[mg/dl]$ is the pancreatic "target glycemia" (pancreas produces insulin as long as $G(t) > p_5$);
- p_6 $[min^{-1}]$ is the first order decay rate constant for plasma insulin;
- $p_7 = \mu UI/ml$ is the theoretical plasma insulin concentration at time 0, above basal insulinemia, immediately after the glucose bolus.

Parameters p_0, p_1, p_4, p_5, p_6 and p_7 are usually referred to in the literature as G_0, S_G, γ, h, n and I_0, respectively, while the insulin sensitivity index S_I is computed as p_3/p_2.

The Minimal Model was conceived as composed of two parts. The first one, made of eq.'s (11)-(12), describes the time course of plasma glucose concentration as a function of the circulating insulin, treated as a forcing function, known from measurements. The second part consists of eq. (13) and describes the time course of plasma insulin concentration accounting for the dynamics of pancreatic insulin release in response to the glucose stimulus regarded as a forcing function, known from measurements. In this way the problem of model parameter fitting can be separated into two separate subproblems. However, some stability problems of the Minimal Model have been revealed in [7]. The main reason of instability is a term in the third equation that linearly grows with time.

The injection into the bloodstream of a subject of a bolus of glucose during the IVGTT induces an *impulsive* increase in the plasma glucose concentration $G(t)$ and a corresponding increase of the plasma concentration of insulin $I(t)$, secreted by the pancreas. These concentrations are measured during a three hour time interval beginning at the bolus injection. Note that in eq. (13) the positive part of $G(t) - p_5$ multiplies the time t to model the hypothesis that the effect of circulating hyperglycemia on the rate of pancreatic secretion of insulin is proportional both to the hyperglycemia and to the time elapsed from the glucose stimulus [16]. However the multiplication by t in (13) introduces an origin for time, making the model non stationary and binding the model to the IVGTT experimental procedure.

The application of the observer (8) to the Minimal Model can be worked out by considering the time t, explicitely appearing in (13), as an external input, and verifying that the relative degree is 3. Before deriving the observer equations, it is convenient to put system (11)–(13) in a suitable form, defining:

$$\begin{aligned} x_1 &= G - p_5, && \text{(glucose conc. exceeding the \textit{target glycemia})} && (14)\\ x_2 &= X + p_1, && \text{(rate of tissue glucose uptake)} && (15)\\ x_3 &= I - I_b, && \text{(insulin conc. exceeding the \textit{baseline insulinemia})} && (16)\\ y &= G - p_5. && \text{(measured variable: } x_1) && (17) \end{aligned}$$

With these definitions the Minimal Model can be written as

$$\dot{x}_1(t) = -x_2(t)\big(x_1(t) + p_5\big) + p_1 G_b, \tag{18}$$

$$\dot{x}_2(t) = -p_2\big(x_2(t) - p_1\big) + p_3 x_3(t), \tag{19}$$

$$\dot{x}_3(t) = -p_6 x_3(t) + p_4 x_1^+(t) t, \tag{20}$$

$$y(t) = x_1(t), \tag{21}$$

with state domain $\mathcal{X}$ and initial conditions $x(0)$:

$$\mathcal{X} = \left(\mathbb{R}^+\right)^3, \qquad \begin{bmatrix} x_1(0) \\ x_2(0) \\ x_3(0) \end{bmatrix} = \begin{bmatrix} p_0 - p_5 \\ p_1 \\ p_7 \end{bmatrix} \tag{22}$$

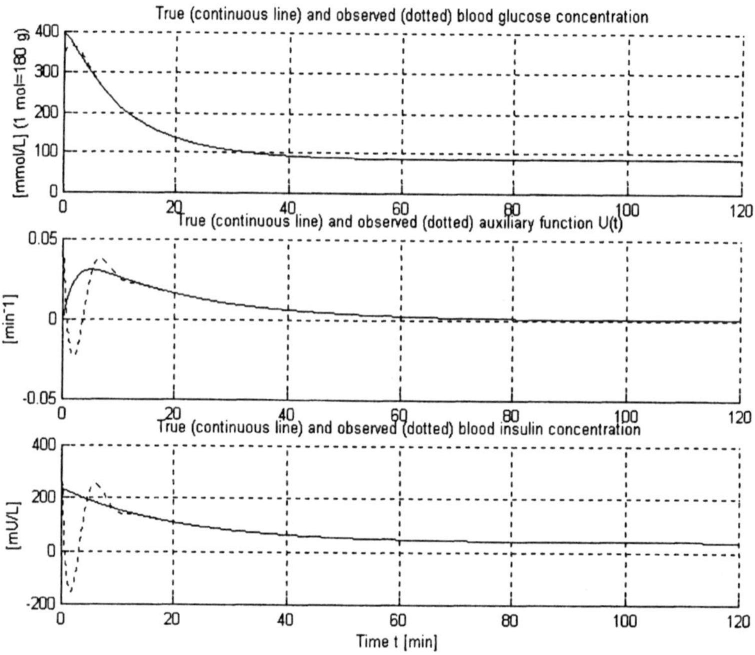

Figure 1. **Minimal Model: state observation**

The drift-observability matrix of system (18)–(21) is

$$Q(x) = \begin{bmatrix} 1 & 0 & 0 \\ -x_2 & -(x_1 + p_5) & 0 \\ Q_{3,1}(x) & Q_{3,2}(x) & -p_3(x_1 + p_5) \end{bmatrix} \tag{23}$$

$$Q_{3,1} = x_2^2 + p_2 x_2 - p_3 x_3 - p_1 p_2 \tag{24}$$

$$Q_{3,2} = (2x_2 + p_2)(x_1 + p_5) - p_1 G_b. \tag{25}$$

All simulations of the observation algorithm revealed a very good tracking capability of the observer. Fig. 1 reports simulation results using a gain vector K that assigns eigenvalues $(-1, -1.2, -1.4)$ to the matrix $A_b - KC_b$. The values of model parameters used in the reported simulation are $G_b = 87$, $I_b = 37.9$, $p_0 = 398$, $p_1 = 0.05$, $p_2 = 0.5$, $p_3 = 10^{-4}$, $p_4 = 10^{-5}$, $p_5 = 150$, $p_6 = 0.05$, $p_7 = 199$.

3. The Fisher Model

In this section we investigate the behavior of the observer (9) applied to the model used by Bergman *et al.* [1]–[2] and by Fisher *et al.* [9]–[10] for the development of control strategies of plasma glucose levels in diabetic individuals. Two main approaches are currently followed in the development of insulin infusion programs: open-loop methods, devoted to the computation of a predetermined amount of insulin to be delivered to a patient, and closed-loop methods, often referred to as *artificial beta cells* or *artificial pancreas.* Closed-loop methods require continuous monitoring of blood glucose levels and can involve quite sophisticated and costly apparatus. An intermediate approach is followed by semi closed-loop methods ([5], [9], [10]), based on intermittent blood glucose sampling. Optimization techniques are used to calculate insulin infusion programs for the correction of hyperglycemia. The semi closed-loop algorithm proposed in [11] is based on three hourly plasma glucose samples and combines a single injection with continuous infusion of insulin.

The model of insulin-glucose homeostasis used in the cited works is a suitable modification of the Minimal Model. In particular, as long as severe diabetic individuals are being considered, in the third equation of the Minimal Model the time-varying term that models the stimulus on the insulin production given by the glucose concentration is removed. In the first equation a glucose infusion term is introduced, representing the effect of glucose intake resulting from a meal. The resulting model is as follows:

$$\dot{G}_\Delta(t) = -p_1 G_\Delta(t) - X(t)\big[G_\Delta(t) + G_b\big] + P(t), \tag{26}$$

$$\dot{X}(t) = -p_2 X(t) + p_3 I_\Delta(t), \tag{27}$$

$$\dot{I}_\Delta(t) = -N\big[I_\Delta(t) + I_b\big] + u(t)/V_I, \tag{28}$$

The same meaning of the parameters used in the Minimal Model is retained in this model, except that $G_\Delta(t)$ and $I_\Delta(t)$ represent the differences of plasma glucose concentration and free plasma insulin concentration from their basal values G_b and I_b (i.e. $G(t) = G_b + G_\Delta(t)$, $I(t) = I_b + I_\Delta(t)$. $P(t)$ and $u(t)$ are the rates of infusion of exogenous glucose and insulin, respectively, V_I is the insulin distribution volume and N is the fractional disappearance rate of insulin. Note that for diabetic patients the basal value of plasma insulin concentration I_b is not a *natural value* but should be interpreted as a *target value* for the insulin infusion program.

The model parameters p_1, p_2 and p_3 are estimated by Bergman *et al.* in [3] in a study of diabetic and normal human subjects. Values they use for normal subjects are $p_1 = 0.028$, $p_2 = 0.025$, $p_3 = 0.000013$. For diabetic (glucose resistant) subjects the value of p_1 is significantly reduced and can be set to zero. The other parameters for a subject of average weight can be set as:

$$V_I = 12\ l, \quad N = 5/54\ min^{-1}, \quad G_b = 4.5 \cdot 18\ mg/dl, \quad I_b = 15\ mU/l.$$

The value of I_b is typical of free insulin levels of controlled diabetic subjects under steady-state conditions.
The steady-state in the model corresponds to a constant insulin infusion rate of $u = NV_I I_b$ $[mU \cdot min^{-1}]$. This is consistent with observations of the infusion rates that are required to maintain steady-state plasma glucose levels of severe diabetics at the basal values of normal subjects.

In the design of an observer for the model (26)–(28) we assume that oral glucose infusion starts at $t = 0$ prior to which plasma glucose and insulin are at their fasting levels. The term $P(t)$ in (26) represents the rate at which glucose enters the blood from intestinal absorption following a meal. In oral glucose tests it is observed that the plasma glucose level rises from the rest level to a maximum in less than 30 min. In normal subjects the glucose level falls to the base level after about 2 – 3 hours. A function that produces this kind of desired behavior in the model (26)–(28) is

$$P(t) = Be^{-kt}, \qquad t \geq 0, \tag{29}$$

where B depends on the amount of glucose ingested during the meal, and k is the rate constant of glucose delivery to the blood circulatory system. A good value for k in normal subjects is $k = 0.05\ min^{-1}$, while for a medium meal $B = 0.5\ mg/(min \cdot dl)$. The introduction of the term (29) in the model (26)–(28) requires an additional differential equation

$$\dot{P}(t) = -kP(t), \qquad P(0) = B. \tag{30}$$

The state space dimension now has dimension $n = 4$. The state variables considered for the observer design are

$$x_1(t) = G_\Delta(t) = G(t) - G_b, \tag{31}$$
$$x_2(t) = X(t), \tag{32}$$
$$x_3(t) = I_\Delta(t) = I(t) - I_b, \tag{33}$$
$$x_4(t) = P(t). \tag{34}$$

With respect to these variable the Fisher Model takes the form (1)–(2) with

$$f(x) = \begin{bmatrix} -p_1 x_1 - x_2\big[x_1 + G_b\big] + x_4 \\ -p_2 x_2 + p_3 x_3 \\ -n\big[x_3 + I_b\big] \\ -k x_4 \end{bmatrix}, \qquad g(x) = \begin{bmatrix} 0 \\ 0 \\ 1 \\ 0 \end{bmatrix}, \tag{35}$$

$$h(x) = x_1, \qquad U = \frac{u}{V_I}. \tag{36}$$

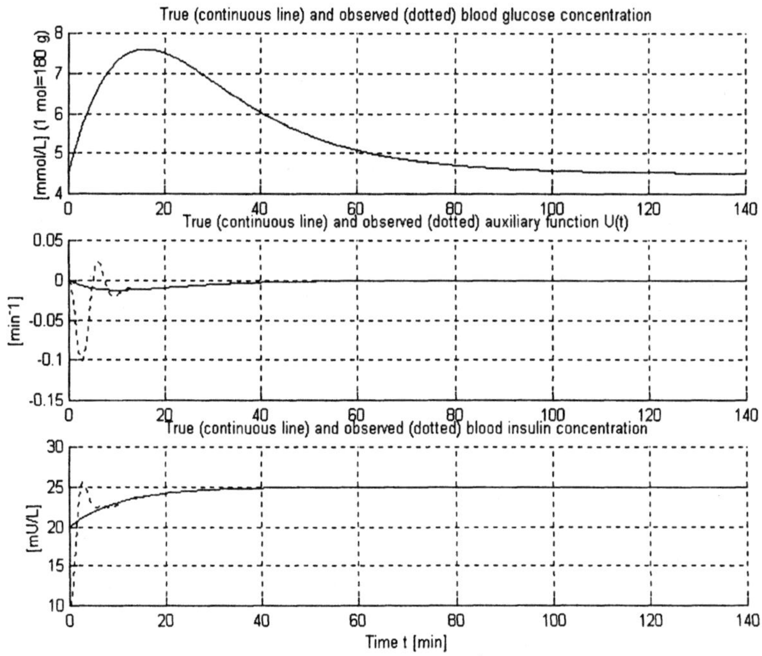

Figure 2. **Fisher Model: state observation (n=4)**

For this system the observation relative degree is smaller than 4, the dimension of the state space. As a consequence, the observer (9) should be preferred to observer (8). The expressions of the map $\Psi(x,u)$ and of the Jacobian $\bar{Q}(x,u)$ are quite long and are not reported here due to lack of space. The values of the parameters used in the simulations are $G_b = 4.5$, $I_b = 25$, $p_1 = 0.05$, $p_2 = 0.1$, $p_3 = 6.5 \cdot 10^{-4}$, $k = 0.05$, $B = 0.5$, $N = 5/54$, $V_I = 12$. The set of eigenvalues chosen for the computation of the observer gain is $(-1, -1.05, e^{+3/4\pi j}, e^{-3/4\pi j})$.

4. Glucose Feedback Model

One disadvantage of the Minimal Model (18)–(21) is its intrinsic non stationarity, due to the presence of a term that grows linearly with time in eq. (20) and affects the system stability. We propose a stationary model with a behavior similar to the Minimal Model:

$$\dot{x}_1(t) = -x_2(t)\big(x_1(t) + p_5\big) + p_1 G_B \tag{37}$$

$$\dot{x}_2(t) = -p_2\big(x_2(t) - p_1\big) + p_3 x_3(t) \tag{38}$$

$$\dot{x}_3(t) = p_4 x_1^+(t) u(t) - p_6 x_3(t) \tag{39}$$

$$y(t) = x_1(t) \tag{40}$$

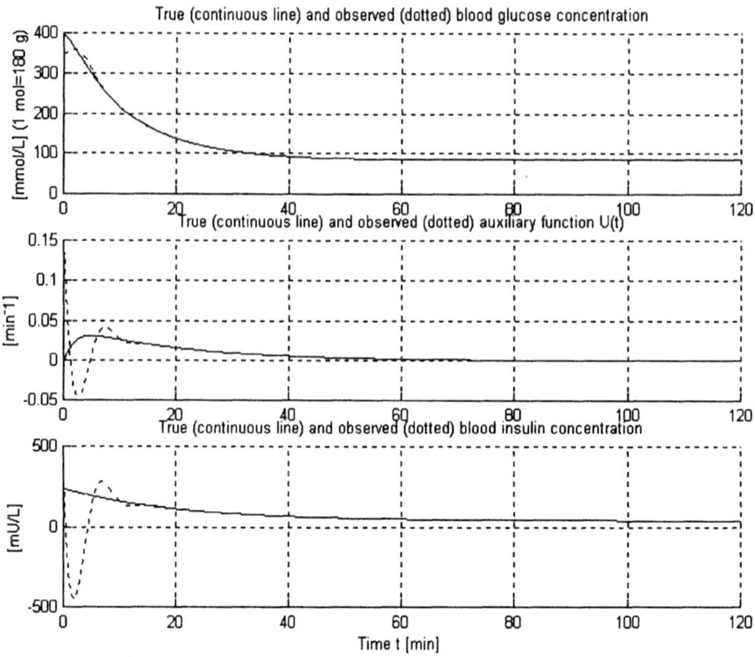

Figure 3. **Output feedback model: state observation**

where the auxiliary variable $u(t)$ is computed as

$$\dot{u}(t) = \frac{e^{a_1 y(t)}}{1 + e^{a_1 y(t)}} - a_2 u(t), \qquad u(0) = 0. \tag{41}$$

This model differs from the Minimal Model in the third equation, where the explicit appearance of time t has been substituted with an auxiliary variable u that approximates the unit ramp only for high values of the measured glucose concentration.

For low glucose concentrations, $u(t)$ decays to a steady state value. The parameters a_1 and a_2 can be adjusted so to give the desired behavior. We found a good behavior with $a_1 = 0.1$ and $a_2 = 1$.

From a control system perspective, equations (37)–(41) describe an output feedback system and therefore we name such model of the insulin-glucose homeostasis the "Glucose Feedback Model". This model has relative degree 3 and therefore admit the observer equation (8). The drift-observability matrix coincides with (25), because the pair $(f(x), h(x))$ is the same of the Minimal Model.

Fig. 3 reports the simulation results using a gain vector K that assigns eigenvalues $(-0.5,\ 0.5 \cdot e^{+3/4\pi j},\ 0.5 \cdot e^{-3/4\pi j})$ to the matrix $A_b - KC_b$.

The values of the model parameters are the same used in the simulation of the Minimal Model.

5. Conclusions and Future Developments

This work explores the use of nonlinear state observers for real-time monitoring of the insulin blood concentration using only measurements of blood glucose concentration. Three models of the glucose-insulin homeostasis have been presented here, on which asymptotic observers have been constructed. The clinical validation of the proposed observers using experimental data will be the object of a future research. In future work, also the delay-differential models presented in [7] will be considered for state observation, using the observer developed in [14].

References

[1] R. N. Bergman, D. T. Finegood, and M. Ader. Assessment of insulin sensitivity **in vivo**. *Endocrine Rev.*, 6:45–86, 1985.

[2] R. N. Bergman, Y. Z. Ider, C. R. Bowden, and C. Cobelli. Quantitative estimation of insulin sensitivity. *Amer. Journal of Physiology*, 236:E667–677, 1979.

[3] R. N. Bergman, L. S. Phillips, and C. Cobelli. Physiological evaluation of factors controlling glucose tolerance in man: measurement of insulin sensitivity and β-cell glucose sensitivity from the response to intravenous glucose. *Journal Clin. Invest.*, 68:1456–1467, 1981.

[4] B. Candas and J. Radziuk. An adaptive plasma glucose controller based on a nonlinear insulin/glucose model. *IEEE Transactions on Biomedical Engineering*, 41:–, 1994.

[5] D. J. Chisolm, E. W. Kraegen, D. J. Bell, and D. R. Chipps. A semi-closed loop computer-assisted insulin infusion system. *Med. Journal Aust.*, 141:784–789, 1984.

[6] M. Dalla Mora, A. Germani, and C. Manes. Design of state observers from a drift-observability property. *IEEE Transactions on Automatic Control*, 45:1536–1540, 2000.

[7] A. De Gaetano and O. Arino. Mathematical modelling of the intravenous glucose tolerance test. *Journal of Mathematical Biology*, 40:136–168, 2000.

[8] R.A. Defronzo, J.D. Tobin, and R. Andreas. Glucose clamp technique: a method for quantifying insulin secretion and resistance. *Am. Journal of Physiology*, 237:E214–E223, 1979.

[9] M. E. Fisher and Kok Lay Teo. Optimal insulin infusion resulting from a mathematical model of blood glucose dynamics. *IEEE Transactions on Biomedical Engineering*, 36:479–485, 1989.

[10] M.E. Fisher. A semiclosed-loop for the control of blood glucose levels in diabetics. *IEEE Transactions on Biomedical Engineering*, 38:57–61, 1991.

[11] S. M. Furler, E. W. Kraegen, R. H. Smallwood, and D. J. Chisolm. Blood glucose control by intermittent loop closure in the basal mode: computer simulation studies with a diabetic model. *Diabetes Care*, 8:553–561, 1985.

[12] J.P. Gauthier, and G. Bornard, Observability for any $u(t)$ of a class of nonlinear systems. *IEEE Transactions on Automatic Control*, 26:922–926, 1981.

[13] A. Germani and C. Manes. State observers for nonlinear systems with slowly varying inputs. 36th IEEE Conf. on Decision and Control (CDC'97), S. Diego, Ca., 5:5054–5059, 1997.

[14] A. Germani, C. Manes, and P. Pepe. An asymptotic state observer for a class of nonlinear delay systems. *Kybernetyka*, 37:459–478, 2001.

[15] R.S. Parker, F.J. Doyle III, and N.A. Peppas. A model-based algorithm for blood glucose control in type i diabetic patients. *IEEE Transactions on Biomedical Engineering*, 46:–, 1999.

[16] G. Toffolo, R.N. Bergman, D.T. Finegood, C.R. Bowden, and C. Cobelli. Quantitative estimation of beta cell sensitivity to glucose in the intact organism: a minimal model of insulin kinetics in the dog. *Diabetes*, 29:979–990, 1980.

CONVERGENCE ESTIMATES OF POD-GALERKIN METHODS FOR PARABOLIC PROBLEMS

Thibault Henri
INSA de Rennes, IRMAR
CS 14315, 35 043 Rennes Cedex
Thibault.Henri@insa-rennes.fr

Jean-Pierre Yvon
INSA de Rennes, IRMAR
Jean-Pierre.Yvon@insa-rennes.fr

Abstract Proper orthogonal decomposition (POD) is a Galerkin method which has been introduced in a fluids mechanics context. It is also known as Karhunen-Loeve decomposition and principal component analysis. The idea of POD consists in using a priori known information on the solution u of PDE, for example snapshots $u_i = u(t_i)$, to determine a set of functions which are the eigenfunctions of an Hilbert-Schmidt operator. This basis can be used to solve the PDE with a smaller amount of computations. Convergence estimates have been proved recently in the parabolic case starting from a particular discretization scheme [7]. Moreover it has been proved that the method converges independently from the scheme [6]. We consider the case of a linear parabolic equation. We give a first convergence estimate in a case where u is regular. However classical POD does not look satisfactory and an improvement consists in considering a POD which takes into account the derivative of u. We will also present some insights into the control of the approximation by introducing what will be called a good order of approximation.

Keywords: Proper orthogonal decomposition, Karhunen-Löve decomposition, model reduction.

1. Principle of Proper Orthogonal Decomposition (POD)

1.1 Proper Orthogonal Decomposition of a Function $u \in L^2(0,T;X)$

Let X be a real separable Hilbert space endowed with the scalar product $(.,.)_X$, let $T > 0$ be a positive real and let $u \in L^2(0,T;X)$ be a (class of) function depending on time $t \in [0,T]$ with values in X. We define the POD operator $K(u) : X \to X$:

$$K(u) \;:\; \varphi \longmapsto \frac{1}{T}\int_0^T (u(t),\varphi)_X \; u(t)\; dt. \tag{1}$$

We consider the kernel $\widetilde{k}(s,t) = \frac{1}{T}(u(s),u(t))_X$ and we define the auxiliary POD operator $\widetilde{K}(u) : L^2(0,T) \to L^2(0,T)$:

$$\widetilde{K}(u) \;:\; v \longmapsto \int_0^T v(s)\widetilde{k}(s,.)\; ds. \tag{2}$$

By concern of clarity, we denote $K = K(u)$ and $\widetilde{K} = \widetilde{K}(u)$. The operators K and $\widetilde{K}$ are self-adjoint semi-definite. Moreover $\widetilde{K}$ is a Hilbert-Schmidt operator since the kernel $\widetilde{k}$ is $L^2([0,T]^2)$. We can therefore index the eigenvalues of $\widetilde{K}$ in a non-increasing sequence: $\widetilde{\lambda}_1 \geq \widetilde{\lambda}_2 \geq \cdots \geq \widetilde{\lambda}_k \geq \cdots \geq 0$. We assume that u is not the null function so that the spectrum of $\widetilde{K}$ is not zero.

LEMMA 1 *The operators K and $\widetilde{K}$ have the same eigenvalues with same multiplicity.*

Proof. Let $\widetilde{\lambda}$ be an eigenvalue of the operator $\widetilde{K}$ with multiplicity $r \geq 1$. We consider a set $(v_k)_{1\leq k\leq r}$ of orthonormal eigenvectors in $L^2(0,T)$ and we define the set $(\psi_k)_{1\leq k\leq r}$ by posing $\psi_k = 1/\sqrt{\widetilde{\lambda}T}(u,v_k)_{L^2(0,T)}$. Then the set $(\psi_k)_{1\leq k\leq r}$ is a set of orthonormal eigenvectors of the operator K in X for the eigenvalue $\widetilde{\lambda}$. Conversely, if λ is an eigenvalue of the operator K and if $(\psi_k)_{1\leq k\leq r}$ is a set of orthonormal eigenvectors of K in X, the set $(v_k)_{1\leq k\leq r}$ defined by $v_k = 1/\sqrt{\lambda T}(u,\psi_k)_X$ is a set of orthonormal eigenvectors of $\widetilde{K}$ in $L^2(0,T)$ for the eigenvalue λ.

DEFINITION 2 *The non zero eigenvalues of the operator K, indexed in a non increasing order, are called the POD eigenvalues associated with the function u. A set $(\psi_k)_{k\geq 1}$ of orthonormal eigenvectors of K in X corresponding to these eigenvalues is called a set of POD eigenvectors associated with the function u.*

Remark. The POD eigenvectors depend on the space X. If the fonction u is in $L^2(0,T;X_1)\cap L^2(0,T;X_2)$, the POD eigenvectors in X_1 differ from the POD eigenvectors in X_2.

Remark. We have not uniquess of the POD eigenvectors, although we have uniquess of the eigenspaces of the operator K. We define the approximated subspaces $Y_\ell = \text{span}(\psi_1, \cdots, \psi_\ell)$ which do not depend of the POD vectors in the case when $\lambda_\ell > \lambda_{\ell+1}$ and we denote $Y = \overline{\text{span}(\psi_k)_{k\geq 1}}^X$ the space spanned by the POD eigenvectors in X.

Remark. The POD eigenvectors correspond to eigenvalues which are indexed in a non increasing order: the order of indexation is significant.

Example. *The method of snapshots.* Let us consider real numbers $\sigma > 0$ and $0 = t_0 < \cdots < t_N = T$ such that $t_{i+1} - t_i = \sigma$ for $i = 0, \cdots, N-1$. Let u be a piecewise constant function defined by posing $u(t) = u_{i+1} \in X$ for $t \in]t_i, t_{i+1}]$. The values $(u_i)_{1\leq i\leq N}$ are called snapshots. The proper orthogonal decomposition of u corresponds to the method of snapshots. For any $\varphi \in X$, we have: $K(u)\varphi = 1/N\sum_{i=1}^N (u_i,\varphi)_X u_i$. The method of snapshots consists in considering the correlation matrix K_{cor} defined by:

$$K_{cor} = \left(\frac{1}{N}(u_i, u_j)_X\right)_{1\leq i,j\leq N}. \tag{3}$$

Then we denote $v_k = (v_{i,k})_{1\leq i\leq N} \in \mathbf{R}^N$, for $k = 1, \cdots, N$, a set of orthonormal eigenvectors of the matrix K_{cor}. Orthonormal eigenvectors of the operator $K(u)$ are given by posing: $\psi_k = 1/\sqrt{N\lambda_k}\sum_{i=1}^N v_{i,k}u_i$. In practice, we compute the proper orthogonal decomposition of a function u by approaching u with a piecewise constant function. The method of snapshots allows us to compute the proper orthogonal decomposition of this piecewise constant function.

Let us mention at last two characterizations of the POD eigenvectors.

THEOREM 3 *Let $(\psi_k)_{k\geq 1}$ be a set of POD eigenvectors associated with u. Then for any $\ell \geq 0$, the vector $\psi_{\ell+1}$ satisfies:*

$$\frac{1}{T}\int_0^T \frac{(u(t),\psi_{\ell+1})_X^2}{(\psi_{\ell+1},\psi_{\ell+1})_X}dt = \max_{\varphi\in span(\psi_1,\cdots,\psi_\ell)^\perp\setminus\{0\}} \frac{1}{T}\int_0^T \frac{(u(t),\varphi)_X^2}{(\varphi,\varphi)_X}dt. \tag{4}$$

Conversely, if $(\psi_k)_{k\geq 1}$ is a set of orthonormal vectors in X which satisfy equality (4), then $(\psi_k)_{k\geq 1}$ is a set of POD eigenvectors associated with u.

Remark. In case when the function u is piecewise constant, we still call the values of u snapshots. The previous proposition characterizes the

POD eigenvectors as the best correlated to the snapshots in a quadratic mean sense.

THEOREM 4 *Let $(\psi_k)_{k\geq 1}$ be a set of POD eigenvectors associated with u. For any integer $\ell \geq 0$ and for any orthonormal set $(\varphi_k)_{k\geq 1}$ in X, we have the following inequality:*

$$\int_0^T \left\| u(t) - \sum_{k=1}^{\ell} (u(t), \psi_k)_X \psi_k \right\|_X^2 dt \leq \int_0^T \left\| u(t) - \sum_{k=1}^{\ell} (u(t), \varphi_k)_X \varphi_k \right\|_X^2 dt \tag{5}$$

Moreover we have:

$$\frac{1}{T} \int_0^T \left\| u(t) \ - \ \sum_{k=1}^{\ell} \ (u(t), \psi_k)_X \psi_k \right\|_X^2 dt \ = \ \sum_{k=\ell+1}^{\infty} \ \lambda_k \xrightarrow[\ell\to\infty]{} 0. \tag{6}$$

In particular for $\ell = 0$, the sum of the POD eigenvalues is equal to the energy of u. Conversely, if $(\psi_k)_{k\geq 1}$ is a set of orthonormal vectors in X which satisfy equations (5) and (6), then $(\psi_k)_{k\geq 1}$ is a set of POD eigenvectors associated with u.

1.2 Proper Orthogonal Decomposition of a Set of Functions

Let $n \in \mathbf{N}^*$ be a positive integer and let $(u_i)_{1\leq i\leq n}$ be a set of (class of) functions in $L^2(0,T;X)$. We define the POD operator $K : X \to X$ by posing:

$$K \ : \ \varphi \longmapsto \sum_{i=1}^{n} \frac{1}{T} \int_0^T (u_i(t), \varphi)_X \ u_i(t) \ dt. \tag{7}$$

DEFINITION 5 *The non zero eigenvalues of the operator K, indexed in a non increasing order, are called the POD eigenvalues associated with the set $(u_i)_{1\leq i\leq n}$. A set $(\psi_k)_{k\geq 1}$ of orthonormal eigenvectors of K in X corresponding to these eigenvalues is called a set of POD eigenvectors associated with $(u_i)_{1\leq i\leq n}$.*

The theorems analogous to theorems 3 and 4 hold.

Example. Let us consider a function $u \in L^2(0,T;X)$. If the time derivative du/dt is in $L^2(0,T;X)$, we can consider the proper orthogonal decomposition associated with $(u, du/dt)$.

2. Problem Formulation

Let V and H be real separable Hilbert spaces. We assume that the embedding $V \subset H$ is dense continuous. So there exists a constant $\alpha > 0$

such that $\|.\|_H \leq \alpha\|.\|_V$. We identify the space H with the dual space H'. Then the space $H = H'$ is identified with a dense subspace of the dual V' of the space V, with continuous embedding.

Let $a : V \times V \to \mathbf{R}$ a bilinear continuous elliptic form. So there exist real numbers $\beta, \kappa > 0$ such that for any $\varphi, \psi \in V$ we have $|a(\varphi, \psi)| \leq \beta\|\varphi\|_V\|\psi\|_V$ and $\kappa\|\varphi\|_V^2 \leq a(\varphi, \varphi)$. Let $\phi \in H$ and $f \in L^2(0, T; V')$, with $T > 0$. We consider the following parabolic problem:

$$(\mathcal{P})\left\{\begin{array}{rcll} \frac{d}{dt}(u(t), \varphi)_H + a(u(t), \varphi) &=& \langle f(t), \varphi\rangle_{V'\times V} & t \in [0, T],\ \varphi \in V, \\ (u(0), \varphi)_H &=& (\phi, \varphi)_H & \varphi \in V, \end{array}\right. \tag{8}$$

where $(.,.)_H$ is the scalar product in H and $\langle .,.\rangle_{V'\times V}$ is the duality V', V. We denote $W(0, T; V) = \{u \in L^2(0, T; V) | du/dt \in L^2(0, T; V')\}$ the space of (class of) functions in $L^2(0, T; V)$ the time derivative of which is in $L^2(0, T; V')$. We denote $C(0, T; H)$ the space of continuous functions with values in H. The space $W(0, T; V)$ is identified with a subspace of $C(0, T; H)$ in the following sense: any (class of) function in $W(0, T; V)$ admits a continuous representative with values in H. We have the following result ([4], theorems 1 and 2, p. 619-620):

THEOREM 6 *The problem $(\mathcal{P})$ admits a unique solution u in $W(0, T; V)$.*

Let u be the solution of problem $(\mathcal{P})$. Then $u \in L^2(0, T; V) \cap L^2(0, T; H)$ and we can consider the proper orthogonal decomposition of u in both cases $X = V$ and $X = H$. Let $(\psi_k)_{k\geq 1}$ be a set of POD eigenvectors associated with u in one of both cases $X = V$ or $X = H$. Let $\ell \geq 1$ be an integer and let $Y_\ell = \mathrm{span}(\psi_1, \cdots, \psi_\ell)$ be the approximated subspace of order ℓ. We consider the following problem ;

$$(\mathcal{P}_\ell)\left\{\begin{array}{rcll} \frac{d}{dt}(U_\ell(t), \varphi)_H + a(U_\ell(t), \varphi) &=& \langle f(t), \varphi\rangle_{V'\times V} & t \in [0, T],\ \varphi \in Y_\ell \\ (U_\ell(0), \varphi)_H &=& (\phi, \varphi)_H & \varphi \in Y_\ell \end{array}\right. \tag{9}$$

which admits a unique solution $U_\ell \in C(0, T; Y_\ell)$. The following theorem is now well known [6]:

THEOREM 7 *In both cases $X = V$ and $X = H$, the sequence $(U_\ell)_{\ell\geq 1}$ of the solutions of the problems $(\mathcal{P}_\ell)$ converges towards the solution u of the problem $(\mathcal{P})$ as $\ell \to +\infty$ in $L^2(0, T; V)$ strong.*

3. Estimates of the Error of POD-Approximation in a Regular Case

Now we want to give an estimate of the error $\|U_\ell - u\|_{L^2(0,T;V)}$. We consider a case when the function u is regular. We make the following assumption:

Assumption 1. Still denoting u the solution of the problem $(\mathcal{P})$ in $W(0,T;V)$, we assume that the time derivative du/dt is in the space $L^2(0,T;V)$.

This assumption is sensible because of the following result ([12]theorem 3.2, p. 70):

THEOREM 8 *If f and df/dt are in $L^2(0,T;H)$ and if the initial condition ϕ is in V, then du/dt is in $L^2(0,T;V)$.*

Under assumption 1, we can also consider the POD approximation of the problem $(\mathcal{P})$ by defining the proper orthogonal decomposition associated with $(u, du/dt)$. In this case we will obtain an estimate of the error of approximation according to the POD eigenvalues.

3.1 Case of the POD Associated with u

We consider the case $X = V$ and we still denote $(\psi_k)_{k\geq 1}$ a sequence of POD eigenvectors associated with u. The theorem 4 allows us to write the following equality in V for $t \in [0,T]$ and for any $\ell \geq 1$:

$$u(t) = \sum_{k=1}^{\ell} (u(t), \psi_k)_V \psi_k + \sum_{k=\ell+1}^{\infty} (u(t), \psi_k)_V \psi_k \tag{10}$$

We define two fonctions u_ℓ and $\widetilde{u}_\ell$ by posing for $t \in [0,T]$:

$$u_\ell(t) = \sum_{k=1}^{\ell} (u(t), \psi_k)_V \psi_k, \qquad \widetilde{u}_\ell(t) = \sum_{k=\ell+1}^{\infty} (u(t), \psi_k)_V \psi_k. \tag{11}$$

As the function U_ℓ, which is the solution of the problem $(\mathcal{P}_\ell)$, is in the space $Y_\ell = \operatorname{span}(\psi_1, \cdots, \psi_\ell)$, we have $U_\ell(t) = \sum_{k=1}^{\ell} (U_\ell(t), \psi_k)_V \psi_k$. By definition, the set $(\psi_k)_{k\geq 1}$ is orthonormal in V, so the Pythagore theorem gives:

$$\|u - U_\ell\|^2_{L^2(0,T;V)} = \|u_\ell - U_\ell\|^2_{L^2(0,T;V)} + \|\widetilde{u}_\ell\|^2_{L^2(0,T;V)}. \tag{12}$$

We observe the following equality:

$$\|\widetilde{u}_\ell\|^2_{L^2(0,T;V)} = T \sum_{k=\ell+1}^{\infty} (K\psi_k, \psi_k)_V = T \sum_{k=\ell+1}^{\infty} \lambda_k, \tag{13}$$

and we know from theorem 4 that the rest $\sum_{k=\ell+1}^{\infty} \lambda_k$ tends towards zero as $\ell \to \infty$. We only have to estimate the term $\|u_\ell - U_\ell\|^2_{L^2(0,T;V)}$. We denote $z_\ell = U_\ell - u_\ell$. The function u (resp. U_ℓ) is the solution of problem $(\mathcal{P})$ (resp. $(\mathcal{P}_\ell)$) so the function z_ℓ satisfies the following equality for $\varphi \in Y_\ell$:

$$\frac{d}{dt}(z_\ell(t),\varphi)_H + a(z_\ell(t),\varphi) = \frac{d}{dt}(\widetilde{u}_\ell(t),\varphi)_H + a(\widetilde{u}_\ell(t),\varphi),\ t \in [0,T], \quad (14)$$

as well as $(z_\ell(0),\varphi)_H = (\widetilde{u}_\ell(0),\varphi)_H$. If we take $\varphi = z_\ell(0)$, we obtain $\|z_\ell(0)\|_H \leq \|\widetilde{u}_\ell(0)\|_H$ and if we take $\varphi = z_\ell(t)$ in equality (14), we obtain the following equality after integration on $[0,T]$:

$$\begin{aligned} &\tfrac{1}{2}\|z_\ell(T)\|_H^2 + \int_0^T a(z_\ell(t),z_\ell(t))dt \\ = &\int_0^T \left(\tfrac{d\widetilde{u}_\ell}{dt}(t), z_\ell(t)\right)_H dt + \int_0^T a(\widetilde{u}_\ell(t),z_\ell(t))dt + \tfrac{1}{2}\|z_\ell(0)\|_H^2. \end{aligned} \quad (15)$$

Then we get:

$$\begin{aligned} \kappa\|z_\ell\|^2_{L^2(0,T;V)} \leq\ & \tfrac{\varepsilon}{2}\left\|\tfrac{d\widetilde{u}_\ell}{dt}\right\|^2_{L^2(0,T;H)} + \left(\tfrac{\alpha}{2\varepsilon}+\tfrac{\beta}{2}\right)\|z_\ell\|^2_{L^2(0,T;V)} \\ & + \tfrac{\beta}{2}\|\widetilde{u}_\ell\|^2_{L^2(0,T;V)} + \tfrac{1}{2}\|\widetilde{u}_\ell(0)\|^2_H, \end{aligned} \quad (16)$$

where $\varepsilon > 0$ is a real to be chosen below. Let us recall that the definition of the form a gives for any $\varphi \in V$: $\kappa\|\varphi\|_V^2 \leq a(\varphi,\varphi) \leq \beta\|\varphi\|_V^2$. Moreover we assume $\kappa - \beta/2 > 0$. Now we choose $\varepsilon > 0$ such that $\kappa - \alpha/2\varepsilon - \beta/2 > 0$ and we obtain the following estimate:

$$\left(\kappa - \frac{\alpha}{2\varepsilon} - \frac{\beta}{2}\right)\|z_\ell\|^2_{L^2(0,T;V)} \leq \frac{\varepsilon}{2}\left\|\frac{d\widetilde{u}_\ell}{dt}\right\|^2_{L^2(0,T;H)} + \frac{\beta}{2}\|\widetilde{u}_\ell\|^2_{L^2(0,T;V)} + \frac{1}{2}\|\widetilde{u}_\ell(0)\|^2_H \quad (17)$$

The term $\|\widetilde{u}_\ell\|^2_{L^2(0,T;V)}$ tends towards zero as $\ell \to \infty$ by definition of $\widetilde{u}_\ell$ and because of equality (6) in theorem 4. Moreover equality (10) holds in V for almost any $t \in [0,T]$, so in H for any $t \in [0,T]$ because $u \in C(0,T;H)$, in particular for $t = 0$, and we obtain that the term $\|\widetilde{u}_\ell(0)\|^2_H$ tends towards zero as $\ell \to \infty$. However we have in general $du/dt \in L^2(0,T;V')$, which does not ensure the convergence of the term $\|d\widetilde{u}_\ell/dt\|^2_{L^2(0,T;H)}$. That is why we are led to make assumption 1.

THEOREM 9 *Under assumption 1, we choose $\varepsilon > 0$ such that $\kappa - \alpha/2\varepsilon - \beta/2 > 0$ and we set:*

$$\rho_\ell = \frac{1}{\kappa - \alpha/2\varepsilon - \beta/2}\left(\frac{\varepsilon}{2}\left\|\frac{d\widetilde{u}_\ell}{dt}\right\|^2_{L^2(0,T;H)} + \frac{1}{2}\|\widetilde{u}_\ell(0)\|^2_H\right). \quad (18)$$

We choose $X = V$ and we consider the proper orthogonal decomposition associated to the function u. We get the following estimate:

$$\|u - U_\ell\|^2_{L^2(0,T;V)} \leq \rho_\ell + \frac{(\kappa - \alpha/2\varepsilon)T}{\kappa - \alpha/2\varepsilon - \beta/2}\sum_{k=\ell+1}^{\infty}\lambda_k \underset{\ell\to\infty}{\longrightarrow} 0, \quad (19)$$

on the error between the solution u of the problem $(\mathcal{P})$ and the approximation U_ℓ, which is the solution of the problem $(\mathcal{P}_\ell)$.

Proof. According to inequality (17), it suffices to proove that the term $\|d\widetilde{u}_\ell/dt\|^2_{L^2(0,T;H)}$ tends towards zero as $\ell \to \infty$. As the subspace $Y = \overline{\text{span}(\psi_k)_{k\geq 1}}^V$ is closed in V and as $u \in L^2(0,T;Y)$ according to equality (10), we obtain, under assumption 1, that $du/dt \in L^2(0,T;Y)$. We can then write the following equality in V for $t \in [0,T]$: $du/dt(t) = \sum_{k=1}^{\infty} (du/dt(t), \psi_k)_V \psi_k$. Then we have $\|d\widetilde{u}_\ell/dt\|^2_{L^2(0,T;V)} = \|\sum_{k=\ell+1}^{\infty} (du/dt(t), \psi_k)_V \psi_k\|^2_{L^2(0,T;V)}$ which tends towards zero as $\ell \to \infty$. On the other hand, we have the bound $\|.\|_H \leq \alpha\|.\|_V$ and we get the expected result from equalities (12) and (13) as well as from inequality (17).

3.2 Case of the POD Associated with $(u, du/dt)$

The definition of the term ρ_ℓ in theorem 9 let the term

$$\|d\widetilde{u}_\ell/dt\|^2_{L^2(0,T;H)}$$

arise, which represents the energy of the rest of the time derivative of u. It seems quite natural to consider the proper orthogonal decomposition associated with the set $(u, du/dt)$ so that the POD eigenvalues also take into account the energy of du/dt.

We still consider the case $X = V$ and we denote $(\psi_k)_{k\geq 1}$ a set of POD eigenvectors associated with $(u, du/dt)$. In this case we have the following equality, according to theorem 4, in particular according to equality (6):

$$\sum_{k=\ell+1}^{\infty} \int_0^T \left(\frac{du}{dt}(t), \psi_k\right)_V^2 dt + \sum_{k=\ell+1}^{\infty} \int_0^T (u(t), \psi_k)_V^2 \, dt = T \sum_{k=\ell+1}^{\infty} \lambda_k. \quad (20)$$

We set $c = \max(\varepsilon\alpha^2/(2\kappa - \alpha/\varepsilon - \beta), (\kappa - \alpha/2\varepsilon)/(\kappa - \alpha/2\varepsilon - \beta/2))$ and we can now express the following theorem:

THEOREM 10 *Under assumption 1, we choose $X = V$ and we consider the proper orthogonal decomposition associated with $(u, du/dt)$. With the above notations, we obtain the following estimate:*

$$\|u - U_\ell\|^2_{L^2(0,T;V)} \leq cT \sum_{k=\ell+1}^{\infty} \lambda_k + \frac{1}{2\kappa - \alpha/\varepsilon - \beta}\|\widetilde{u}_\ell(0)\|^2_H \underset{\ell\to\infty}{\longrightarrow} 0. \quad (21)$$

4. Choosing the Order of Approximation

When proper orthogonal decomposition is utilized for model reduction, the question arises of the choice of the order $\ell \geq 1$ of approximation. The integer ℓ must be large enough for the approximation to be good and small enough for the model to be reduced enough. The usual criterion (cf. for example [1, 5, 8, 9, 10, 11]) consists in setting a percentage $\delta \in [0, 1]$ then in choosing the smaller integer $\ell \geq 1$ such that:

$$\sum_{k=1}^{\ell} \lambda_k \Big/ \sum_{k=1}^{\infty} \lambda_k \ \geq \ \delta, \tag{22}$$

where the λ_ks are the POD eigenvalues. As the sum of the POD eigenvalues is equal to the energy of u, this criterion consists in projecting the studied system onto the modes which capture the largest portion of the energy of u. We are going to see that this criterion can be irrelevant and we will propose alternative ideas.

4.1 A Counter-example to the Usual Criterion of the Choice of the Order of Approximation

We still consider the parabolic case of the problem $(\mathcal{P})$. We set $\Omega =]0, 1[$, $V = H_0^1(\Omega)$, $H = L^2(\Omega)$ and $X = V$. We define the bilinear form a by posing $a(.,.) = (.,.)_V$ so that the problem $(\mathcal{P})$ is the variationnal form of the heat equation with homogeneous Dirichlet boundary conditions. Let ψ_1 and ψ_2 be orthonormal vectors in V but not in H defined by:

$$\begin{array}{rcll} \psi_1(x) & = & \left| \begin{array}{ll} x & \text{if} \quad x \in [0, 1/2] \\ 1 - x & \text{if} \quad x \in [1/2, 1] \end{array} \right. \\ \psi_2(x) & = & \left| \begin{array}{ll} \frac{1}{\sqrt{2}\pi} \sin(2\pi x) & \text{if} \quad x \in [0, 1/2] \\ \frac{1}{\sqrt{2}\pi} \sin(2\pi(x - 1/2)) & \text{if} \quad x \in [1/2, 1] \end{array} \right. \end{array} \tag{23}$$

Then we have

$$\|\psi_1\|_H^2 = 1/12, \ \|\psi_2\|_H^2 = 1/(4\pi^2) \text{ and } (\psi_1, \psi_2)_H = (2\pi)^{-3/2}$$

We set $T = 1$ and we define two orthogonal functions h_1 and h_2 in $L^2(0, T)$ by posing for $t \in [0, T]$: $h_1(t) = 1$ and $h_2(t) = 1.2 \ \sin(20\pi t)$. We set at last for $x \in \Omega$ and $t \in [0, T]$: $u(x, t) = h_1(t)\psi_1(x) + h_2(t)\psi_2(x)$. We assume that the function u is the solution of the problem $(\mathcal{P})$ for a well chosen right-hand term f. We still denote K the POD operator associated with u and we observe that $K\psi_1 = \psi_1$ and $K\psi_2 = 0.72 \ \psi_2$ so the vectors (ψ_1, ψ_2) are the POD eigenvectors associated with u and the POD eigenvalues are $\lambda_1 = 1$ and $\lambda_2 = 0.72$. We denote y_1

(resp. y_2) the solution of the projection of the problem $(\mathcal{P})$ onto span(ψ_1) (resp. span(ψ_2)). As $\lambda_1 > \lambda_2$ we expect the projection onto span(ψ_1) to be better than that onto span(ψ_2), that is:

$$\|u - y_1\|_{L^2(0,T;V)} \leq \|u - y_2\|_{L^2(0,T;V)}. \tag{24}$$

In fact, this inequality is not satisfied. Indeed a computation with *Maple* allows us to obtain:

$$\|u-y_1\|^2_{L^2(0,T;V)} \simeq 1.2641, \quad \|u-y_2\|^2_{L^2(0,T;V)} = 1+(1-e^{-8\pi^2})/4\pi, \tag{25}$$

that is:

$$\|u - y_1\|_{L^2(0,T;V)} > \|u - y_2\|_{L^2(0,T;V)}, \tag{26}$$

instead of the expected inequality (24). If we set the percentage $\delta = 58\%$ and if we apply criterion (22), we find $\ell = 1$ and we compute y_1, whereas the function y_2 is a best approximation of the solution u of the problem $(\mathcal{P})$. We could conclude that the POD set (ψ_1, ψ_2) is not well indexed and that the indexation must be modified to consider the set (ψ_2, ψ_1). However, if the number of POD eigenvectors is infinite, it is difficult to reindex them. We rather consider the point of view which consists in keeping the same order of indexation as that of the eigenvalues and in going deeper into the calculation of the approximation.

4.2 Definition of Some Criteria for Choosing the Order of Approximation

Let us first mention two natural criteria for choosing the order of approximation. The first natural criterion is the absolute value $\ell \geq 1$ of the order of approximation: we can decide not to go over a certain value, for instance in order to limit the computation time. This is not the usual point of view: in general, one prefers to choose the smallest integer $\ell \geq 1$ which satisfies a certain condition, for example the inequality (22), which amounts to favouring the precision of the approximation rather than the speed of computation ; we will follow this line by defining criteria for the order of approximation. The second natural criterion consists in utilizing the bounds (19) and (21) in theorems 9 and 10 to ensure a given precision.

We still make the regularity assumption 1 and we set $X = V$. We consider the case of the proper orthogonal decomposition associated with the solution u of the problem $(\mathcal{P})$. We propose the following definition:

DEFINITION 11 *We define the real numbers $\varepsilon, \rho_\ell > 0$ as theorem 9. The integer $\ell \geq 1$ is a* good order *of approximation if we have the following* *inequality:*

$$\rho_\ell + \frac{(\kappa - \alpha/2\varepsilon)T}{\kappa - \alpha/2\varepsilon - \beta/2} \sum_{k=\ell+1}^{\infty} \lambda_k \leq T \sum_{k=1}^{\ell} \lambda_k. \tag{27}$$

THEOREM 12 *If the integer $\ell \geq 1$ is a good order of approximation, the Galerkin projection of the problem $(\mathcal{P})$ onto the subspace Y_ℓ is better than the Galerkin projection onto the subspace Y_ℓ^φ, for any subspace Y_ℓ^φ orthogonal to Y_ℓ in Y, in the following sense:*

$$\|u - U_\ell\|_{L^2(0,T;V)} \leq \|u - U_\ell^\varphi\|_{L^2(0,T;V)}, \tag{28}$$

where U_ℓ^φ is the solution obtained by Galerkin projection onto Y_ℓ^φ.

DEFINITION 13 *We define real numbers $\varepsilon, \rho_\ell > 0$ as in theorem 9. The integer $\ell \geq 1$ is a* very good order *of approximation if the following inequality holds:*

$$\rho_\ell + \frac{(\kappa - \alpha/2\varepsilon)T}{\kappa - \alpha/2\varepsilon - \beta/2} \sum_{k=\ell+1}^{\infty} \lambda_k \leq T\lambda_\ell. \tag{29}$$

THEOREM 14 *If the integer $\ell \geq 1$ is a very good order of approximation, then for any integer $k \geq \ell + 1$ the Galerkin projection of the problem $(\mathcal{P})$ onto the subspace $Y_\ell = span(\psi_1, \cdots, \psi_{\ell-1}, \psi_\ell)$ is better in the sense of the norm $L^2(0,T;V)$ than the projection onto the subspace $span(\psi_1, \cdots, \psi_{\ell-1}, \psi_k)$.*

Analogous definitions can be expessed in the case of the POD associated with $(u, du/dt)$.

5. Conclusion

We have recalled the principle of proper orthogonal decomposition and we have considered POD-Galerkin methods for a parabolic problem. If the solution u of the parabolic problem $(\mathcal{P})$ is regular, i.e. u satisfies assumption 1, we can obtain bounds of the error of the POD-Galerkin approximation in the case $X = V$. We have considered the POD associated with u and the POD associated with $(u, du/dt)$. These bounds do not depend on the discretization scheme and allow us to define some criteria for choosing the order of approximation.

References

[1] H. Banks, L. Joyner, B. Wincheski, and W. Winfree. Nondestructive evaluation using a reduced order computational methodology, 2000. Nasa/CR-2000-209870, ICASE Report No. 2000-10.

[2] G. Berkooz, P. Holmes, and J. Lumley. The proper orthogonal decomposition in the analysis of turbulent flows. *Annu. Rev. Fluid Mech.*, 25:539–575, 1993.

[3] J. Bonnet, L. Cordier, J. Delville, M. Glauser, and L. Ukeiley. Examination of large-scale structures in a turbulent plane mixing layer. Part 1. Proper orthogonal decomposition. *J. Fluid Mech.*, 391:91–122, 1999.

[4] R. Dautray and J.L. Lions. *Analyse mathématique et calcul numérique pour les sciences et les techniques, tome 3.* Masson, Paris, 1985.

[5] A. Glezer, Z. Kadioglu, and A. Pearlstein. Development of an extended proper orthogonal decomposition and its application to a time periodically forced plane mixing layer. *Phys. Fluids A*, Vol. 1, No. 8:1363–1373, 1989.

[6] T. Henri and J.-P. Yvon. Stability of the POD and convergence of the POD-Galerkin method for parabolic problems, 2002. preprint IRMAR 02-40.

[7] K. Kunisch and S. Volkwein. Galerkin proper orthogonal decomposition methods for parabolic problems. *Numer. Math.*, 90:117–148, 2001.

[8] S. Lall, J.E. Marsden, and S. Glavaski. Empirical model reduction of controlled nonlinear systems, 1999. Proceedings of the IFAC World Congress.

[9] B.C. Moore. Principal component analysis in linear systems : controllability, observability and model reduction. *IEEE Transactions on Automatic Control*, vol. AC-26, no 1, 1981.

[10] S.S. Ravindran. Proper orthogonal decomposition in optimal control of fluids, 1999. Nasa/TM-1999-209113.

[11] S.Y. Shvartsman, C. Theodoropoulos, R. Rico-Martinez, I.G. Kevrekidis, E.S. Titi, and T.J. Mountziaris. Order reduction for nonlinear dynamic models of distributed reacting systems. *J. of Process Control*, 10:177–184, 2000.

[12] R. Temam. *Infinite-dimensional dynamical systems in mechanics and physics.* Spinger-Verlag, New York, 1988.